BIOLOGY
An Uncommon Introduction

Contributing Editors

Michael G. Barbour
Richard H. Falk
Seymour M. Gold
James A. Harding
Thomas E. Ragland
University of California, Davis

Diane Hersh

BIOLOGY
An Uncommon Introduction

Robert McNally

Canfield Press
San Francisco

A Department of Harper & Row, Publishers, Inc.
New York Evanston London

To Terry

Library of Congress Cataloging in Publication Data

McNally, Robert, date
Biology: an uncommon introduction.

1. Biology. I. Title. [DNLM: 1. Biology.
QH308.2 M169b 1974]
QH308.2.M35 574 73-22277
ISBN 0-06-380435-2

BIOLOGY: AN UNCOMMON INTRODUCTION

74 75 76 10 9 8 7 6 5 4 3 2 1

Interior design and cover by Russell Leong
Biological art by Heather Kortebein
Other line art by Charlotte Kay and Ayxa Art
Cartoons by Sam Daijogo

Credits

Chapter 2 *p. 14:* Courtesy of the Hale Observatories. *Fig. 2-4:* Richard H. Gross.

Chapter 3 *Fig. 3-3:* Richard H. Gross. *Fig. 3-4:* Adapted from Annette Tussing, "The Fight to Save the Snake," *Field and Stream*, October 1971, pp. 22, 126-127. *Fig. 3-8:* Adapted from data supplied in *Newsweek*, January 22, 1973, p. 54. *Fig. 3-11:* EPA-DOCUMERICA, Gene Daniels. *Fig. 3-13:* Adapted from Clarence Hylander, *Wildlife Communities*, Boston, Houghton Mifflin, 1966. *Fig. 3-14:* Richard H. Gross. *Fig. 3-15:* Steve H. Kratka. *Fig. 3-16:* Adapted from Clarence Hylander, *Wildlife Communities*, Boston, Houghton Mifflin, 1966, p. 72. *Fig. 3-17:* Adapted from W. A. Niering, *The Life of the Marsh*, New York, McGraw-Hill, 1966, pp. 100-101.

Chapter 4 *Fig. 4-1:* The Bettmann Archive, Inc. *Fig. 4-4:* A. Photo Derrick Witty, Photocopyright: George Rainbird Ltd. B. Courtesy Oregon Regional Primate Research Center. *Fig. 4-6:* From the experiments of Dr. H. B. D. Kettlewell, University of Oxford. *Fig. 4-8:* A. U.S. Navy. B. Carl Roessler. *Fig. 4-12:* Richard H. Gross. *Fig. 4-13:* Richard H. Gross. *Fig. 4-14:* Courtesy of the American Museum of Natural History. *Fig. 4-15:* Courtesy of the American Museum of Natural History. *Fig. 4-16:* Courtesy of the American Museum of Natural History.

Chapter 5 *Fig. 5-2:* Eel, California Academy of Sciences, Steinhart Aquarium; bird, Richard H. Gross; athletes, Claus Meyer, Black Star; bacteria, Frank H. Johnson, Princeton University; bear, Richard C. Simmonds.

Chapter 6 *Fig. 6-2:* California Academy of Sciences, Steinhart Aquarium. *Fig. 6-3:* Richard H. Gross. *Fig. 6-9:* A. Richard H. Gross. B. Courtesy of Dr. M. F. Perutz. *Fig. 6-11:* Adapted from CRM Books, *Biology Today*, Del Mar, Calif., CRM, 1972, p. 386. *Fig. 6-13:* From K. L. Johnson et al., *Science*, vol. 160, no. 67, 1968. *Fig. 6-14:* Courtesy of *New York Journal American* and the Environmental Protection Agency.

The credits are continued on p. 516.

Contents

A Word from the Editors

In researching and planning this book, the editors convened a meeting of concerned two- and four-year college biology instructors to discuss the needs, interests, and aims of their nonmajor introductory biology students. At this meeting a consensus was reached on the need for a fresh approach to biology, one that would emphasize essential principles and concepts within a context immediately accessible to the student.

We also outlined some of the instructors' objectives in teaching the nonmajors course. Among them, the participants listed the desire to convey their own personal excitement and enthusiasm about biology; the crucial need to provide an understanding of biology to all college-educated students, irrespective of their future plans; to provide an appreciation of the problems and methods of science; and in general, to make an important and lasting contribution to the student's education in the brief time available. What it all came down to was the concept of biology literacy—an asset surely as important as an understanding of history, a foreign language, or the social sciences.

The question was: Are available texts going in the same direction as these innovative and exciting courses? The reluctant conclusion was that some other—different—approach to the writing of a text was needed. It was concluded that an attractive alternative would be—from the outset—to develop a book with the following goals: to put man back into the biological context from which his technology seems to have removed him; to place the priority on comprehension rather than encyclopedic comprehensiveness; to present the material in a style which was both engaging and comfortably informative, and which contributes to, rather than interferes with, an appreciation of biological concepts.

Canfield Press has taken an active role in the development of this book because it believes that the publishing community shares with faculty and administrators a responsibility for the quality and success of our higher education. In that a demonstrable need for this book exists, we welcomed the opportunity to assist in providing it. We would particularly like to thank those instructors who gave of their time and ideas to get us started in this project: Jean Chapman, Paul Hansen, Patrick Brunnelle, Stephen Barnhart, Alan Holbert, Carlo Vecchiarelli, Robin Lenn, Alyce Fiedler, and Laurence Fulton.

We very much hope that you will enjoy your reading and review of this book, and that you will then be able to judge whether the aforementioned goals have been met. With that hope, and with a good deal of personal enthusiasm, we offer you BIOLOGY: AN UNCOMMON INTRODUCTION.

R. Wayne Oler
Malvina Hindus

San Francisco
January, 1974

About This Book

The British biologist Sir Julian Huxley has predicted that, in the course of human history, the twentieth century will be remembered more for its biology than for its physics or its chemistry. Let me offer two possible explanations for this. First, in recent decades, man has come very far along in his understanding of his favorite subject—himself. His heredity, his medicine, his evolutionary heritage, the functions of his body, the physical basis of the way he thinks and feels are being explored in ways and with results inconceivable seventy years ago. Second, in this century, biology has become the imperative science. It used to be that an understanding of biology's principles, like knowledge of Herodotus and Pliny, was the mark of an educated elite. Today it is a necessary tool for all the occupants of this fragile planet.

Indeed, it has become a truism that biology is profoundly important to the layman, that the news media pay close attention to advances in biological research and encourage, even implore, the "average citizen" to deal with their moral and social consequences. However, the important question is, how many people are actually aware of what is reported? Fewer still, undoubtedly, can and do respond from an enlightened, responsible point of view. The majority of people are put off by technical jargon, do not recognize the potential importance of biological discoveries to their own lives, or lack the background needed to assess the significance of scientific developments.

For many college students, the introductory biology course will be their only formal opportunity to remedy such problems. Unfortunately this need often gets lost in the academic shuffle. In too many survey courses, the student plows through several hundred pages of a standard textbook, processing data and facts in preparation for examinations. Rarely does the introductory student have the opportunity for reflection, which is what the best education is or ought to be about.

The nonmajors course is designed to meet the special needs of the student who doesn't have to meet the professional requirements made of those pursuing further study in science. At its best, the nonmajors course is not a "brief" version of anything, nor is it a variation of another approach. It is a unique view of an exciting field with an emphasis on concept and application, on perspective and context. *Biology: An Uncommon Introduction* lets biology speak for itself, free of much of the classifying, labeling, and excessive detail that fast extinguish the student's initial enthusiasm for the field.

This textbook builds on two central purposes, each dependent on the other. The first is to teach not the accumulated data of biology in their most condensed and categorical form, but to present the concepts of the science as clearly as possible. We have attempted to portray biology as a mode of thinking about the living world, a way of giving it order and form. The second purpose is to select those topics that aid the student's understanding of himself, his species, his living neighbors, and his environment.

The book is organized into three parts. In Part One, *The Design of Life,* we explore the basic physical, chemical, and biological laws that shape life. The focus is on the elemental ideas about matter and energy and on the concepts of evolution and ecology. Part Two, *The Processes of Life,* takes a close-up look at how organisms work, with an emphasis on man. Here we look at respiration, circulation, digestion, excretion, genetics, reproduction, development, motion, and internal and external communication. Part Three, *The Community of Life,* examines the biology of populations, the peculiar biological dilemmas of man, and some predictions about the shape of the human future.

Each chapter begins with an introduction that serves to show the reader what will be covered and how the topic fits into the overall plan of the book. A summary at the end of the chapter recapitulates all major concepts presented. Following the summary is a list of questions for the student to use in checking his comprehension and organizing his review. All new terms are italicized at first appearance, and their definitions appear in the glossary at the end of the chapter.

Two publications designed for use with this text should be mentioned. The Instructor's Manual by Carlo Vecchiarelli provides helpful outlines of the material as well as additional lecture topics. The accompanying laboratory manual, *Exploring Biology in the Laboratory* by Alyce Fiedler and Laurence Fulton presents experimental biology in a style parallel to that of this text—a nonquantitative, fundamentals approach.

Many people play a significant part in bringing any new text to light, and this one is no exception. In addition to the contributing editors, several instructors provided important assistance throughout the planning and writing. Alyce Fiedler and Laurence Fulton of American River College contributed to the early development of the book as did Abraham Flexer of the University of Colorado at Boulder and Steven Wolfe of the University of California, Davis. In its first and later drafts, the manuscript was carefully read and criticized by Robin Lenn, Sacramento City College; Paul Hansen, Napa College; and Carlo Vecchiarelli, Chabot College. Richard Gross of West Palm Beach Junior College contributed his talents as a biologist and photographer. Phyllis Keenan and Ann Hundley accurately and swiftly transformed heaps of penciled sheets into typescript.

For many books, the publisher's role begins and ends with production and sales, but the people of Canfield Press have been deeply involved in this project. R. Wayne Oler, editor-in-chief, recognized the need for a book of this type and provided a steadying and encouraging hand throughout the long process of translating idea into reality. Patricia Brewer, production editor, capably handled the complex task of producing a book from a manuscript. Special thanks is due Malvina Hindus, developmental editor. Everything an editor might do, from evaluating copy through providing ideas to planning and selecting artwork, she did well and then some. I am indebted to her.

Although many people have contributed to this text in one way or another, any errors are my sole responsibility. I welcome the comments, criticisms, and suggestions of teachers who use the book.

Whatever else may be said, a text exists only to serve the needs of students. I hope my readers find this effort useful.

Robert McNally
San Anselmo, California
January, 1974

1 Starting Out

In the course of the day-to-day—or test-to-test—pressure of a college education, it's all too easy to be swamped by the flood of facts that have to be learned and to forget why one is learning them, to be so concerned with memorizing the names of the trees that you forget the way in and out of the forest. This is unfortunate. One of the marks of an educated person is the ability to tell the significant from the insignificant, to be able to separate important concepts from trivial details. An encyclopedia contains a wealth of facts, but memorizing its 20-odd volumes isn't the same as getting an education. Oftentimes, though, science courses do seem like memorizing the encyclopedia, or at least a large part of it. There seem to be so many facts and so little time that one can quickly forget what the science is concerned with and how the facts add up to a whole.

Before we begin to look at the facts and principles of biology, let's put this particular science in perspective, indicating what it is and why it's important. What precisely is biology? What is the subject matter of this

An encyclopedia contains a wealth of facts, but memorizing its 20-odd volumes isn't the same as getting an education.

science? Is there anything distinguishing about science as a form of knowledge? Finally—and perhaps most importantly—why should you spend your time studying this science?

What is biology?

Take yourself as an example of a living organism, a fit subject for biology.

In the widest sense, biology is the science of life. More specifically, biology studies the structures, functions, and relationships of living organisms. Take yourself as an example of a living organism, a fit subject for biology, and think about what these words mean, what they imply. First, consider the number of structures, or parts, there are to your body: brain, teeth, hair, skin, eyes, stomach, feet, and so on. Second, think of the many different functions, or activities, your body performs every day: sleeping, thinking, pumping blood, growing, digesting. Then consider the relationships you have as an organism, both internal and external: the integration of all the structures and functions of the body into a working whole and the social relationships you have with other human beings. Multiply these activities by the number of kinds of living organisms—there are more than 700,000 varieties of insects and 250,000 kinds of flowering plants, to give two numerous examples—and you'll see that a wide range of phenomena falls within the purview of biology.

Because biology encompasses such a huge portion of the natural world, biologists researching different questions work in very different ways. A biologist who specializes in the chemical activities of cells comes closest to the movie image of the scientist as a white-coated, bespectacled technician who spends all day in a shining laboratory filled with glass instruments, flickering signal lights, and many-gauged control panels. Such a biologist works very much as a chemist does. A very different image of the science comes from a biologist who studies the behavior of wolves or chimpanzees. This kind of researcher spends his days outdoors, living where the animals do and observing them carefully. His tools are not test tubes and Bunsen burners, but camera equipment for photographing the animals, notebooks for keeping accurate records of their daily activities, and well-made game blinds for observing without being observed. This biologist's methods of work are much more like those of the anthropologist studying a primitive tribe than they are like those of the cell chemist.

Even though these two biologists are studying very different phenomena, both studies are contributing to a better understanding of life—what it is, how it works. But how do the scientists know that they are studying life? How can they tell the living from the nonliving?

The qualities of life

Although many scientists have tried, no one has come up with a simple definition of life. However, it is possible to list the special characteristics that are shared by organisms, from the very simplest to the most complex, and that distinguish them from nonliving things.

First of all, living things are organized quite differently from their environment on two levels. All living things are composed of one or more cells, a highly complex structural unit found only among living forms. On the other level, organisms are chemically distinct from their surroundings. They are composed of the same chemical elements as the nonliving environment, but in different proportions.

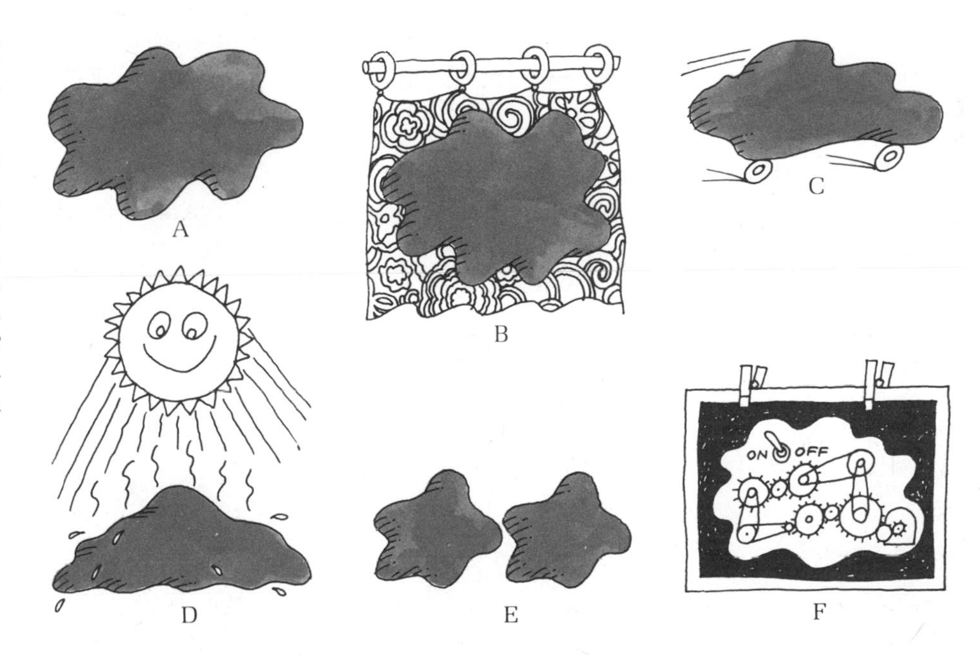

A. How do we know if it's alive? B. First, is it composed and organized differently from its environment? C. Does it take in energy and put it to work? D. Does it respond to its environment? E. Can it reproduce itself accurately? F. Does each of its component parts serve a definite function?

The second characteristic of all organisms is their ability to take in energy and put it to work. Plants, for example, use the energy of sunlight to build food molecules from carbon dioxide and water. Fireflies use some of their energy to give off light signals, and certain eels are able to give off an electric shock in order to catch prey. When you take in food and the energy stored in it, your body puts the energy to such uses as building the protein molecules needed in growth or providing the power for running the 100-yard dash or for reading a book.

Third, all living things respond to their environments. A plant will grow in the direction of its light source; a trout will strike at the shadow of an insect on the surface of the water; a seed put into the right kind of soil with the right amount of water at the right temperature will sprout.

The fourth characteristic of life is the most distinguishing: living things reproduce themselves with considerable accuracy. However, reproduction doesn't produce exact copies—no one looks exactly like either of his parents. The mechanics of reproduction have allowed enough alteration that over the last 3 billion years, the many different organisms populating the world today have all derived from the very first, primitive cells.

The final characteristic of life forms is somewhat related to the first. Any sample of the nonliving environment—a bucket of sterile soil or sand or water—is a collection of material found together at random. Just the opposite is true of something alive. Everything in and about that life form, from such complex behaviors as thinking to microscopic anatomical structures like the cell membrane, serves a purpose: it contributes to the survival of the organism. Much of biology thus consists of asking: "What function does this serve?" It's a question we'll be asking time and again throughout this book.

The role of the scientific method

With a sense of what biology studies, we can ask how biology studies its material. The standard answer is that biologists, as scientists, adhere strictly to the canons of the scientific method and that this method ensures accuracy and truth.

What we now think of as science began with the careful observation and description of the natural world—the types of animals and plants, the

Science is more than a catalog of observations or a list of phenomena.

apparent movements of the planets and stars, the structure and composition of the nonliving world. But, a few hundred years ago, various scientists realized that to learn more about the world they had to do more than just watch, they also had to manipulate the world in a controlled way and evaluate the results. These manipulations are what we call experiments, and the scientific method is an outgrowth of the use of experiments to further scientific knowledge.

According to the classical formulation of the scientific method, the scientist begins by observing a certain phenomenon. Working from these observations, the scientist formulates a hypothesis, which is a tentative explanation of the phenomenon. The hypothesis allows for a prediction. That is, if the system under observation is manipulated or controlled as directed, then it should behave in a certain fashion. The scientist then performs an experiment that manipulates the system in the way the prediction directs, and observes the results. If the system behaves as expected, then the hypothesis is supported. If not, the scientist discards it and tries another, continuing this process until he finds one that fits. A hypothesis supported by many experiments gains the status of a theory, which can be used to predict the behavior of other, similar systems. If the theory successfully explains every system it is applied to, it becomes a law. As might be expected, science has only a few laws, some theories, and a great many hypotheses (Figure 1–1).

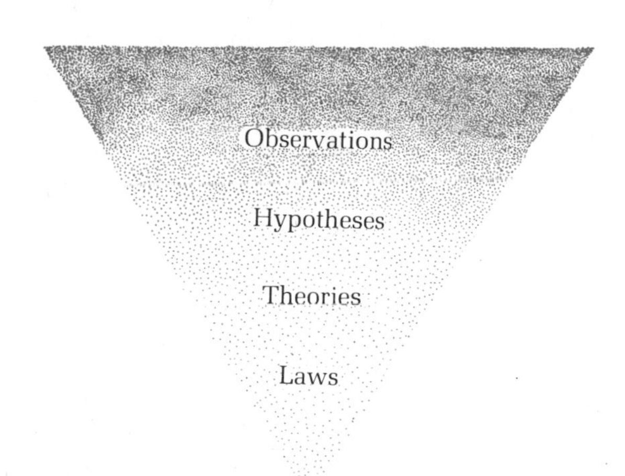

Figure 1–1 Science is a selective process in which a great number of observations leads to fewer hypotheses, still fewer theories, and an even smaller number of basic laws.

The most important qualities ascribed to the scientific method are its objectivity and skepticism. Theories and laws are never conclusively proved; they are valid only so long as they successfully predict and explain every system they are applied to. Once a theory fails in a particular case, it must be discarded, and a new one formulated that accounts for the problematical case. Thus the scientist must be objective about what he sees and skeptical about what he accepts as true.

There is a measure of validity to this textbook view of the scientific method. No scientist can, for example, let loose a wild idea unsupported by evidence and expect to be regarded as anything other than unbelievable. However, too much credit is given to the scientific method as the only road to truth, with the result that a very false view of the scientist and his creativity emerges.

The formulation of the scientific method we have just given was not drawn up by working scientists explaining how they go about their research but by philosophers and historians of science seeking to distinguish science from other forms of intellectual activity. It is a portrait drawn at a distance. In truth, no scientist, not even the best, follows the scientific method religiously. Some day-to-day, routine investigative work does indeed adhere to the classical canon of observation–hypothesis–experiment, but none of the great scientific discoveries and ideas came exclusively from the scientific method.

The scientific method, as an abstraction, necessarily discounts the role of the individual creative mind. In truth, random bumbling, accidents, hunches, lucky finds, and curious poking around have played as great a role in all the great leaps forward of modern science, and many of the shorter ones, as has the scientific method. Positing a hypothesis or designing the definitive experiment to test a theory are acts of creative insight, not the result of the careful application of a preconceived method.

What distinguishes good technical work in the laboratory from great science? Why do we recall, for example, the work of Louis Pasteur, Albert Einstein, Isaac Newton, and Charles Darwin, without paying very much

Great science is the insight with which a person looks at the same set of facts that hundreds of other people have noticed, but sees something special within them.

attention to the hundreds of thousands of other people who contributed to their work? What makes a great scientific mind?

It's an intangible thing really. Partially, it's the insight with which a person looks at the same set of facts that hundreds of other people have noticed, but sees something special within them—insight to draw an important unifying conclusion from seemingly unrelated facts.

Usually we tend to separate the arts and the sciences into two distinct areas that share nothing with one another. The scientist is cast as the diligent and careful worker who proceeds as the scientific method demands, while the artist is thought to rely mainly on intuitive flash and little on discipline. These distinctions are just not accurate. Formulating a hypothesis to account for a particular phenomenon or designing a good experiment are intuitive, creative acts much like the sort of insight that gives a novelist the structure of his book or shows a painter the outline of his next canvas. And the patient, hard work characteristic of scientific investigation is also true of art, for a good novel or sculpture is the product of months or years of hard, painstaking labor.

Science attempts to explain a wide range of phenomena in the simplest possible way.

A basic goal of science is to describe the greatest number of phenomena in the briefest possible way.

Thus the body of knowledge that we think of as science is the product of the combined interplay of the scientific method, a goodly number of accidents, and a very healthy dose of individual creativity. It would be wrong to think of the concepts and explanations in this book as cold and disembodied truths arrived at by equally cold and disembodied researchers. These concepts are creative explanations of the living world based on the best available evidence. Nor are these ideas permanent and absolute truths. Change best describes scientific ideas, for as better formulations and new information appear, old ideas are dropped in favor of new ones. A basic goal of science is to describe and explain the greatest number of phenomena in the briefest possible way. Science is the child of our ability to see, to judge, and to think, and not of our slavish devotion to a method.

What good is it?

Even with answers given to the what and how of biology, there remains a set of important questions, ones that have to do with the worth of the science. Why should anyone want to study it? What benefit can one derive from knowing what evolution means or from tracing the chemical pathways of photosynthesis?

Some people have to know biology in order to work at their chosen professions. Obviously, college and university teachers fall into this category, as well as a great number of other professionals who need to be conversant with at least certain aspects of the science: physicians,

Why should anyone want to study biology? What benefit can one derive from knowing what evolution means or from tracing the chemical pathways of photosynthesis?

agronomists, pathologists, nurses, dentists, dieticians, veterinarians, forest and range managers, science writers and journalists. However, these professions represent only a small proportion of all working people. Most students who spend a term in a biology course do so not because they need the science for their subsequent professional work but because they have to fulfill a graduation requirement. What good can they hope to derive from such a course?

Making the world understandable

The hypotheses, theories, and laws of biology serve to explain the occurrences of the living world. They provide a way of ordering the welter of information we have about organisms and providing it with meaning. Biology, like all sciences, makes the world more understandable.

Biology helps explain many of the seeming mysteries of the living world.

If you reflect about your own experiences with other organisms, you'll quickly see that a wide range of phenomena seems mysterious and unexplainable. Why does a growing plant always bend toward the light? Why is the hummingbird the only bird able to hover under its own power? Why do ants walk in a line, head to tail, like soldiers marching in single file? Why do migrating geese fly in a V-formation? For that matter, how do the geese know when to migrate, and how do they know where they are going? Why do animals breathe? If you say they do it because they need oxygen, what do they—and we—need the oxygen for? Can plants breathe? What about fish?

On and on this list could go. In fact, there is no real end to it, for with every answer, a new question arises. But with each new bit of knowledge, we have learned something important about the way the world works. Even for the nonbiologist, biology offers a better understanding of life, an understanding that applies to ourselves as well as to other organisms. The principles and theories of biology deal with man as with all living things, and they offer us a novel way of examining ourselves. Throughout this book, we'll refer again and again to man and see how he fits into the scheme of things.

Making the world livable

At this time, we Americans are faced with a long list of political and social issues that are either rooted in biological facts or have biological implications. We can only understand the social and moral implications of such developments as sperm banks, heart transplants, kidney machines, and "test-tube babies" if we know the biological principles behind them. In recent years, the word *ecology* has become a sort of rallying cry for people worried that man's growing population and technology are ruining the environment in a way that could eventually spell the end of the human race. Newspapers, magazines, and television have carried story after story about legal and political fights between developers who want to build houses or freeways and conservationists who want the land left as is. While some people spread warnings about poisons in the air and water, others call them misinformed Jeremiahs. Timber companies maintain that their cutting merely speeds nature's work, but their opponents accuse them of raping the land for their own profit.

Acting on biological knowledge alone, we cannot decide whether a forest should be cut, but we can know the consequences of cutting it or of letting it be.

Although each of these typical issues deals with biological facts, biology can't decide which side is right. The reason is that these issues involve human values outside the realm of science. The conflict between a timber company and a conservation group can only be settled by comparing the values each side wants to realize: the economic value of the cut timber as opposed to the aesthetic and ecological value of letting the forest stand. Biology has only the value of seeking truth. However, biology can often tell us something of the consequences of our actions. Acting on biological knowledge alone, we cannot decide whether a forest should be cut, but we can know the consequences of cutting it or of letting it be. And that kind of information is crucial to making an informed and intelligent decision.

As man's population continues to grow at a stupendous rate and as we exhaust an increasing number of our resources, issues centered about the way man should handle the environment are bound to become more common and more difficult. Any American seeking to decide how these issues should be settled will need to know at least something of biology. Setting air pollution standards without knowing the effects of inhaled lead would be a bit like trying to buy shoes without knowing the size of your feet!

The themes of the book

The subject matter of this book is organized around these two benefits of the study of biology. On the one hand, we'll be looking for those patterns of order that characterize the living world. At the same time, we'll see how the principles and ideas of biology add to our understanding of environmental problems.

Part One

The Design of Life

The extraordinary phenomenon known as life is closely dependent upon its nonliving setting. Despite the unique qualities of living organisms, they are affected and controlled by the same laws that regulate the rest of the natural world. In this first part, we will look at life from this perspective—how life is shaped by its chemical and physical components. Chapter 2, *Earth*, describes our planet, the only known site of life in the universe. The major link between the living and the nonliving is at the chemical level, and we will discuss some basic ideas about matter—how it is organized and how it behaves. Chapter 3, *Cycles*, charts the path of important chemical substances between living things and their environments. It then describes how biologists have organized these relationships through the concepts of ecology. In Chapter 4, *Evolving*, we roll time back and see how the process of evolution has structured the adaptation of living things to their environments. Chapter 5, *Energy*, presents the basic laws of energy and how energy is applied by living things.

A portion of our galaxy, the Milky Way, photographed by five cameras.

2

Earth

One of the curious characteristics of life, one that we rarely think about, is the rarity of living organisms. In terms of the space we humans occupy, life seems commonplace, but from any wider perspective, the picture quickly changes. Of the nine planets that orbit the sun, only the earth is known to support life as we know it. The other planets are either too hot or too cold for water to exist as a liquid, and they have either no atmosphere at all or one so thick that the light of the sun cannot penetrate (Figure 2–1). Even on Mars and Venus, the planets most often touted as possible sites of life, conditions are so harsh that only the most primitive organisms can exist. But the earth is more than just a setting, an inert stage with just enough air and the right temperature for life to act out its drama. It is also a storehouse of all the materials on which life depends.

To understand the nature of life, we have to get a clear idea of the earth's structure and its contribution to life.

Taking the earth apart

For many centuries the insides of the earth were a great mystery. The few shallow mines in existence around the time of Christ and for centuries thereafter were mere scratches in the earth's surface that told man only

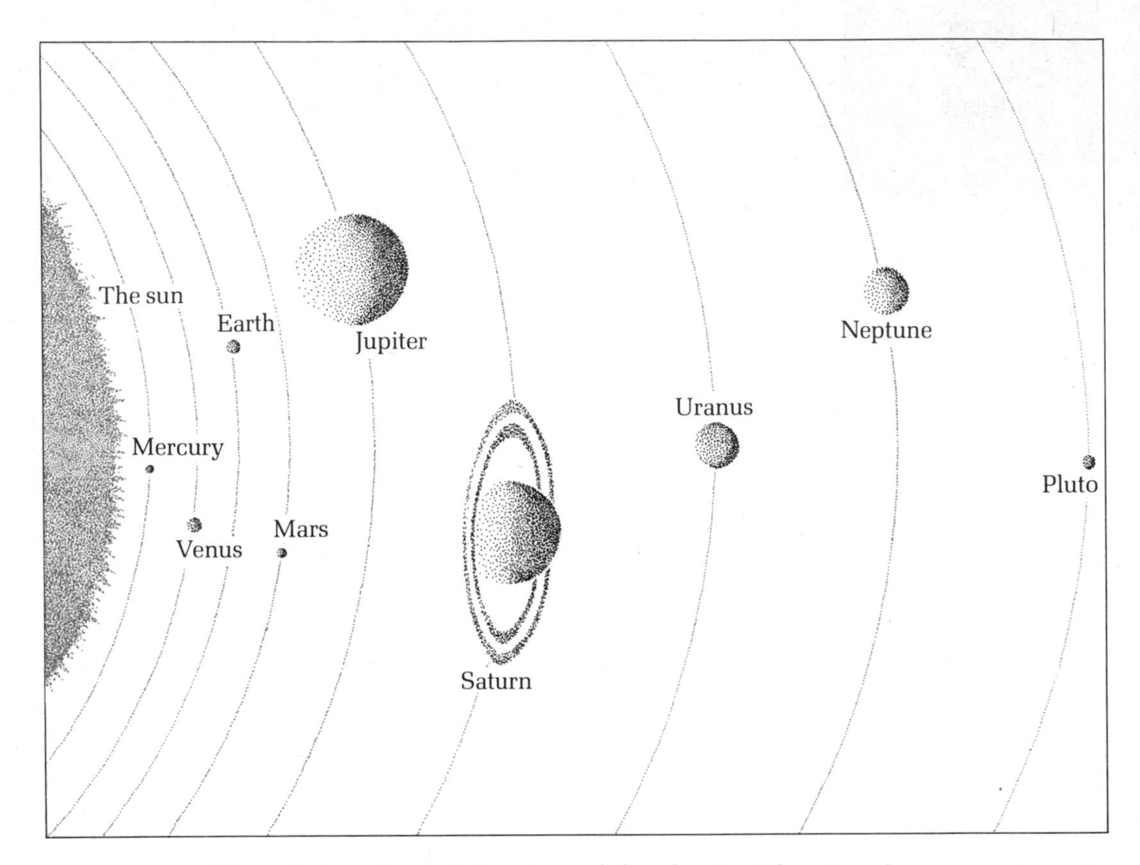

Figure 2–1 The relative sizes of the planets. (The nine planets are drawn to scale; distances from the sun are not drawn to scale.)

that there was quite a bit of rock down there. As mining technology improved, man was able to go deeper into the earth, but even today the very deepest shafts go only seven or eight miles into a planet that is nearly 8,000 miles in diameter. How do we know what is under our feet?

Most of our information about the earth's interior has come from the study of earthquakes. The violent slipping of rock surfaces that is an earthquake sends waves like sound waves through the earth. How these waves behave depends on the nature of the materials through which they pass. Perhaps you've noticed that a noise sounds one way in the air and quite different under water. An outboard motor sounds pretty much the way you would expect it to when the noise travels through the air, but if you have your head under water when a powerboat passes by, it sounds as if you're being buzzed by an angry bee. Likewise, an aluminum bar and an iron bar of the same length and diameter struck with the same force give off different tones. On the basis of this principle, the behavior of earthquake waves provides a clue to what the waves have passed through. With extremely sensitive instruments known as seismographs, scientists monitor the waves produced by earthquakes and use them to draw a picture of the earth's interior. After many earthquakes and many such pictures, we have a pretty good idea of how the inside of the planet looks.

The layers of the earth

Figure 2–2 shows a schematic representation of what you would get if you could accomplish the impossible task of cutting the earth in half. Like an onion, the earth is composed of layers laid one on top of the other, but there the similarity ends. While each of the layers of an onion is of the same material and of the same thickness as all the others, the earth's layers vary in composition and size.

The centermost layer, often referred to as the *core*, takes up only one-eighth of the total volume of the earth but accounts for nearly a third of its weight. The core's great weight is due to its two main ingredients, the heavy metals iron and nickel. Besides being very heavy, the core is

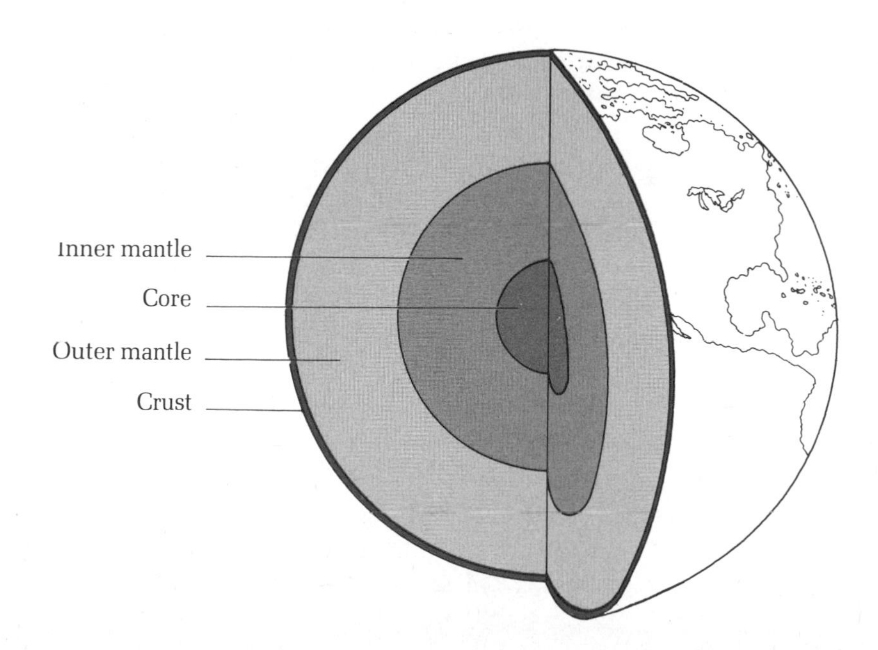

Figure 2–2 The internal structure of the earth. The crust, the only portion of the earth where we find life, is only a small fraction of the earth's diameter.

also very hot, about 5,000° Fahrenheit (F), well over the melting points of iron and nickel. However, we do not know for sure whether the core is liquid or solid. Although at sea level the core's temperature would certainly be hot enough to melt the iron and nickel and keep them molten, the outer layers of the earth exert great pressure on the core—much as deep water does on the hull of a submarine. Such pressure raises the metals' melting points, but no one knows just how far.

The layer beyond the core is known as the *mantle* and is the largest of the layers. It makes up 82 percent of the earth's volume and accounts for 68 percent of the weight. Like the core, the mantle is both very dense and very hot, although it is composed of materials that we think of as rock. The heat and pressure of the mantle have occasionally been the cause of great disasters. Sometimes parts of the mantle become so hot that they melt away the rock of the outermost layer of the earth, opening a fissure to the outside. Then the great pressures to which the mantle is subjected force molten material known as *magma* up and out. This is the way all volcanoes are created.

On top of the mantle sits the one layer we have been able to examine directly and not by educated seismographic guess: the *crust*. Of all three layers, the crust is the hands-down lightweight. It is merely 20 miles thick, and represents only 1.5 percent of the earth's volume and less than 1 percent of its weight. The crust is not perfectly smooth, but a little uneven. The high portions of the crust make up the great land masses we call continents, and in the hollows collect the waters of the oceans. These waters are as much a part of the crust as are the great layers of rock forming the foundations of the continents. In fact, water covers about three-quarters of the earth's surface, and over nine-tenths of that water is found in the oceans. All the earth's water taken together is sometimes given the name *hydrosphere*.

Although it is invisible, the gaseous envelope surrounding the earth is as much a part of the planet as are the various rocky layers. This envelope is known as the *atmosphere*, and it in turn is made up of a number of layers.

Where life is

Despite the bulk and size of the earth, life is limited to a thin theoretical region called the *biosphere* (Figure 2–3). It comprises the outermost layers of the crust, portions of the hydrosphere, and the lowest levels of the atmosphere. If the earth were an orange, life would exist only as a nearly invisible film on the skin of the fruit.

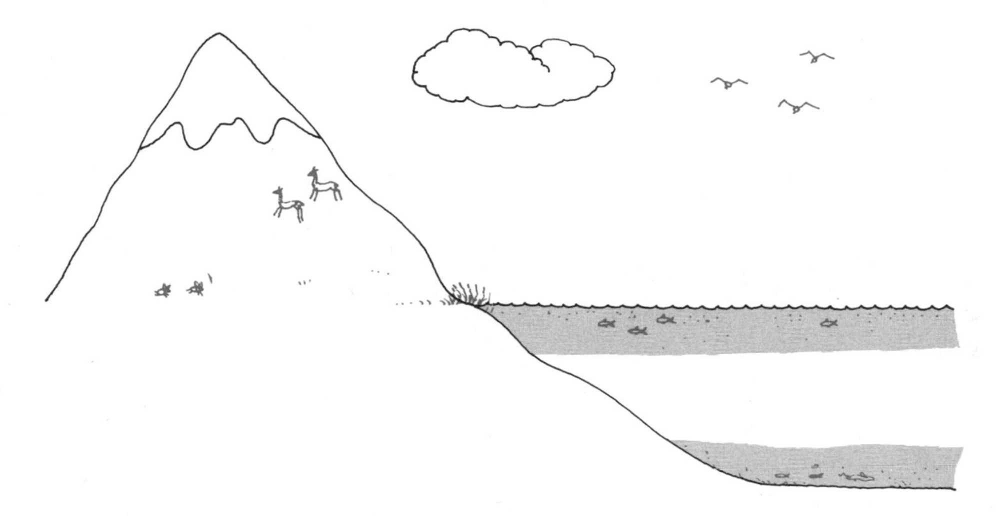

Figure 2–3 The range of the biosphere. Living things are found only within a relatively small portion of the earth's crust, the atmosphere, and the hydrosphere.

This statement gives the impression that life can survive in relatively few places. To a certain extent, that's true. There are definite physical limitations on life. All living things, both plant and animal, contain liquid water. Obviously, life cannot survive in those places where relatively high or low temperatures change all liquid water into steam or ice. Likewise, practically all living things need oxygen and can exist only where oxygen is available.

Within these physical limits, however, life has been able to make its home almost everywhere. Think of the wide variety of different environments included in the phrase "outermost layers of the crust, portions of the hydrosphere, and the lowest levels of the atmosphere." The biosphere includes everything from the dark, cold, and salty waters of the ocean at a depth of 1,500 feet to the warm and sunlit fresh water of a farm

If the earth were an orange, life would exist only as a nearly invisible film on the skin of the fruit.

pond; from the scorching desert of Death Valley to the frozen tundra of the Alaskan north slope; from the thick rain forests of Central Africa to the barren steppes of Central Asia. In fact, in the Canadian ice fields—as bleak and inhospitable a setting as can be imagined—the entire living population consists of small ice worms and the windblown pollen on which they feed. In every one of these places life is found.

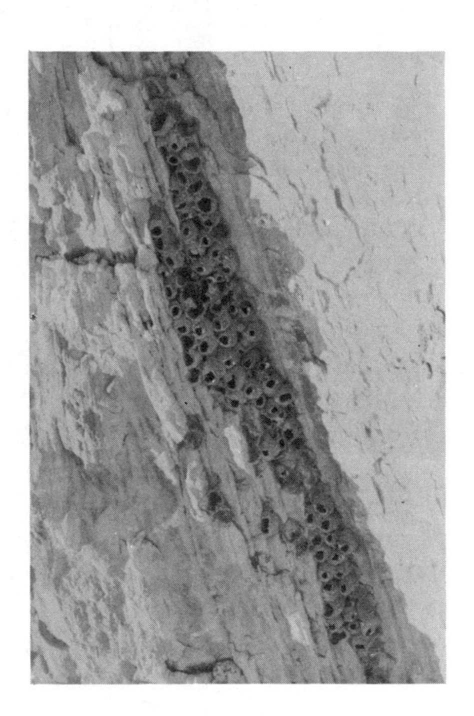

Figure 2–4 Living things have found ways of getting along in some extraordinary harsh environments. Figure 2–4a shows the nests of desert cliff swallows. The nests are built deep in the crevices to avoid the strong winds. Figure 2–4b shows alpine tundra in Roosevelt National Forest, Colorado. Although also a harsh environment, it is the home of many animals and plants, all of which have adapted to the special demands such an environment makes.

In terms of the whole solar system, living things occupy an infinitesimally small area. But measured in more human-sized terms, living things have been able to settle and to maintain themselves in an amazing variety of places.

Basic ideas about matter

But the earth is something more than simply a stage. Living things depend on the earth for the essential materials of life. All organisms are

Atoms are so small that 100 million of them stacked end to end would be a little less than an inch long.

composed of materials found in the biosphere, and they are dependent on these materials as nutrients. To understand how the living world makes use of the materials found in the nonliving environment, we have to first explore a few basic ideas about matter. And talking about matter means learning a few things about chemistry.

Chemistry is simply a language for talking about matter—how matter is constructed, how different kinds of matter combine with one another, and how they break apart. Only a little over a hundred years ago, chemists and biologists alike assumed that the processes of life, even the simplest ones, could not be duplicated in the laboratory and that the theories and laws of chemistry did not apply to living things. But we now know that chemistry's concepts of matter apply to the living as well as to the nonliving.

We can begin to analyze matter by disassembling it into its building blocks and then by seeing how these building blocks are cemented together.

The ABC's of matter

If you took any one of the many kinds of matter around you—solid or liquid or gas—and analyzed it for its component parts, you would find that it's composed of one or more *elements*. Water, for example, is a combination of the two elements hydrogen and oxygen, or, in chemist's shorthand, H and O. The "lead" in a pencil is a form of the element carbon, or C. Elements are simply the basic types, or kinds, of matter. The list of elements, known as the periodic table, gives 105 elements, but only 92 are found in nature. The other 13 were synthesized under special laboratory conditions (Figure 2–5).

The smallest unit of an element that still behaves like that element is an *atom*. Atoms are so small that 100 million of them stacked end to end would be a little less than an inch long. But the atom itself is composed of still smaller units. The *nucleus*, or center, contains two kinds of

Fundamental elements ▪ Secondary elements ▪

H 1																	He 2
Li 3	Be 4											B 5	C 6	N 7	O 8	F 9	Ne 10
Na 11	Mg 12											Al 13	Si 14	P 15	S 16	Cl 17	Ar 18
K 19	Ca 20	Sc 21	Ti 22	V 23	Cr 24	Mn 25	Fe 26	Co 27	Ni 28	Cu 29	Zn 30	Ga 31	Ge 32	As 33	Se 34	Br 35	Kr 36
Rb 37	Sr 38	Y 39	Zr 40	Nb 41	Mo 42	Tc 43	Ru 44	Rh 45	Pd 46	Ag 47	Cd 48	In 49	Sn 50	Sb 51	Te 52	I 53	Xe 54
Cs 55	Ba 56	La 57	Hf 72	Ta 73	W 74	Re 75	Os 76	Ir 77	Pt 78	Au 79	Hg 80	Tl 81	Pb 82	Bi 83	Po 84	At 85	Rn 86
Fr 87	Ra 88	Ac 89	Ku 104	Ha 105													

Ce 58	Pr 59	Nd 60	Pm 61	Sm 62	Eu 63	Gd 64	Tb 65	Dy 66	Ho 67	Er 68	Tm 69	Yb 70	Lu 71
Th 90	Pa 91	U 92	Np 93	Pu 94	Am 95	Cm 96	Bk 97	Cf 98	Es 99	Fm 100	Md 101	No 102	Lr 103

Figure 2–5 The periodic table of the elements. Among the 105 elements, only 19 are needed in any significant quantity by living systems.

particles, *protons* and *neutrons* (Figure 2–6). Protons carry a positive electrical charge, while neutrons, as their name implies, are electrically neutral. Because all the protons in a nucleus are positive, they repel each other, just as if they were the positive poles of two magnets. So-called *nuclear forces* overcome this repulsion and bind the nucleus into a stable whole. The positive charge of the protons attracts *electrons*, which are considerably smaller than protons and electrically negative. One electron

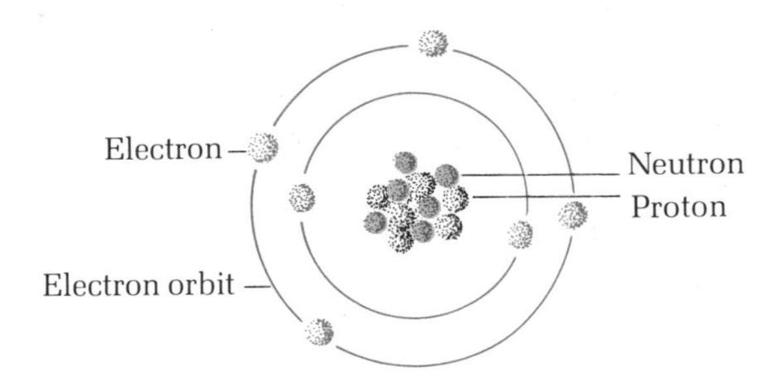

Figure 2–6 The carbon atom. Within the densely packed nucleus are the protons (shown in black) and the neutrons. Orbiting around the nucleus are the six electrons, arranged in shells.

is attracted for each proton, so that the numbers of both are equal. As a result, the atom is electrically neutral.

The reason that the various elements differ is that each contains a characteristic number of protons and electrons. (Neutrons are another matter altogether, which we need not be concerned with here.) Hydrogen (H), the very simplest element, contains only one proton and one electron; helium (He) contains two protons and two electrons; lithium (Li) has three of each. This addition of particles continues right on up to the 105th element, hahnium, characterized by 105 protons and 105 balancing electrons.

Because of the conventional ways that atoms are depicted in books, people often think of an atom as a tight bunch of little dots. In fact, most of the volume of an atom is nothing more than empty space. Take hydrogen, for example, with its single proton and single orbiting electron. The distance between these two particles is so great that if the proton were the size of a period on this page of print, the electron would be an invisible fleck of dust some 40 feet away! Even in the case of hahnium, the atom is still almost 100 percent open space.

The atom is mostly empty space. If a hydrogen nucleus were a period on this page, its lone orbiting electron would be forty feet away.

10ft.

Even with all that open space though, the electrons orbiting the nucleus follow definite paths, called *electron shells*. A precise number of electrons fits into each shell. The first may contain one or two electrons; every shell after that can have as many as eight (except for the third shell, which may be home for as many as eighteen electrons once the fourth shell gets its allotment of eight). Atoms can have as many as seven shells, but it takes a great many protons to attract enough electrons to fill them all. Most of the elements crucial to living things are simple, so only the first two or three shells are of importance.

As an example of how the shell distribution works, consider oxygen, with its eight protons and eight electrons (Figure 2–7a). There are two electrons in the first shell and six in the second. In the case of sodium (Na), which has eleven protons and eleven electrons, the first shell has two electrons and the second shell a full complement of eight, resulting in one electron left over for the third shell (Figure 2–7b).

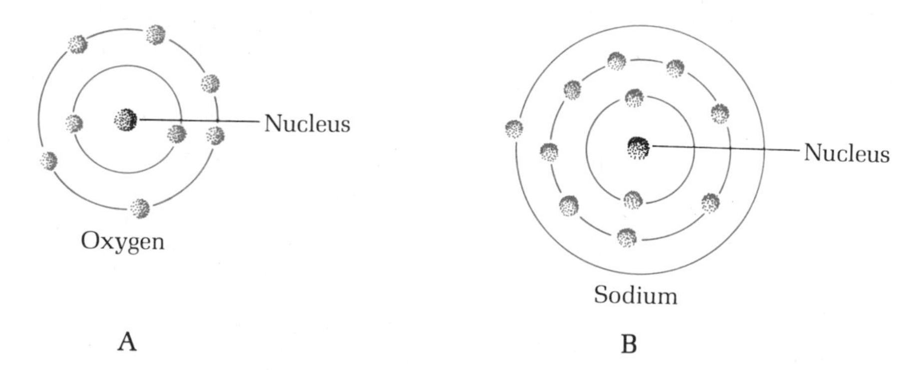

Figure 2–7 The electron shells of oxygen (A) and sodium (B).

Putting atoms together

Single, isolated atoms are generally not found in nature. Instead, atoms may be held to one another by forces known as *chemical bonds*, and a group of atoms so bound to each other is called a *molecule*. The atoms may be of the same element, such as the oxygen molecule, written O_2, where the subscript indicates the number of atoms involved. When the molecule is composed of different elements, the substance is known as a *compound*. Water molecules are composed of two atoms of hydrogen and one of oxygen: H_2O. Thus, water is a compound.

Why do atoms stick together to form molecules? The answer lies in the outermost electron shell. For reasons as yet unknown, all atoms share a tendency to stabilize, or complete, that outer shell. This can be accomplished by either gaining or losing electrons. Whether a particular atom tends to be reduced or oxidized depends on the number of electrons in the outer shell. For example, when an atom bonds with oxygen, in the process known as oxidation, it tends to lose electrons to the oxygen atom. The opposite process is known as reduction, which is equivalent to the gaining of electrons, or bonding with hydrogen.

Note that after either oxidation or reduction, the atom is no longer electrically neutral. The atom has become charged, and it is called an *ion*. A gain of electrons results in a negative ion, while a loss produces a positive one. An ion is designated by a + or a − sign depending on the net gain or loss of charge; a number is added if the gain or loss involves more than one electron. Thus, after sodium (Na) loses its one outermost electron, it is Na^+. Chlorine (Cl) adds one electron to its outer shell of seven, becoming Cl^-. Beryllium (Be) loses two outer electrons, becoming Be^{2+}, or Be^{++}.

If an atom likely to give up electrons—that is, likely to be oxidized—comes into close contact with an atom likely to gain electrons—that is, likely to be reduced—an electron or electrons will be passed from one atom to the other. If Na comes close to Cl, Cl takes one electron from Na:

$$Na \rightarrow Na^+$$
$$Cl \rightarrow Cl^-$$

The arrow indicates the change, or *reaction*. Since the two resulting ions have opposite charges, they attract each other strongly and form a compound:

$$Na^+ + Cl^- \rightarrow NaCl$$

Such a bond between ions is known as an *ionic bond*. NaCl is sodium chloride, or common table salt.

Atoms can also complete their outermost electron shells by sharing electrons with other atoms, so that the electron does double or triple or quadruple duty in the outermost shells of two or more atoms. Shared electrons provide a binding force known as a *covalent bond*. Covalent bonds are particularly important for biological molecules.

Water is a good example of a covalently bonded compound. Since the hydrogen atom contains only one electron, it needs an additional

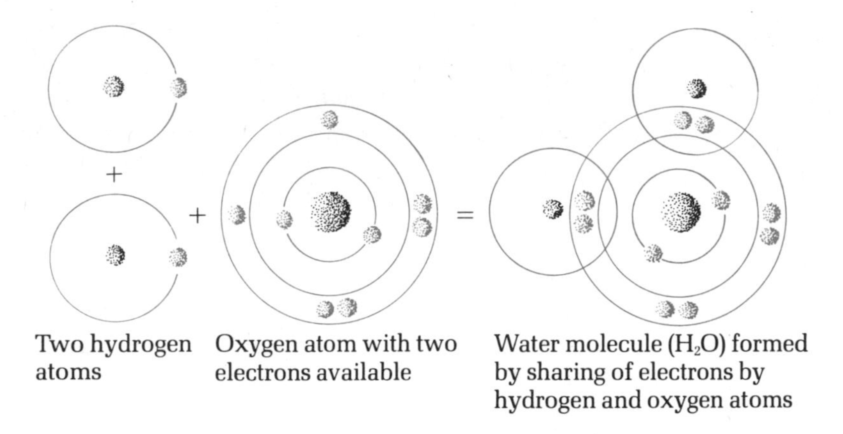

Figure 2–8 The formation of a water molecule.

electron to complete its shell. The oxygen atom, with its eight protons and eight electrons, has two electrons in its first shell and six in the second. The second shell is shy two electrons of the eight needed to complete it. If two hydrogen atoms share their electrons with one oxygen atom, then the electron requirements of all three atoms are satisfied, and a molecule of water results (Figure 2–8). The equation for this reaction is

$$2H + O \rightarrow H_2O$$

Because the force an atom exerts on electrons varies from one element to another, the electrons of a covalently bonded compound will bunch toward the more attractive element. This shift gives the atom with the greater electron affinity a small negative charge and the atom with the lesser affinity a small positive charge. Even though the molecule is electrically neutral as a whole, it has negative and positive poles, much like a magnet, and is therefore called a *polar molecule*. Water is a good example. Oxygen has a much greater affinity for electrons than does hydrogen. The water molecule is polar, with two negative and two positive "corners" (Figure 2–9).

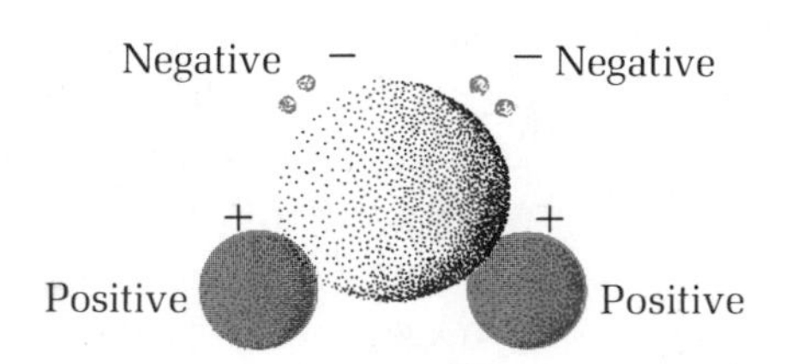

Figure 2–9 The polar water molecule.

By no means are all covalently bonded molecules polar. In the cases of O_2 and H_2, each of the component atoms has the same affinity, and the whole molecule is nonpolar. Compounds of carbon and hydrogen, the hydrocarbons like coal, gasoline, and natural gas, are also nonpolar.

Both ionic and covalent bonds can vary greatly in strength, depending on the characteristics of the atoms involved. Some bonds are so weak and so unstable that they last only for the tiniest units of time, while others can last for a seeming eternity.

Putting molecules together

Molecules stick to each other, just as atoms do, because of a number of special forces. In the case of polar molecules like water, the pull of opposites will bring the negative and positive poles of molecules together. Each water molecule can form such a bond, often called a *hydrogen bond*, with four other water molecules (Figure 2–10).

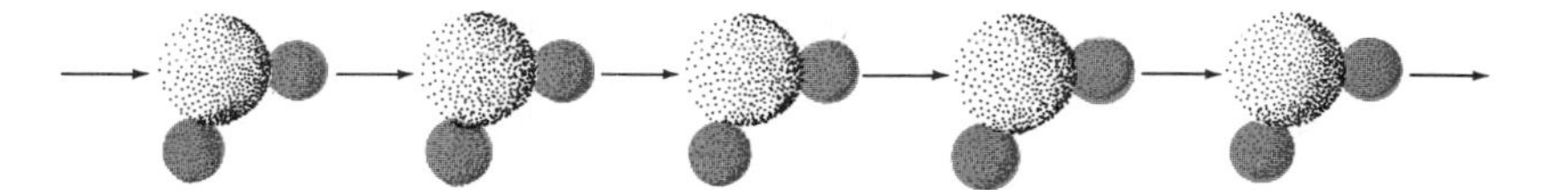

Figure 2–10 Hydrogen bonding in water molecules. The positive end of one water molecule is attracted to the negative end of another, forming a "linkage" between them.

Nonpolar molecules like those in petroleum also adhere to each other, but much more weakly than do polar molecules. As the electrons move in their shells, simultaneous electron displacement in opposite directions in two molecules creates a short-lived polarity. The oppositely charged parts of the molecules are then attracted to each other. These forces are only about one-tenth as strong as hydrogen bonds but, in the case of large molecules, they provide sufficient holding force to keep the molecules together.

Since we now have a basic idea of the vocabulary used to talk about matter and of some of the concepts used to describe its behavior, we can take the earth and the biosphere apart and see what they're made of.

The earth

The elements contained in our bodies are exactly the same as those contained in the nonliving sections of the earth, but the proportions are

The elements contained in our bodies are exactly the same as those contained in the nonliving sections of the earth, but the proportions are different.

different (Figure 2–11). As we said when describing the earth's layers, the core and the mantle are made mostly of iron and nickel. Of course, since these portions of the earth are not part of the biosphere, the elements contained there are of relatively little significance to living things. The crust, obviously, is of extreme importance. Table 2–1 gives the relative composition of those parts of the earth accessible to man.

A glance at the table reveals several intriguing contrasts. The crust is much richer in carbon, calcium, potassium, silicon, magnesium, aluminum, sodium, and iron than either the hydrosphere or the atmosphere. Curiously, all these elements, except silicon and aluminum, are very important to living systems.

Table 2–1 Relative composition, in percentages, of the earth's surface

Element and symbol	*Crust*	*Hydrosphere*	*Atmosphere*
Oxygen (O)	60.4	33	21
Hydrogen (H)	2.92	66.4	trace
Carbon (C)	0.16	0.0014	0.03
Nitrogen (N)	trace	trace	78.3
Calcium (Ca)	1.88	0.006	trace
Potassium (K)	1.37	0.006	trace
Silicon (Si)	20.5	trace	trace
Magnesium (Mg)	1.77	0.034	trace
Phosphorus (P)	0.08	trace	trace
Sulfur (S)	0.04	0.017	trace
Aluminum (Al)	6.2	trace	trace
Sodium (Na)	2.49	0.28	trace
Iron (Fe)	1.90	trace	trace
Chlorine (Cl)	trace	0.33	trace
Boron (B)	trace	0.0002	trace

Adapted from E. S. Deevey, Jr., "Mineral Cycles," in *The Biosphere*, San Francisco: Freeman, 1970, p. 85.

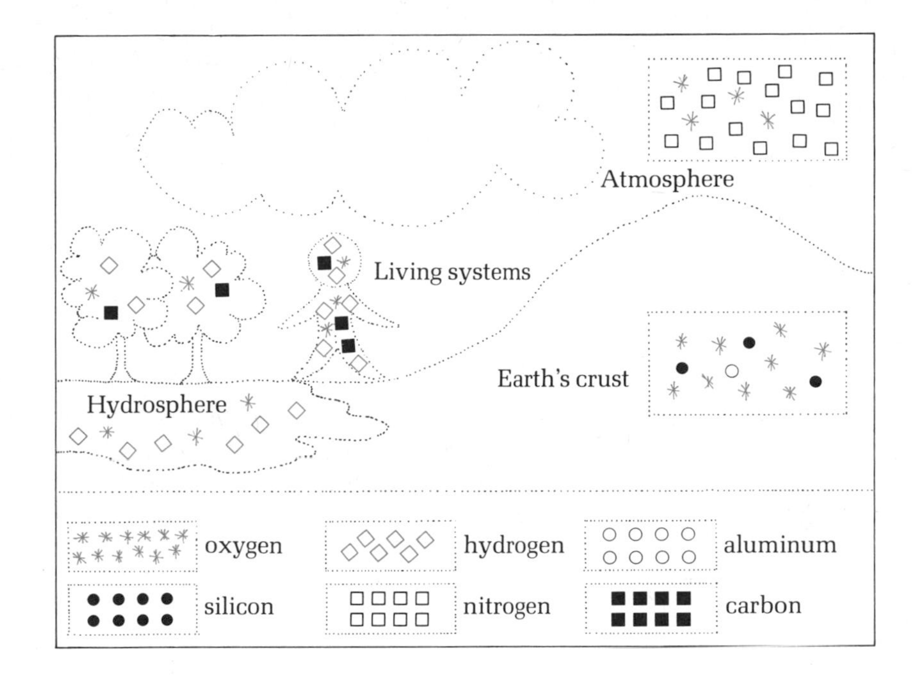

Figure 2–11 The gross chemical composition of the living and nonliving worlds. Just six chemical elements compose the greatest proportion of the hydrosphere, the atmosphere, the biosphere, and the earth's crust. Note the similarity between the chemical proportions of the hydrosphere and living systems.

Since the hydrosphere is more than 99 percent water, it's easy to understand why hydrogen and oxygen are so abundant. Besides serving as a welcome thirst quencher on a hot day in summer, water is the major source of hydrogen for living organisms. The saltiness of the oceans accounts for the sodium and chlorine content of the hydrosphere. Plants and animals both need chlorine. Sodium is necessary to the animal world, but some plants can get along without it.

In comparison to the crust and the hydrosphere, the atmosphere is a simple collection of elements. Nitrogen and oxygen predominate, and both are essential to plant and animal life. Carbon's relative scarcity (it exists in the atmosphere mostly in the form of carbon dioxide, CO_2) belies its importance to life, an importance we will explore soon.

This table includes only the naturally occurring elements; it fails to list the many pollutants man has introduced to the soil, water, and air. In fact, man's effect on the balance of elements has affected their relative proportions. For example, the CO_2 content of the air has been rising for the past 80 years. Man burns great amounts of fuel to cook food, light cities, and power automobiles, and CO_2 is produced by burning. Soon we will see the tremendous biological importance of carbon and learn why seemingly small changes may have large-scale effects.

The biosphere

Table 2–2 lists those chemical elements that make up living organisms. As you can see, living organisms resemble the hydrosphere much more than either the crust or the atmosphere. Does this fact suggest anything to you about the nature and origin of living things? In Chapter 4, we'll look at a likely answer to this question.

Table 2–2 The composition of living organisms

Element	*Percent of the total*
Hydrogen (H)	49.8
Oxygen (O)	24.9
Carbon (C)	24.9
Nitrogen (N)	0.27
Calcium (Ca)	0.073
Potassium (K)	0.046
Silicon (Si)	0.033
Magnesium (Mg)	0.031
Phosphorus (P)	0.030
Sulfur (S)	0.017
Aluminum (Al)	0.016
Sodium (Na)	trace
Iron (Fe)	trace
Chlorine (Cl)	trace

Adapted from E. S. Deevey, Jr., "Mineral Cycles," in *The Biosphere*, San Francisco: Freeman, 1970, p. 85.

This table of the elements of living organisms is not just a simple list of chemicals. It's also much like a parts list for our own bodies. Because elements, by definition, are individual and quite different from each other, they are suited for certain very specific and important jobs within an organism. The first three elements in the table—carbon, hydrogen, and oxygen—comprise about 96 percent of the total weight of the human body. The body's high water content, about 70 percent, accounts for much of the hydrogen and oxygen. Water is perhaps the most important compound for keeping living systems living; it makes up a large proportion of the structure of most organisms, and it is the basic means for transporting necessary chemicals from one part of an organism to another. Carbon is also a very versatile element. Because of its unique properties, carbon is the essential building block in nearly all the molecules of living systems, ranging from very simple compounds like sugar to complex compounds formed of tens of thousands of atoms, such as the protein molecules of muscle tissue.

By comparison with the first four elements, such relative lightweights as sulfur and magnesium seem hardly worth bothering about, but appearances are often deceiving. Sulfur serves as one of the building blocks in those complicated protein molecules. And magnesium is the kingpin of the chlorophyll molecule, the pigment that gives plants their green hue. Chlorophyll is much more than another pretty color—without it, life as we know it would not exist. We'll see why when we talk more about carbon. All of these "lesser" elements, with the exceptions of silicon, sodium, aluminum, and boron, are essential to all forms of life, and the exceptions are essential to certain specific organisms. For example, silicon is an important structural element in some of the tiny, one-celled aquatic plants known as phytoplankton and in some sponges.

The essential essentials

All the elements listed in Table 2–2 are essential to life, but some are "more essential" than others, at least in terms of amount. These particular elements are given the name *macronutrients*. The prefix *macro-*, derived from the Greek word for large, indicates that living things require large amounts of these elements. To remember the macronutrients, imagine the world as a kind of restaurant, and the restaurant's spot advertisement as a code for the menu. The ad is: "C. HOPKiNS CaFe, Mighty good." The letters of the ad are the chemical symbols of the macronutrients: carbon (C), hydrogen (H), oxygen (O), phosphorus (P), potassium (K), nitrogen (N), sulfur (S), calcium (Ca), iron (Fe), and magnesium (Mg, from Mighty good).

But the macronutrients aren't the whole story. A number of other elements essential for plants or animals or both are required in smaller amounts, sometimes as small as a few atoms a year. These elements are called *micronutrients*, or *trace elements*. Although these nutrients are needed in only small quantities, they are no less crucial to life than the macronutrients. The micronutrients are often building blocks in biological molecules that are important for a chemical process necessary to the organism's survival. Table 2–3 lists the micronutrients.

Table 2–3 Micronutrients required by plants and animals

	Required by			
Element and symbol	*all plants*	*some plants*	*all animals*	*some animals*
Manganese (Mn)	X		X	
Zinc (Zn)	X		X	
Copper (Cu)	X		X	
Chlorine (Cl)	X		X	
Molybdenum (Mo)		X	X	
Cobalt (Co)		X	X	
Sodium (Na)		X	X	
Vanadium (V)		X		X
Silicon (Si)		X		X
Boron (B)	X			
Iodine (I)				X

Note: Sodium and chlorine are macronutrients for animals and micronutrients for plants.

A missing micronutrient can have about the same effect as an absent production worker on an automobile assembly line. As long as each man does his job, the cars come off the line properly assembled. But if one worker, say the man responsible for putting on the distributor cap or for bolting down the right rear fender, walks out and isn't replaced, then the finished cars either won't start or their right rear fenders will fall off when they do.

A missing micronutrient can have about the same effect as an absent production worker on an automobile assembly line.

The turning wheel of matter

After looking at these tables of elements, one could get a very static idea of the chemical composition of living things and their environment. In fact, change is the major happening. Elements are constantly forming bonds to make compounds, even as other compounds are broken down into their component elements. The very same thing—building and breaking down—is happening to your own body at this very moment. Saying that you are composed of certain kinds of elements assembled into complex molecules is much like contemplating only one frame of a very long motion picture. You may look pretty much the same as you did five years ago, but the particular atoms and molecules you have now aren't the same ones you had back then. There is a constant turnover of molecules, much like the turnover of merchandise in a department store. Old molecules change and are discarded; new ones are built to take their place. This constant change is the reason that all organisms need a steady supply of nutrient molecules.

However, there's one other point to keep in mind: regardless of how many compounds are being built up or broken down or how fast the process occurs, the total amount of matter remains the same. This principle, called the *conservation of matter*, means that matter isn't created and that it isn't used up. It may go through many combinations, breakdowns, and recombinations, but the total amount of matter remains the same. If you carefully measured and weighed every nutrient entering a plant and performed the same measurements on all wastes leaving it and on any growth in the plant, and then added both groups of figures, the totals would come out equal, just like a bookkeeper's balance of credits and debits. The earth's supply of matter is constant. It is almost as if the earth were a storeroom filled with a certain amount of merchandise that is neither added to nor used up but simply passed around.

The earth's supply of matter is constant. It is almost as if the earth were a storeroom filled with a certain amount of merchandise that is neither added to nor used up but simply passed around.

Living organisms have a never-ending need for nutrients, but at the same time the earth's supply of nutrients is limited and constant. How can these two seemingly opposite forces be reconciled?

Now, consider the two facts just pointed out. Living organisms have a never-ending need for nutrients, but, at the same time, the earth's supply of nutrients is limited and constant. On the one hand, there is a constant need for matter in the variety of chemical forms represented by the macro- and micronutrients, and, on the other, the supply of these nutrients is fixed. How can these two seemingly opposite forces—limitless consumption and limited supply—be reconciled? If all life needs calcium, and if there is a limited supply of calcium, won't the calcium eventually run out and all life pass away from calcium starvation?

The answer to this apparent dilemma is that the various nutrients are recycled like bottles and tin cans (Figure 2–12). All the elements essential for life pass from the crust, the hydrosphere, and the atmosphere into the living matter of the biosphere. The living organisms then use, but do not destroy, these elements. The nutrients are returned to the nonliving world through various waste products and finally by the decay of the organisms themselves into their component elements.

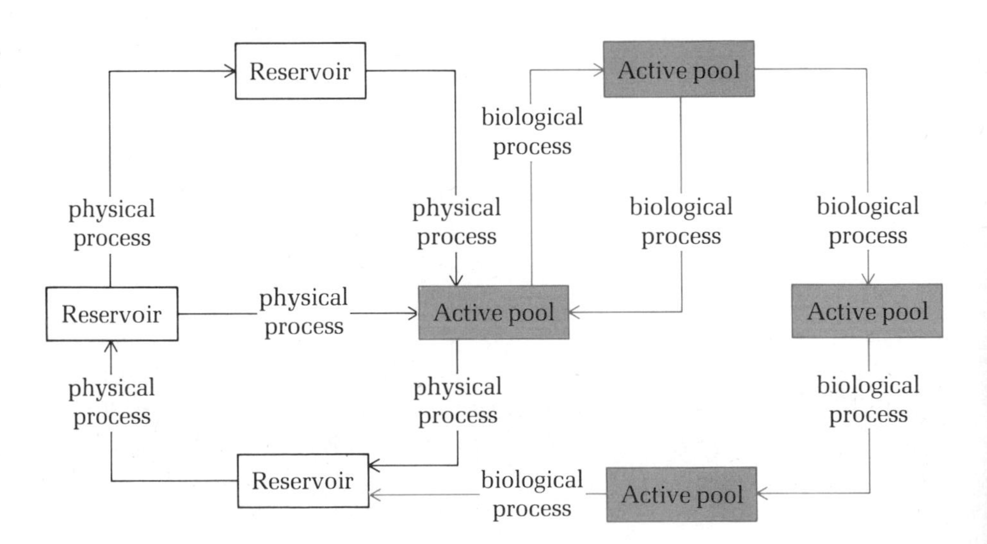

Figure 2–12 A generalized nutrient cycle.

In the next chapter, we'll trace these cycles in detail and see how they add to our understanding of living things.

Summary

The nature of living things depends absolutely on the nonliving world.

The earth itself is composed of layers, the centermost portion of which is the core, a very heavy and very hot ball of iron and nickel. Outside the core is the mantle, a solid layer of rock, which is lighter and cooler than the core but still quite hot and heavy. The thinnest and outermost layer, the crust, sits atop the mantle. About three-quarters of the crust is covered with water, and this area is called the hydrosphere. The atmosphere is the gaseous envelope surrounding the earth. Those portions of the crust, hydrosphere, and atmosphere that can support life are called the biosphere.

All matter is composed of combinations or mixtures of elements, which are the basic varieties of matter. The unit of all elements is the atom. The nucleus of the atom comprises protons, which have a positive charge, and neutrons, which carry no charge. Negatively charged electrons orbit the nucleus in distinct electron shells. Elements differ from one another in the number of protons and electrons.

Two or more atoms can bind together to form a molecule. Certain kinds of atoms complete their outermost shells by gaining electrons. This process of reduction results in a negative ion. Other kinds of atoms tend to oxidize, or lose electrons, and become positive ions. Two ions of opposite charge can be held together by an ionic bond. Other kinds of atoms complete their outer electron shells by sharing electrons with other atoms in what is known as a covalent bond. Some covalently bonded molecules have regions of local positive or negative charge that make the molecule polar. Whole molecules are held together by intermolecular forces.

The chemical composition of nonliving matter differs considerably from that of living organisms. Living things need some of these elements, known as macronutrients, in relatively large quantities. Micronutrients are required in trace amounts, and vary greatly among the various kinds of organisms. Both macro- and micronutrients move in cycles from the nonliving world through the living world and back to the nonliving world again, a reflection of the law of conservation of matter.

Questions to consider

1. a. What are the layers of the earth?
   b. How can each one be described?
2. a. What are the component parts of the atom?
   b. How are they assembled?
3. a. What happens when an atom is oxidized?
   b. What happens when it is reduced?
   c. The ion resulting from oxidation or reduction is electrically charged. Why is this so?
4. a. What is an ionic bond?
   b. What happens when an atom of Na comes into close contact with an atom of Cl?
5. a. What is a covalent bond?
   b. How does a covalent bond differ from an ionic bond?
6. What is a polar molecule?
7. a. Why do water molecules adhere to each other?
   b. What name is given to the bonds between these molecules?
8. a. How do macronutrients differ from micronutrients?
   b. Which group is more important to the survival of an organism?
9. What is the law of conservation of matter?
10. How are the limited resources of the earth available to supply the constant needs of living systems? Doesn't this seem like a violation of the law of conservation of matter?

Glossary

atmosphere the layer of gases surrounding the earth

atom the extremely small units of which all matter is composed *biosphere* the regions of the earth's surface, the atmosphere, and the hydrosphere in which life is found

chemical bond the force holding atoms together to form a molecule

compound a molecule formed of atoms of two or more elements

conservation of matter the principle that the amount of matter in the universe is fixed and that it cannot be destroyed

core the innermost layer of the earth

covalent bond a bond between two atoms in which electrons are shared rather than transferred

crust the very thin, exterior layer of the earth

electron the negatively charged particle that "orbits" the nucleus of the atom

electron shell one of several definite paths that electrons follow in their "orbit" of the atomic nucleus

element one of the 105 distinct types of matter of which the universe is composed

hydrogen bond a force between molecules in which a hydrogen atom in one molecule is attracted to a fluorine, oxygen, or nitrogen atom in another molecule

hydrosphere the region of the crust and atmosphere in which salt or fresh water is found

ion an atom which has gained or lost electrons, becoming electrically charged

ionic bond a bond between atoms in which electrons are transferred from one atom to another

macronutrient element required by organisms in relatively large amounts

mantle the layer of the earth between the core and the crust

micronutrient element required by organisms in relatively small amounts, also called *trace elements*

molecule a unique cluster of atoms joined by chemical bonds

neutron an uncharged particle, one of the two types of particles found in the atomic nucleus *nuclear forces* the forces which keep the particles of the atomic nucleus together

nucleus (pl., *nuclei*) the central, positively charged portion of the atom, composed of protons and neutrons

oxidation a chemical reaction in which an atom loses electrons from its outer shell, the opposite of *reduction*

polar molecule a molecule that has distinctly positive and negative regions

proton a positively charged particle found in the nuclei of atoms

reaction a process in which atoms or molecules are changed or rearranged to form new atoms or molecules

reduction a chemical reaction in which an atom gains electrons in its outer shell, the opposite of *oxidation*

trace element element required by organisms in relatively small amounts

3 Cycles

At the conclusion of Chapter 2 we saw that the *nutrient cycles* are the product of all organisms' constant need for nutrients and of the conservation of matter. But why are the nutrient cycles of interest as anything other than evidence for the conservation of matter? For the most part, these cycles exist only because life does. There would be no carbon cycle, for example, if there were no life. In examining the nutrient cycles, we are tracing out one of the most basic processes of life: how organisms obtain the matter they are made of.

There are as many nutrient cycles as there are nutrients. We'll limit our study to the four principal cyclic movements of matter: water, carbon, nitrogen, and oxygen. We'll also consider how man has tampered with these cycles and thus affected life. In the concluding section we'll see how the concepts of ecology add to our understanding of the interactions of organisms with each other and with the environment.

The earth's water wheel

Although water is a compound and not an element, it is numbered among the element cycles because of its significance to life. All the oxygen in the atmosphere comes ultimately from water; likewise, water is the source of nearly all the hydrogen found in living organisms. Water serves as an important transport medium for nutrients and other specialized molecules in both plants and animals. It also makes up a large percentage — 50 to 90 percent — of the bulk of all functioning living organisms. Without water, there would be no life.

The uniqueness of water

Why is life very literally based on water? The answer lies in water's unique properties, particularly its hydrogen bonds and its polarity.

All liquids tend to evaporate; that is, the bonds between the molecules break, and the individual molecules escape into the air. Breaking the bonds requires the energy of heat, and the stronger the bond, the more heat it takes to break it. Because water is held together by strong hydrogen bonds, it takes a great amount of heat to vaporize it, and water has a high boiling temperature. Even though evaporation will occur as long as the water is in contact with the air, water evaporates quite slowly, much more slowly than, say, gasoline or alcohol. Because of these characteristics of the hydrogen bond, organisms whose cells contain large amounts of water can withstand a wide variation of temperature without boiling away or evaporating rapidly.

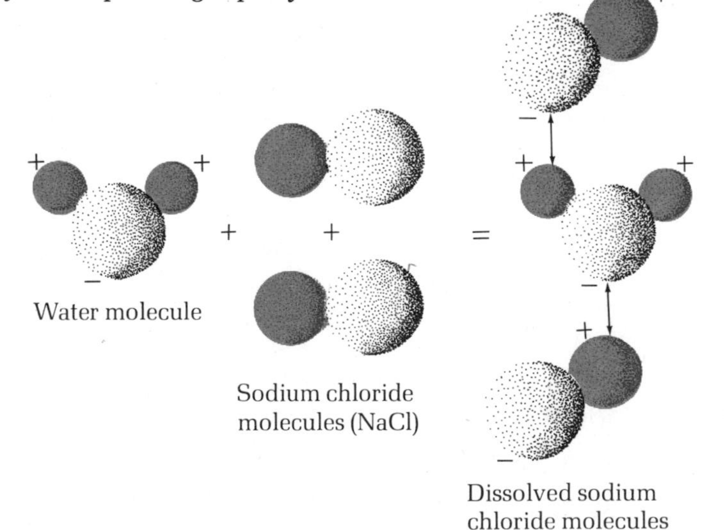

Figure 3–1 Solution of sodium chloride in water. The polar water molecule attracts and aligns the positive and negative ends of the sodium chloride molecules. The sodium chloride molecules will be pulled apart into their component sodium and chloride ions, thus dissolving the salt.

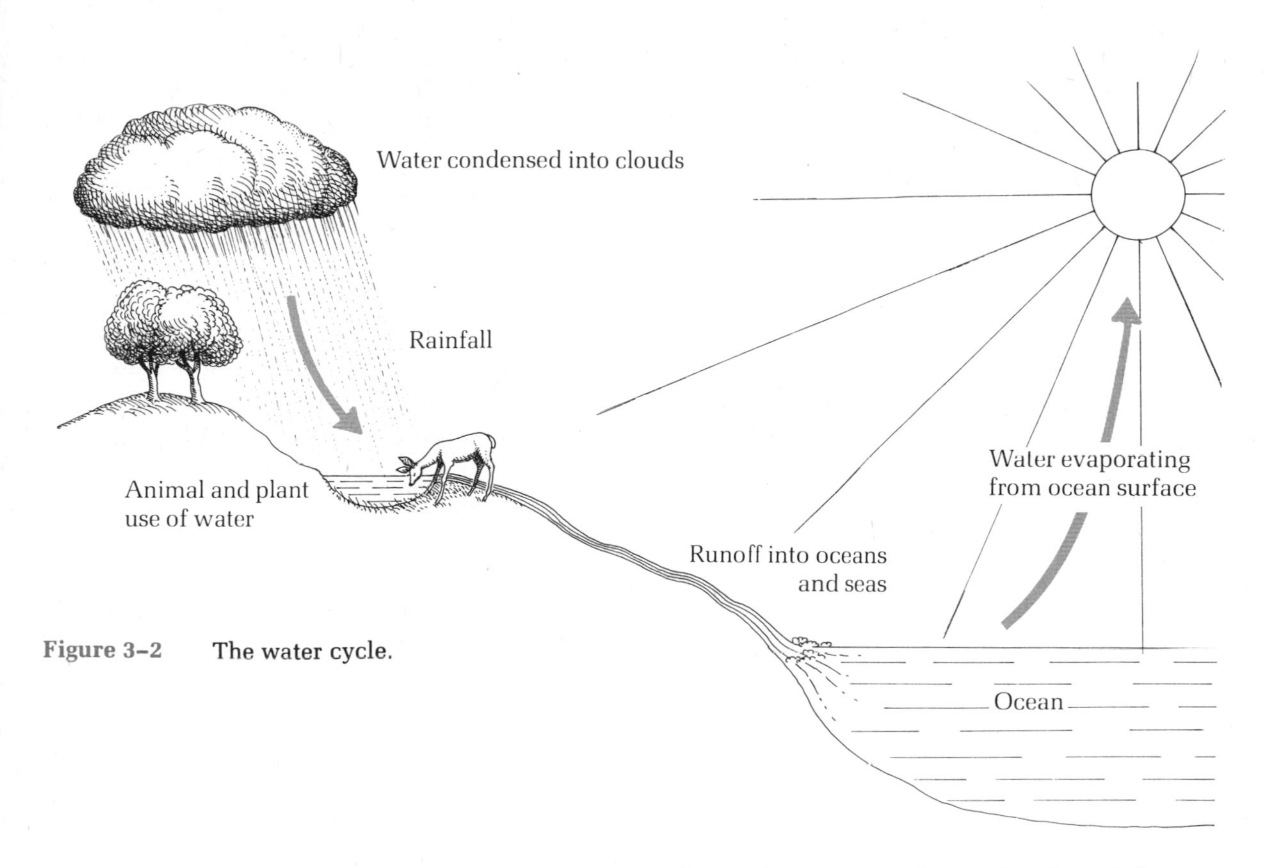

Figure 3–2 The water cycle.

The free ions then orient themselves to the charged poles of the water molecule. The salt dissolves (Figure 3–1). Water's ability to dissolve and hold compounds makes it more suited than any other liquid for carrying nutrients from one part of an organism to another.

The stages of the water cycle

Figure 3–2 provides a diagrammatic scheme of the water cycle. Let's take a closer look at each of the steps.

$$NaCl \rightarrow Na^{+} + Cl^{-}$$

Perhaps even more important than water's hydrogen bonds is its polarity. If an ionic compound is introduced to water, the compound will break apart into its component, charged ions. In the case of NaCl,

The oceans contain over 90 percent of the earth's water, and act as the reservoir from which the water cycle begins and the receptacle in which it ends. For the past several thousand years, the level of the oceans has been the same even though the many mighty rivers of the world, like the Mississippi, the Nile, and the Congo, constantly pour billions and billions of gallons of water into them each year. Obviously, the oceans must be losing as large a volume of water as they gain, or the world's coastal cities would be flooded.

The loss comes from evaporation caused by the heat of sunlight beating on the surface of the oceans. Water vapor is invisible in warm air, but it can be seen in cooler air. Since warm air is less dense than cool air, it rises above the cool air, gradually cooling as it gains altitude, and the previously invisible water vapor becomes visible as clouds.

Once the water vapor cools enough, evaporation is reversed: The gaseous water becomes a liquid again. In the cooling cloud, liquefying water molecules adhere to small bits of dust or debris until a mass of water too heavy to be suspended in the air is formed. Then the water falls, becoming a gentle spring shower or a torrential monsoon downpour.

When the rain stops, the heat of the sunlight evaporates some of the fallen water on the ground and returns it to the atmosphere. Most of the water, though, begins the trip downhill back to the ocean. Part of the water runs off almost immediately into streams and rivers and is back to the sea in no time. In those places where the soil is porous enough for water to penetrate, the rain seeps into rock layers and returns to the ocean underground. Such an underground trip may take a very long time, as long as several centuries, but the water finally gets back to the ocean, only to be evaporated again in its turn and make another trip through the water cycle.

Water that passes through soil and rocks on its way to the sea dissolves and carries with it many ionic compounds. Among these ionic compounds are salts. Salts dissolved from the soil and rock and carried to the sea by returning water over the course of billions of years are what make seawater salty.

Living things and the water cycle

The volume of water used by living organisms pales before the immense amount of water that passes through the water cycle in any given

Figure 3-3 All living things need a way of finding water and retaining it, regardless of their environments. The iguana's tough skin prevents excessive evaporation of body fluids in his warm environment. The cactus is a succulent plant; many of its tissues retain available water for storage. The snook, a fish that can live in both fresh and salt water, has a complex internal system that regulates its internal salt concentration and prevents both flooding and drying of internal tissues.

period of time, but that water is absolutely crucial to the maintenance of life.

All living things have some way of finding water and taking it in (Figure 3-3). The methods vary considerably. Obviously a tiny bacterium living in a farm pond has much less of a problem with its water supply than does a kangaroo rat hopping about on the hot sands of the Mojave

Obviously a tiny bacterium living in a farm pond has much less of a problem with its water supply than does a kangaroo rat hopping about on the hot sands of the Mojave desert.

Desert. But once they take the water in, both organisms put it to roughly the same uses. Most of the water is used to transport materials through the body of the organism; some of it also goes to replace lost water. In both of these uses, the water remains unchanged. However, some 2 percent of the water taken up by plants is broken into the component hydrogen and oxygen atoms. For a while, the water cycle becomes the hydrogen and oxygen cycles, which we will examine soon. In the end, though, the hydrogen and most of the oxygen are recombined to form water molecules that go back into the earth's water cycle.

What man does with water

The water cycle is a precise and balanced thing, something like a finely tuned automobile engine. Changes of even a very small sort can have far-reaching effects. Much of man's technology has affected the delicate water cycle. With any finely tuned and balanced thing, change is rarely beneficial.

Water now carries with it large amounts of materials it never carried before the advent of an industrial civilization. Some of these materials, such as human and industrial wastes, are dumped into the water on purpose. Others end up there because of such accidents as oil spills. Some, like the insecticide DDT and nitrogen fertilizers, get into water quite unintentionally, added to soil and crop plants and then slowly carried away by rainwater running off the land. Originally such pollution mainly affected the rivers, streams, lakes, and ocean coasts near major cities, but over the years the amount of waste has accumulated to such an amount that even the oceans themselves have been affected. Thor Heyerdahl, who in 1970 crossed the Atlantic in a reed boat christened the *Ra II*, found the central Atlantic, far away from the shipping lanes, so laden with filth that he thought he was traveling on the heels of a tramp steamer.

Besides adding pollutants to the water, man also affects the physical or structural aspects of the water cycle—such things as the temperature and the relative amounts of fresh and salt water. Because standing water absorbs heat energy more rapidly than does moving water, damming a rushing river like the Snake River in Idaho to form a lake has the effect of heating the water by a few degrees. This may sound like a small change, but it can have such drastic effects on life that it is called *thermal pollu-*

tion. For example, trout and salmon cannot live in water hotter than 55°F for an extended period of time. The Snake is a migration route for many schools of steelhead trout and salmon, which must pass through regions of lethally warm water trapped by the dams on their journey upstream. As a result, many of the fish and their offspring die (Figure 3–4).

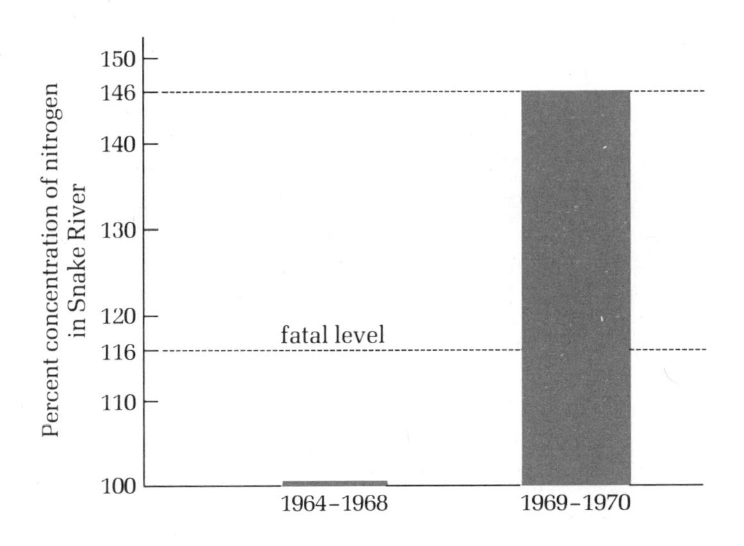

Figure 3–4 The nitrogen threat in the Snake River. Following the damming of the river, the amount of dissolved nitrogen in the waters of the Snake increased substantially. Within five years, it had reached a level equivalent to 146 percent of the earlier base level. However, at a level equivalent to only 116 percent of the base level, many fish died of nitrogen in their bloodstreams, analogous to the human condition known as "the bends." The cause of this massive fish kill was the original change in the temperature of the water.

The water that travels for centuries through layers of rock to the sea is something like a savings account against bad times, a reserve in case of trouble. Man has chosen to use up that reserve rather carelessly by drilling wells and using the groundwater far faster than it is replaced. No one yet knows the eventual effects of tampering with the supply of groundwater.

Carbon: the essential element

If there were an award to be given for that one element most important to the structure of living things, carbon would have no rival. At first glance carbon seems a remarkably unlikely candidate for such a role. The element is found in only two pure forms in nature. One is graphite,

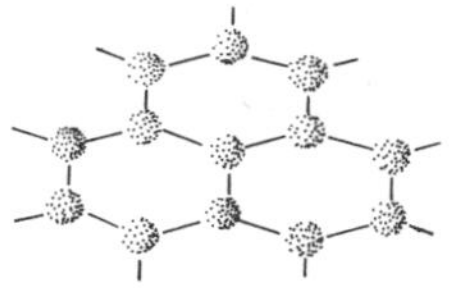

Graphite

A

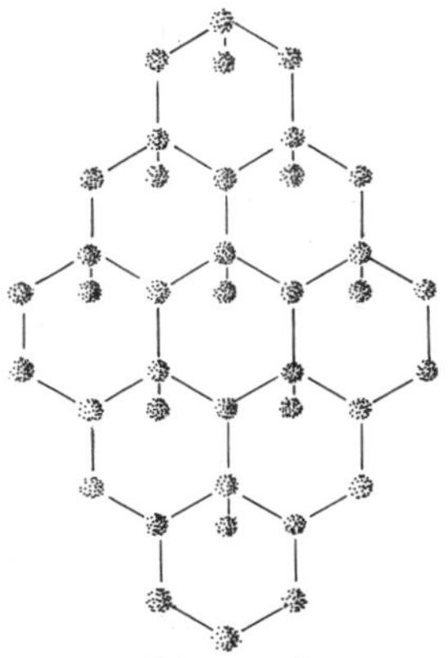

Diamond

B

Figure 3–5 The two naturally occurring forms of carbon. A. Graphite. This slippery substance is formed of two-dimensional hexagonal layers of carbon atoms. B. Diamond. The rigid structure of this substance results from the three-dimensional hexagonal organization of carbon atoms.

the black and slippery material used to make pencil "leads" and powder lubricants (Figure 3–5a). The other is diamond, that incredibly hard substance used for so many tasks, from putting a hard edge on an industrial saw or a stereo needle to acting as a sparkling and very expensive statement of good intentions (Figure 3–5b).

But when you look at carbon the way a chemist does, you begin to see why it plays such an important role: Carbon shows great versatility in forming bonds with itself and other elements. Atoms that ionize lose or gain a characteristic number of electrons, and to form an ionic bond the charges of the ions must be balanced. For example, an atom that loses two electrons must bond either with a single atom that has lost two electrons or with two atoms that have each lost one. Much the same holds for covalent bonds. The atom with "extra" electrons in the outermost shell must find atoms with the right number of "missing" electrons. As a result, each element, whether it bonds covalently or ionically, can bond only with a few other elements.

Not so carbon. Because of the position of its electrons, an atom of carbon can bond covalently with oxygen, nitrogen, phosphorus, hydrogen, and other carbon atoms, and can form four such bonds at any one time. Carbon can make more bonds than any other element. Because of this unusual versatility, the carbon atom is the principal building block of the many kinds of molecules making up living things. In fact, the scientific vocabulary equates *organic compound* with "carbon compound." When we talk about the carbon cycle, we are talking about the most essential stuff of life, the element from which all living things are built.

The three steps

Figure 3–6 provides a schematic summary of the carbon cycle. As you can see, all the carbon begins as carbon dioxide in the atmosphere, runs through the plant and animal worlds, and then returns to the atmosphere.

Plants transform the carbon dioxide of the atmosphere into biologically available carbon compounds through the process known as *photosynthesis*. In the same way that gasoline is needed to power a car, energy is required for photosynthesis. The energy for photosynthesis comes from sunlight, a topic we'll turn to again in Chapter 5. Basically what happens

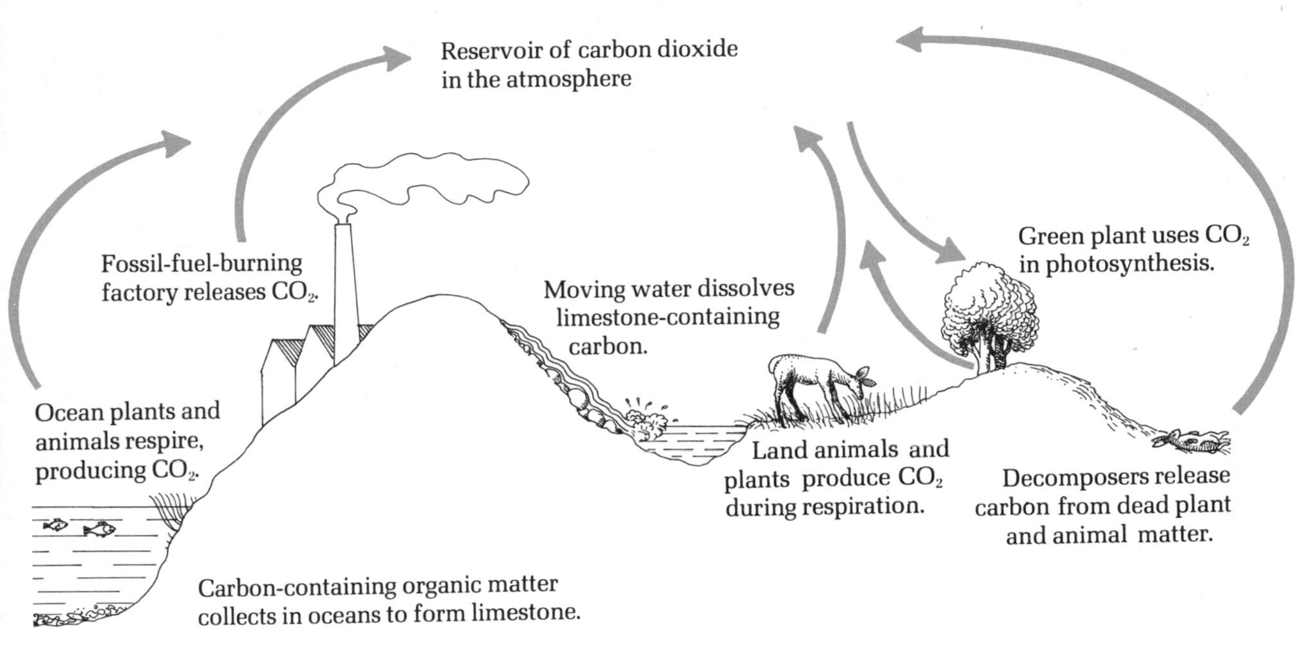

Figure 3–6 The carbon cycle.

in photosynthesis is that the plant takes in carbon dioxide from the air and water from the soil, chemically manipulates them with other molecules to produce *carbohydrate* (a molecule comprising carbon, hydrogen, and oxygen, such as a sugar or starch) and oxygen. In the form of a word equation, photosynthesis is

$$\text{carbon dioxide} + \text{water} + \text{light energy} \xrightarrow{\text{chlorophyll}} \text{carbohydrate} + \text{oxygen}$$

The oxygen is released to the atmosphere as a gas. In fact, all the gaseous oxygen in the atmosphere results from the photosynthetic activities of green plants. The plant retains the carbohydrate molecules. At this point, the carbon that has been removed from the atmosphere is said to be "fixed" into living tissue, and a new part of the carbon cycle has begun.

The plant uses some of this fixed carbon as fuel for its own growth, development, and reproduction. The chemical process by which the carbohydrates are broken down is called *respiration*, which is very nearly the exact opposite of photosynthesis:

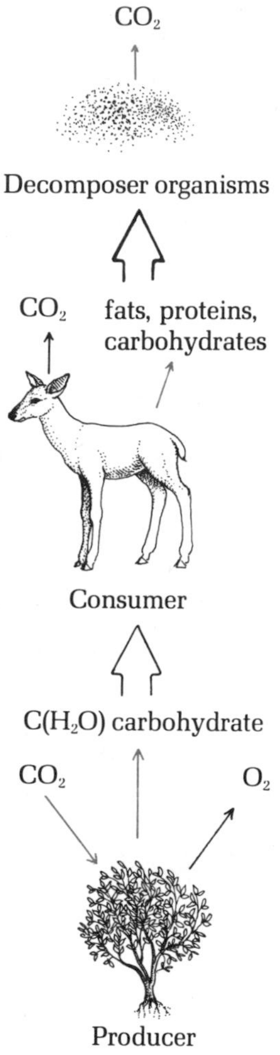

CO_2

Decomposer organisms

CO_2 fats, proteins, carbohydrates

Consumer

$C(H_2O)$ carbohydrate

CO_2 O_2

Producer

Figure 3–7 The path of fixed carbon through the biosphere.

$$\text{carbohydrate} + \text{oxygen} \rightarrow \text{carbon dioxide} + \text{water} + \text{energy}$$

The carbon dioxide produced by the plant's respiration returns to the atmosphere, where it can begin another cycle of photosynthesis. Respiration will also be discussed in Chapter 5.

But the great bulk of fixed carbon, about 90 percent of it, has a longer cycle. It's first stored in the plant's body or seeds. Animals that feed on plants take in the fixed carbon. These animals retain some of the carbon for use in their bodies as structural components of tissues and organs, break some of it down in respiration and return it to the atmosphere as carbon dioxide, and excrete the remainder in urine or feces. The component molecules of these excreted materials as well as the remains of dead plants and animals are consumed by bacteria and other microorganisms, and their respiratory activities return the carbon to the atmosphere as carbon dioxide.

Thus fixed carbon ultimately reverts back to the atmosphere, the place where it began, through the action of *producers*—the plants that fixed the carbon initially—or *consumers*—animals that feed on the plants or on other animals that have fed previously on plants—or *decomposers*—bacteria and fungi that live on excreted or decomposed matter (Figure 3–7). As far as the carbon cycle is concerned, every living organism, plant or animal from the biggest to the smallest, fits into one or another of these categories.

A certain amount of the fixed carbon in dead organisms is not converted back to carbon dioxide. It remains in the soil, and a portion of it combines with oxygen to form carbonates, a class of compounds that do not dissolve in water and are therefore not available to living organisms as nutrients. Also, the remains of some marine organisms drift into the deepest portions of the ocean bottom, which is relatively free of decomposers, and the unused fixed carbon accumulates on the sea floor. Such carbon-rich sediments, deposited some 300 million years ago and altered by great geological pressures, were the raw materials for what we call fossil fuels: coal, oil, and natural gas. When man burns these fuels, the carbon in them is returned to the atmosphere as carbon dioxide.

Since such processes as the formation of carbonates produce a slight but steady deficit in the cycle, is the earth's supply of carbon gradually running down? No, the slight deficit is made up in an appropriately subtle way. Rocks rich in carbon, such as limestone, are slightly soluble in water. Rainwater running over and through beds of such carbon-rich rocks picks up a small amount of that carbon in the form of dissolved carbon dioxide and returns it to the sea, where the molecules can escape into the atmosphere. This slight gain in carbon balances that lost to the depths and to the carbonates, and the amount of carbon dioxide available to the biosphere is held in a constant and delicate balance. Or, is it?

Man's role in the carbon cycle

The answer is that for millions and millions of years the balance was indeed constant and delicate. Since the time of the Industrial Revolution and particularly since the beginning of this century, man has burned an increasing amount of the carbon-rich fossil fuels to provide energy for everything from electric toasters to steel mills. Man has been using fossil fuels for over 800 years. However, most of this use has taken place in just the past 25 years! This accelerating use of fossil fuels is presenting us with a true energy crisis (Figure 3–8).

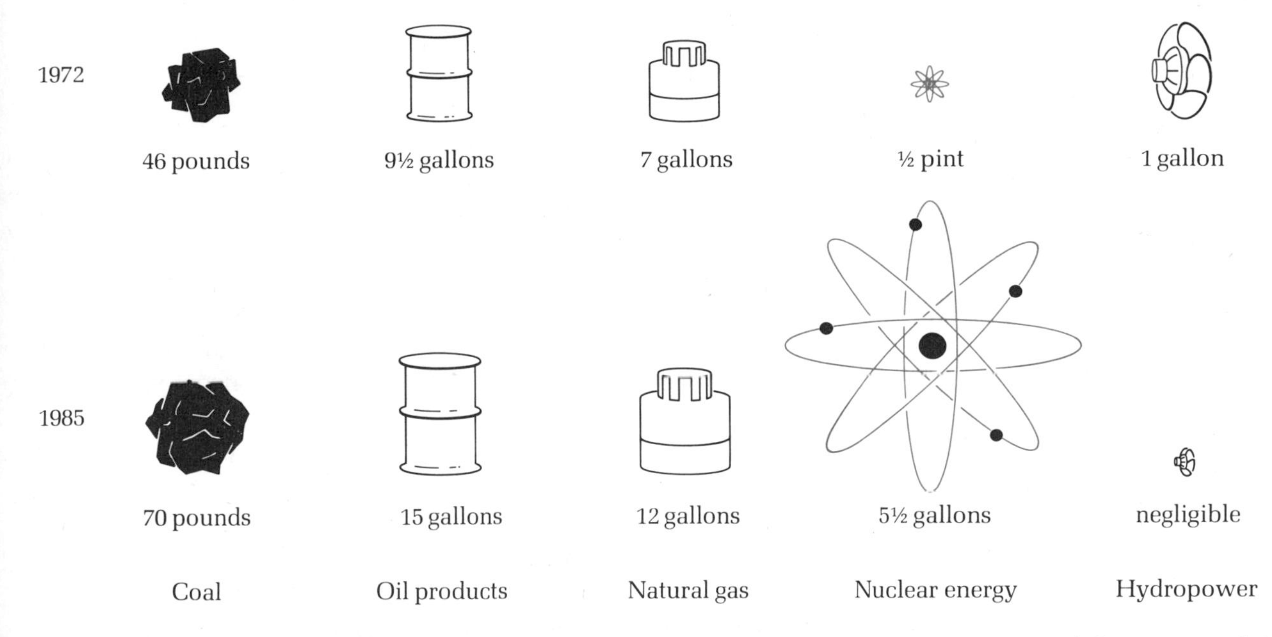

Figure 3–8 A comparison of daily fuel use by the average American household in 1972 and that estimated for 1985. As our use of fossil fuels (natural gas, coal, and oil) increases, we are steadily depleting the fixed and irreplaceable stores of these resources. Water power and nuclear energy are at present relatively small factors in the total energy picture.

Of course, such massive burning produces great amounts of carbon dioxide. Some of this carbon has been dissolved in water or trapped in sediments where it is biologically unavailable, at least for the present. But enough of this excess has remained in the atmosphere so that there is 10 percent more carbon dioxide in the air today than there was 100 years ago. What effect is such an increase having?

It's hard to tell. Since carbon dioxide is so important to plants, the increase may have given the earth's plant population a real shot in the arm, so to speak. However, some authorities feel that the increase is changing the *greenhouse effect* of the atmosphere (Figure 3–9). The

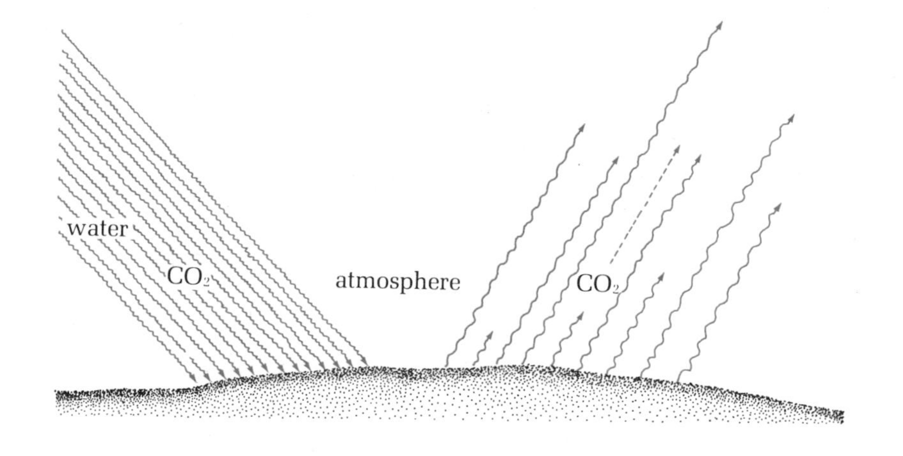

Figure 3–9 The greenhouse effect. The water and carbon dioxide in the atmosphere act to regulate the earth's surface temperature by reducing the amount of escaping heat. If man continues to add carbon dioxide to the atmosphere through his use of fossil fuels, the earth might become dangerously overheated.

carbon dioxide and water vapor of the atmosphere act as a shield, trapping a portion of the sun's heat and warming the surface of the earth in much the same way that a greenhouse roof holds in the heat and produces a warm environment. Without the greenhouse effect, the earth's average temperature would be −40°F rather than its actual average of 60°F. However, adding extra carbon dioxide could warm the earth too much. Such a warming trend might melt the polar ice caps, raise the level of the oceans, and flood the coastal zones of the continents. Other authorities maintain that the increase will have just the opposite effect, increasing the rate of evaporation of the oceans and cooling the surface of the planet, thus causing a disruption of life equally as severe as the flooding of the coastal plains. In either case, such consequences seem rather dire for the seemingly small crime of squandering coal and oil.

Actually, though, it's very hard to predict the results of the increase in carbon dioxide, and for a good reason: This is the first time man has played around with the earth's atmosphere.

Nitrogen: the curious role of the bean

All plants and animals use nitrogen as an important structural component of many necessary organic compounds. As you saw in Table 2–1 (p. 28), nitrogen is the principal ingredient of the atmosphere, and it might seem that all an organism would have to do to get an adequate supply of nitrogen is step outside and take a few deep breaths.

However, the nitrogen in the atmosphere is in the form of nitrogen gas (N_2), which does not tend to form compounds with other molecules and is therefore useless to most living things (Figure 3–10). As in the case of carbon, nitrogen has to be fixed. The task of fixing falls mainly to the class of plants known as *legumes*, which includes alfalfa, beans, and peas.

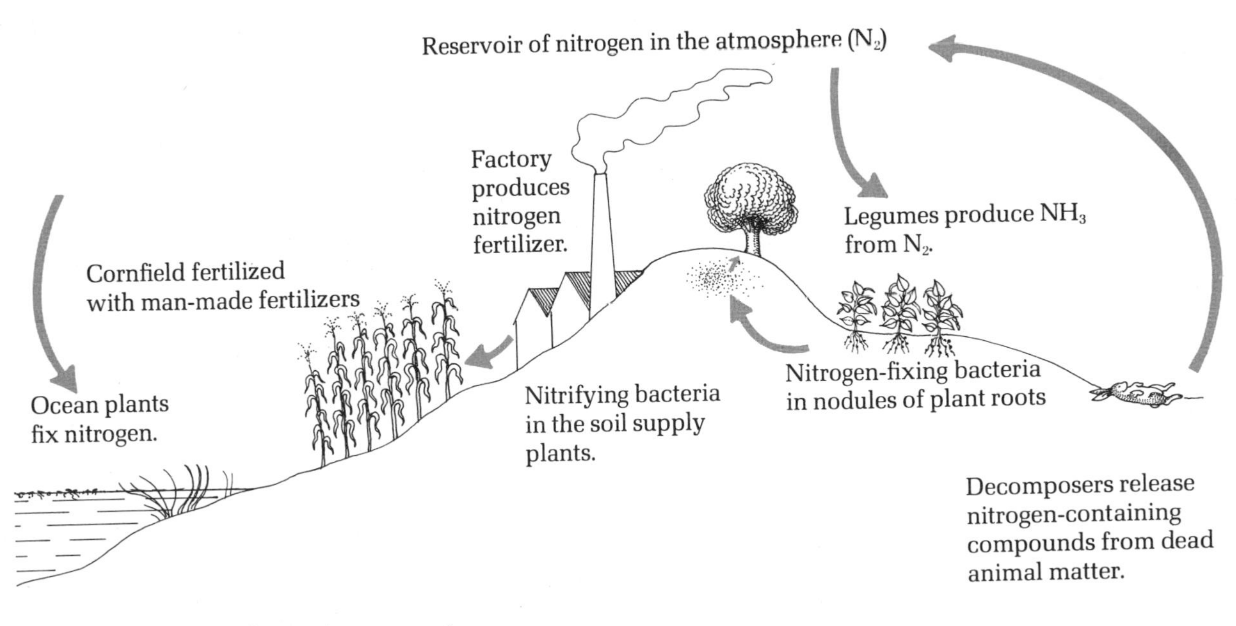

Figure 3–10 The nitrogen cycle.

The legumes' roots are covered with gall-like nodules that are the home of a kind of soil bacterium that can convert the atmospheric nitrogen N_2 to ammonia (NH_3). Ammonia can be used by both plants and animals as a nitrogen source. Other microorganisms convert ammonia to another compound, nitrate (NO_3^-) which can be used by most plants and many animals. In addition, significant amounts of atmospheric nitrogen are converted into biologically useful forms by volcanic eruptions, electrical storms, and fertilizer-manufacturing plants.

The nitrogen incorporated into the tissues of plants and animals is returned to the cycle at the death of the plant or animal. Decomposer bacteria break down the organic nitrogen compounds into ammonia and nitrate, which other organisms can reuse immediately. Another group of bacteria, known as *denitrifiers*, convert the ammonia back to nitrogen gas, thus completing the cycle.

Man's effect on the nitrogen cycle

Before man took control of the earth, the amount of nitrogen removed from the atmosphere by fixation equaled the amount returned by denitrifiers. This balance prevailed until man expanded his agriculture and his use of fertilizers. Cultivated soils planted with a single crop plant such as corn or wheat show a gradual decline in nitrogen content and eventually in plant yield. To boost productivity, and in turn profit, fertilizers rich in nitrogen compounds, particularly ammonia and nitrate, are added to the soil to make up for the lost nitrogen. The total production of such man-made nitrogen has increased rapidly. In 1968, chemical factories produced 30 million tons of nitrogen, an amount equal to one-third of all the nitrogen fixed both naturally and artificially in that year. The man-made total has been doubling every six years, so this amount will certainly increase. Obviously we are tipping the balance of the nitrogen cycle one way or another, but the precise long-run effects can only be guessed at.

But one thing we do know for sure is that nitrogen fertilizers added to the soil run off and eventually end up in streams, rivers, and lakes used for drinking water. Nitrate is relatively innocuous but, in solution, a small amount becomes nitrite (NO_2^-). Nitrite is quite poisonous, particularly to children under the age of five. In parts of the intensively farmed and fertilized San Joaquin Valley of California, all water from wells and reservoirs has been so poisoned with nitrite that it is a hazard to health to drink it.

In addition, nitrogen runoff can drastically alter the nutrient cycles in the streams and lakes where it is deposited. Lake Erie is a prime example. The lake is popularly referred to as a "dead" body of water, but that is not the case at all. In point of fact, it is too much alive. Fertilizer runoff from the heavily cultivated lands around the streams and rivers emptying into the lake greatly increased the amount of nitrogen in the water. Many kinds of waterborne algae (a simple form of plant life) can use this nitrogen as a food source. Presented with such an immense increase in the food supply, the population of algae in the lake doubled and redoubled and doubled again. When the algae died, they were consumed by bacteria. The respiratory activities of the bacteria consumed much of the oxygen dissolved in the water, denying it to the fish in the lake. As a result, the fish literally suffocated, and huge numbers of them were wiped out. This process, in which oxygen is used up and animal life displaced by algae and bacteria, is called *eutrophication* (Figure 3–11).

Figure 3–11 Eutrophication of a river. Dam construction and chemical changes in the Merced River in California have encouraged the thick growth of surface water hyacinths and slowed fish migration.

Oxygen: the significance of breathing

Oxygen is very important to living things. Almost as much as carbon, oxygen forms biologically important compounds with most of the other elements of the biosphere. The cycles of many nutrients—nitrogen, carbon, phosphorus, water, and others—are all tied to the oxygen cycle (Figure 3–12). Perhaps as a reflection of its importance, oxygen can follow

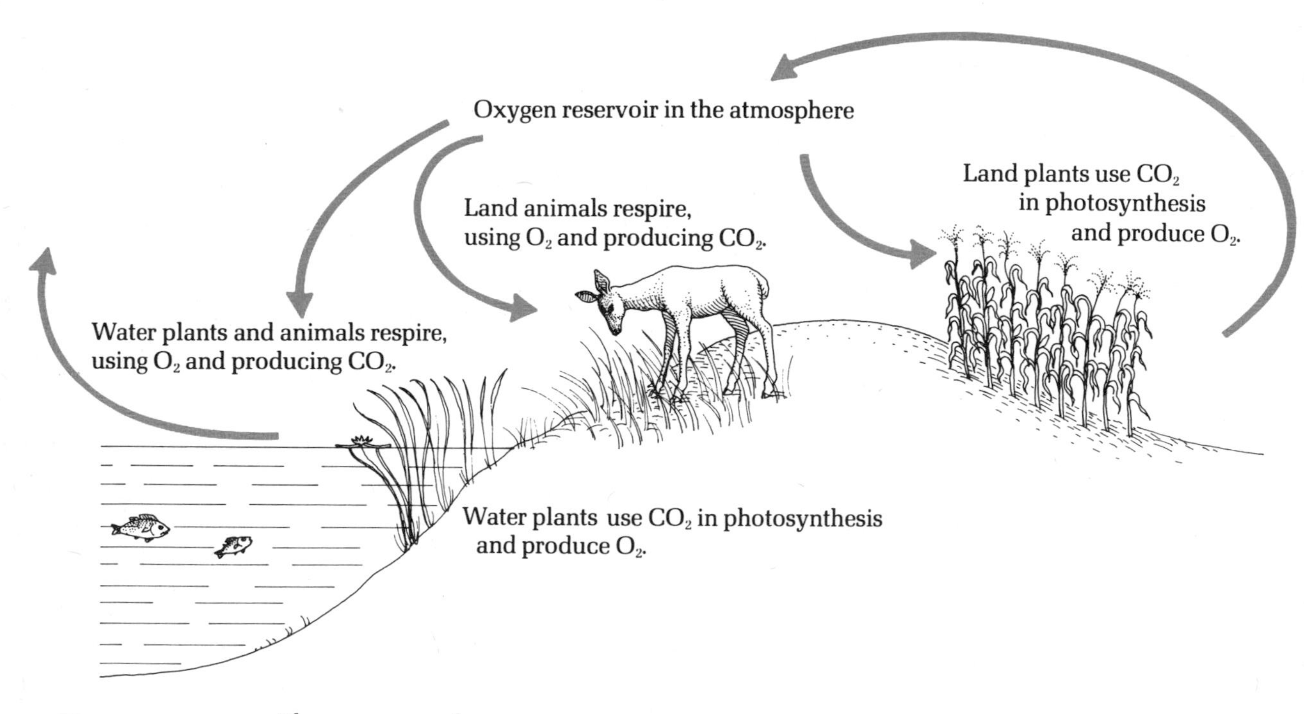

Figure 3–12 The oxygen cycle.

any of a number of pathways through the biosphere. Any given oxygen molecule is something like a hitchhiker standing at a four- or five-way intersection and willing to take a ride with the first car that comes along, no matter what its destination.

During photosynthesis, water is broken into its component hydrogen and oxygen atoms, and the oxygen gas is released to the air. The oxygen in our atmosphere is ultimately the product of photosynthesis. What

The oxygen atom reacts readily with many commonly occurring elements. As a result, the oxygen cycle follows the path that oxygen takes in combination with other chemicals.

happens to the oxygen once it is released into the air? Some of it remains free, and some of it forms compounds with other elements, such as carbon to produce carbon dioxide. Carbon dioxide taken in by plants during photosynthesis is broken down, and the oxygen molecules are used as one of the building blocks of carbohydrate compounds.

The free oxygen—oxygen molecules in the atmosphere that do not form compounds with other elements or dissolve in water—is the "fuel" for energy-releasing respiration in both plants and animals. A simple sugar like glucose is combined with oxygen to produce carbon dioxide, water, and energy. The plant or animal can either use this energy or store it for the future, in ways we shall investigate closer in Chapter 5. Your own need for the energy produced by respiration, and thus for the oxygen needed to release that energy, is the reason that you breathe continuously.

The one outstanding fact about the oxygen cycle is the crucial role of green plants. Plants actually produce the oxygen in the atmosphere, and that oxygen is responsible for the energy-yielding reactions of nearly all forms of living things. If there were no green plants, there would be no oxygen. And without oxygen only certain organisms—yeasts and a few bacteria—could survive.

In addition to its role as a nutrient, oxygen also performs an important life-protecting job in the upper reaches of the atmosphere, outside the realm of the biosphere. When subjected to very high energy, the chemical properties of atoms change, and the atoms behave differently from the way they do under more usual conditions. Such a high-energy situation exists in the upper atmosphere, created by the radiation of sunlight. When exposed to high energy, oxygen's tendency to bond changes, and the usual form of atmospheric oxygen, O_2, becomes O_3, a molecule known as ozone. Ozone is something more than a chemical oddity. The light striking the upper atmosphere contains ultraviolet light, which is lethal to life on earth. Fortunately for us, the ozone in the upper atmosphere effectively filters sunlight, absorbing almost all the ultraviolet light.

Man's role in the oxygen cycle

Even though many of man's industrial and agricultural activities have an effect on the oxygen cycle, there has been very little change in the amount of atmospheric oxygen. Apparently, the amount of oxygen removed from the cycle is roughly balanced by the oxygen put back in. Man's burning of large amounts of fossil fuels and the clearing of all plants from land intended for superhighways or apartment buildings reduce the amount of oxygen available. But rises in agricultural productivity—which are really increases in the total amount of photosynthesis in a specified area—plus the cultivation of previously barren lands like deserts have added to the oxygen supply. There is no definite proof that these activities are in balance at present nor any guarantee that whatever balance there is will last, but at least up till now the level of atmospheric oxygen has remained stable.

Where man has adversely affected the oxygen cycle has been in rivers, streams, and lakes. Take the example of Lake Erie again. Analyzed from the standpoint of nitrogen, the problem with the lake is an increase in that element. The other side of the coin is a decrease in oxygen. The dying algae feed increasing numbers of bacteria, whose respiration depletes the supply of dissolved oxygen. If the algae have enough nutrients, oxygen depletion can continue until the water is effectively oxygenless. Then the water can be home only for a class of bacteria known as *anaerobic,* which can survive only in the absence of oxygen and which produce the foul smells of a stagnant pond. The damage done by adding a given nutrient to a body of water is measured by the *biological oxygen demand,* or BOD. The BOD is an index of the amount of dissolved oxygen used up by the respiration of microorganisms degrading nutrients and wastes dumped into the water.

Bodies of water close to major population centers have suffered the most pronounced depletion of dissolved oxygen. Of all the Great Lakes, Lake Erie contains the smallest volume of water, and any pollution becomes concentrated fairly fast. The great amount of nutrient runoff from the farmlands of Ohio and Michigan and the industrial areas of Detroit, Toledo, and Cleveland has a very high BOD, and the effects have been disastrous. The oceans have always seemed safe from such damage simply because they were so big that such pollutants as excess nitrogen simply seemed to disappear in the vastness of so much water. Even though the BOD of the nutrients that flow into the oceans each year is great, it has not significantly reduced the amount of dissolved oxygen in the sea. But there is always the risk that the store of dissolved oxygen will eventually be depleted and that the seas will eutrophicate just like Lake Erie. In terms of survival, that's a rather chilling prospect.

Putting things together

Thus far, we've been talking in a rather generalized way. For one thing, it's somewhat artificial to separate one nutrient cycle from the others. In reality, all the cycles are tied closely together. In taking a careful look at any one type of plant or animal, the pathways of the various cycles become hard to distinguish. For example, when a rabbit eats fresh grass, it is taking in its carbon, its nitrogen, and a large percentage of its water needs—all at the same time. Also, we have been talking at a high level of generalization by using such terms as "living things." That term covers a wide, wide variety of life forms, from tiny one-celled plants to very complex animals such as birds and mammals. Even though all these forms need some of the same nutrients, they have many different ways of finding and maintaining a steady supply. Obviously an oak tree, which spends its life rooted in one place, meets its needs for water quite differently from a dog, which can move about.

All the cycles are tied closely together. In taking a careful look at any one type of plant or animal, the pathways of the various cycles become hard to distinguish.

Even though all nutrients originate in the nonliving world, not all living things take them directly from that source. Instead, many living things satisfy their needs for certain nutrients from other living things. For example, we humans fill our needs for oxygen and water by breathing and drinking. But we are totally dependent for the energy we need on carbohydrates obtained either directly from photosynthesizing green plants or indirectly from animals that feed on those plants.

Thus nutrient cycles depend on two important kinds of relationships: those between the nonliving environment and living things and those between various kinds of living things. To talk about these relationships, we need to take a closer look at *ecology*.

The dislocated word

In any area charged with controversy and emotion, certain key words have a way of losing their original sense and being invested with a new meaning. This is particularly true of words that become political rallying cries. Such has been the fate of the word *ecology*. For example, if you absentmindedly toss a beer can into a lake, someone may shove the can back in your face and tell you that what you've done isn't very ecological. It is, and then again it isn't.

The word ecology was originally coined in 1869 by a German botanist named Ernst Haeckel. In those days, most biological scientists tended to study individual plant and animal species in isolation. Haeckel maintained that this single-mindedness produced false results, that the only way to understand a species fully was in terms of its relationship to its environment and to the other types of organisms in that environment. Haeckel invented the term ecology from the Greek word *oikos*, meaning household, the word that is also the root for economics. Economics studies the flow of resources among men, and ecology, which since Haeckel's time has become a branch of biology, studies the flow between

Nutrient cycles depend on two important kinds of relationships: those between the nonliving environment and living things and those between various kinds of living things.

In order to survive and reproduce, living things obviously need an orderly, predictable way of dealing with their environment.

various forms of life and the environment. It doesn't judge whether that flow is good or bad. It just analyzes what that flow is.

To return to that example of the beer can. Tossing a beer can into a lake is ecological in the sense that the act, casual trashing though it may be, is part of the relationship between man and his environment. Whether that particular aspect of his relationship is good or bad for the environment or for man's neighbor organisms—and therefore whether it should be allowed or not allowed—is quite another issue.

Organizing the living world

How do these physical and chemical factors combine with living things to form a natural community? What is the relationship between the living members of a community and their *abiotic* (that is, nonliving) supplies and setting?

In order to survive and reproduce, living things obviously need an orderly, predictable way of dealing with their environment. They have to find usable nutrients, sufficient water, oxygen (except for anaerobes), a waste-disposal method, and a suitable way of raising their offspring to independence. They need to know which other plant and animal organisms are worth avoiding, which can be helpful as food or protectors, and which are of no consequence.

It is almost impossible to recognize these relationships if we look at the biotic sphere as a total entity. In order to study the organizing patterns of the natural world, biologists have divided the biotic sphere into practical units that are convenient for study. At each finer step of organization, new and different things are revealed about that particular environment that would not be apparent at a larger, more inclusive vantage point. To demonstrate the usefulness of the organized study of environments and their living residents—the science of ecology—we will look at the Florida Everglades, one of the most complex and extraordinary environments on this planet (Figure 3–13).

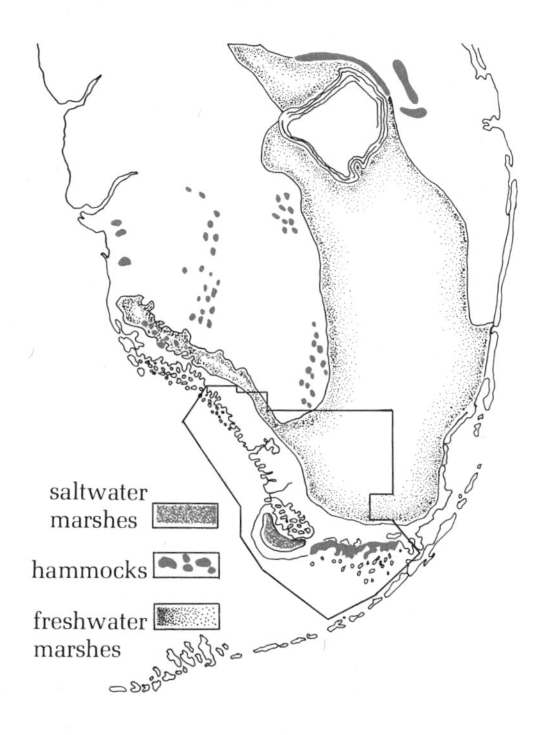

Figure 3–13 The Everglades. The boxed area shows the borders of Everglades National Park.

Why the Everglades of all places? It's a beautiful place—lush with vegetation and staggering in its variety of life forms. It has fascinated every culture which has seen it—the American Indians, the Spanish explorers, and the nineteenth century Americans. Today, it remains a very popular "outdoor museum," and a national park has been established in its southwest corner.

The Everglades are vast—covering 2,500,000 acres of southern Florida. But despite the size of this massive marshland, events in the past 50 years have revealed its frightening delicacy and its vulnerability to poorly planned technological and commercial intrusion.

The only subtropical environment in the continental United States, the Everglades is a freshwater marsh. To a biologist, this statement, though brief, is very informative. It indicates that the Everglades is part of the tropical *biome*, a geographical area whose climate leads us to expect certain types of life forms. Alligators, palms, wading birds, water rats, and dense marsh grasses are examples of the types of organisms adapted to this environment.

The earth has been "divided" into seven major biomes, and each constitutes a handy biological "profile"—tundra, desert, grassland, coniferous forest, deciduous forest, rainforest, and tropics. And thanks to the southern tip of Florida skirting the Tropic of Cancer, North America includes examples of each type of biome.

In addition to providing a capsule climate description of an area, the biome designation permits comparisons with all other biomes of the same type. For example, a tropical area like the Everglades has a good deal in common with tropical marshlands in Indochina, although they are obviously geographically separated.

What do we find in this corner of the tropical biome known as the Everglades? As in every biome, we find *communities*, groups of organisms that have one big thing in common—they have all adapted to that particular environment, and find it possible to survive there. One of the most unusual communities in the United States is the Everglade hammock (from the Arawak Indian word for jungle) (Figure 3–14). This is literally

Figure 3–14 An Everglade hammock. These "islands of trees" are densely covered with plant growth and form a unique environment within the Everglades.

an island of trees, holding rich soil by their roots, and surrounded by the riverlike flow from huge Lake Okeechobee. The Everglade hammocks are dense communities populated by a wide variety of water-adapted animals and plants with established patterns of mutual interest and dependence; that is, the component organisms relate to each other in specific and nonhaphazard ways. Thus, if you look at a community, you will find that the organisms within it "make sense."

How are the living things that comprise a community organized? In the Everglades we find so many kinds of organisms: several dozen types of birds (the anhinga, the limpkin, the purple gallinule, the egret, the heron, the flamingo, and that comical fellow the roseate spoonbill), many kinds of plants (royal palms, the sea grape, the coconut plum, the fish-poison tree, and the wild tamarind), mammals such as deer and opossums, and numerous reptiles (Figure 3–15). How can we keep them

Figure 3–15 A roseate spoonbill. These brilliantly pink birds are characterized by flat, spoon-shaped bills and legs suitable for wading in shallow waters.

all straight and put them into a meaningful context? We are immediately aware that these organisms are doing different things in the Everglades. For example, the flamingo has extremely long, thin legs and spends a good deal of its time wading in shallow water. An anhinga, on the other hand, might be watching the water from its perch high in a tree, a perch to which its own muscle structure is particularly adapted.

Here the concept of *species* comes in. A species is a specific group of organisms (either plant or animal) that normally reproduces by breeding with each other, producing fertile offspring. Thus, by listing the typical species of an environment, we can get a clear picture of the life-supporting potential of the area. Because of the reproductive aspect of the definition of species, we learn that—barring disasters of some kind—there will always be egrets, palms, herons, 'possums, and alligators in the Florida Everglades. So we have species in communities within biomes.

But this is a static view of a living community—all characters and setting, but no plot. It is the relationships between the living and the nonliving that establishes the true "profile" of the environment.

Communities, as we said before, are supported by different organisms performing different roles; the roles relate to energy movement through the environment (more about that in Chapter 5), food production and waste disposal, and reproductive methods. When looked at this way, a community reveals its dependency relationships, which, taken as a whole, describe an *ecosystem*—the balanced, self-supporting juncture of living and nonliving (Figure 3–16).

Here's an example. The role a species population plays within the community is its ecological *niche*, an occupational pigeonhole. If that particular species dies off or leaves the habitat, some other species would have to replace it or the species totally dependent on it would not survive.

One of the most important aspects of the niche concept is that of food relationships. For example, a large land snail used to be a common inhabitant of the Everglades. Drainage changes caused by canal construction in recent decades have almost completely wiped out the snail and left the Everglade kite without its food supply. As a result, the birds' name is ironic; there are almost no Everglade kites left in the area.

Frequently, of course, species adapt to new environmental conditions, and, in a case like this, might find a new food source. So the niche occupied by the snail might be preempted by another, possibly similar, organism.

By listing the typical species of an environment, we can get a clear picture of the life-supporting potential of the area.

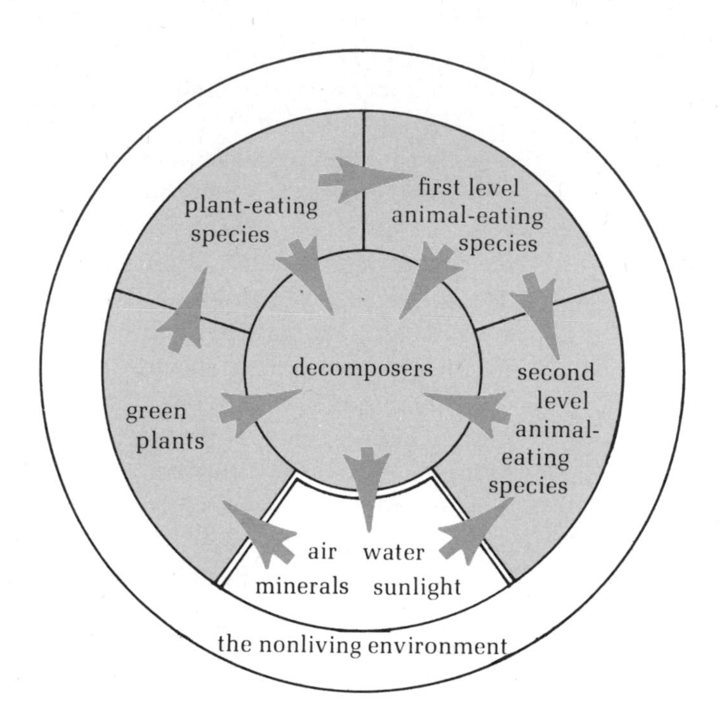

Figure 3–16 The concept of an ecosystem. Within the framework of the nonliving environment, living things interact and are mutually dependent.

The "who-eats-what" of a community constitutes its *food chains*—the energy transfer which occurs between the plants, plant eaters, animal eaters, and the organisms that can eat either plants or animals. Yet, as you know from your own varied diet, most organisms have some variety in their food supply. If an organism is dependent on a single food source—like the kite on the snail—the loss of that supply without immediate replacement is fatal to that consumer.

Thus the stability of a community—its long-range ability to survive environmental changes—is directly related to its complexity. The more options available, the more adaptable the community. One indicator of a community's complexity is its *food web*, the cumulative crisscrosses of its numerous food chains (Figure 3–17). As you can see, several of the organisms play roles as both food sources and predators of other species.

Man and the environment: a case in point

If nature has been so efficient and careful in its "planning," why are we currently so concerned about environmental problems and violations

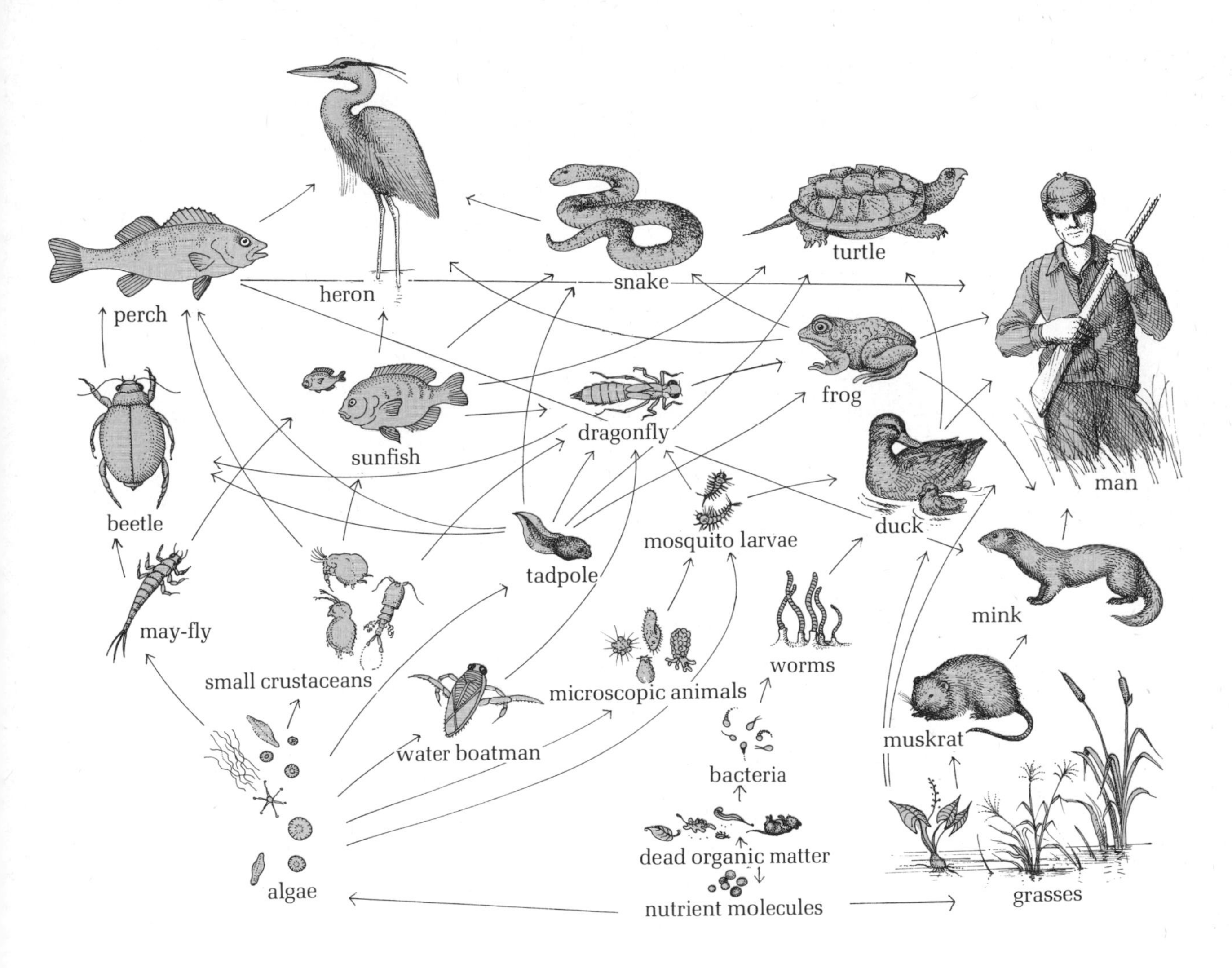

Figure 3–17 The complex food web of a marshland. The interdependent relationships of this community make it very stable.

If nature has been so efficient and careful about its "planning," why are we currently so concerned about environmental problems and violations of ecological principles?

of ecological principles? As we will see throughout this book, the major factor is man and his technology. Although humans clearly fit into the natural way of things, there are some important "exceptions" to their role in the ecosystem.

First of all, man is the only species capable of preying on other species for reasons other than self-defense or food. An example is the case of the snowy egret of the Everglades. During the nineteenth century, these lovely white birds were in danger of being wiped out by plume hunters who supplied the millinery industry. The adult egrets were hunted during the breeding season, leaving the newly laid young without food and protection. It was not until the founding of the Audubon Society that the egrets were protected, and since then, the population of the birds has increased substantially.

Second, man's accelerating population growth has multiplied the environmental effects of his activities. In Chapter 14, we will examine this growth in greater detail.

Third, man has developed technological capacities and needs that exceed his "natural" ability to affect his environment. In the Everglades, man's major impact has been on the natural drainage patterns. Obviously, in an area so dominated by water and water-living organisms, shifts in the level and chemical composition of water can be catastrophic. Today, over 44 percent of the Everglades has been diverted to agricultural or planned residential use, creating a host of water-supply needs. Control of annual flooding from Lake Okeechobee, the constant supply of fresh drinking water, and irrigation have all affected the water level and saltiness of the Everglades.

Large mammals such as deer have been killed by excessive flooding. The increasing concentration of salt in the water has threatened the shrimp fisheries. Dredging and filling for housing construction have eliminated nesting grounds for many species. The most dramatic threat—fortunately headed off by public concern—was the planned construction of a jetport for the city of Miami, a plan that would have shifted water supplies away from the Everglades and greatly compounded the damage already done.

Unfortunately, much of our knowledge about fragile environments occurs after the fact—when a link in a food chain is broken or an aspect of the nonliving world is damaged or otherwise eliminated. An understanding of basic biological principles can help us evaluate potential environmental damage in advance, predicting the consequences of man's economic and political decisions.

Summary

Cycles are the key to life's dependence on the nonliving world.

Water is particularly important to life because of the stability of its hydrogen bonds and because of its polarity, which allows it to dissolve other chemicals and carry them with it. The water cycle begins in the seas with the evaporation of surface water. The cooling vapor condenses and falls to the earth as rain. The fallen water travels back to the sea via rivers or through deep rock layers. Man has affected the water cycle by adding pollutants and extra heat and by tapping groundwater.

Carbon is the most important element for organic compounds because it can form strong bonds with many elements as well as with itself. In the process of photosynthesis, atmospheric carbon dioxide is fixed into plant sugars and starches. Some of these carbon compounds are broken down and released as carbon dioxide to the air by the energy-releasing respiration of the plant itself, of the animal that feeds on the plant, or by bacteria that feed on the dead plant or animal. Man has added greatly to the carbon dioxide of the atmosphere by burning fossil fuels extensively. The exact effect this will have is unknown, but it could possibly either cool or warm the climate.

Nitrogen also begins in the atmosphere. Bacteria fix it into the form of ammonia and nitrate, which can then be used by both plants and animals. Nitrogen released by the death of plants and animals is converted back to gaseous form by denitrifying bacteria. Man adds great amounts of nitrogen to the soil as fertilizer. Runoff carries it to bodies of water, where it provides a ready food source for algae. Bacteria consuming dead algae deplete the supply of oxygen dissolved in the water and cause that body of water to eutrophy.

Oxygen is released to the atmosphere by green plants during photosynthesis. Oxygen is needed by practically all organisms for respiration. Oxygen in the form of ozone in the upper atmosphere shields the earth from lethal ultraviolet rays. Man's principal effect on the oxygen cycle has been through eutrophication.

Ecology is the branch of biology that studies the relationships between living things and the nonliving world and among various living things. Environments can be studied in terms of both their geographic location and the activities of the organisms within them. The species within any community relate to each other in a predictable manner, as shown by food webs. The greater the complexity of a community, the greater its stability. Man is unique in his ability to affect the environment because of three characteristics of his activities: depleting other species for reasons not essential to survival, a rapidly growing population, and technology.

Questions to consider

1. How do water's unique chemical properties explain its importance to living things?
2. a. What are the major stages of the water cycle?
   b. How has man affected the water cycle?
3. Carbon is the most important structural molecule in organisms. Why is this so?
4. a. What are the major steps in the carbon cycle?
   b. How has man affected the flow of carbon between the living and nonliving worlds?
   c. What effect might these changes have?
5. a. What are the steps of the nitrogen cycle?
   b. How has man affected the nitrogen cycle?
6. a. What are the stages of the oxygen cycle?
   b. Why is the ozone of the upper atmosphere important to survival?
   c. What has been man's principal effect on the oxygen cycle?
7. We have seen that chemistry describes and analyzes all forms of matter at the level of atom and molecule. We have also seen that ecology is the study of the interrelationships of living things with each other and with the physical environment. Imagine an ecologist and a chemist both watching untreated sewage pour into a river. What questions would each scientist probably ask about this event?
8. In this example of the dumped sewage, who depends on whom more? That is, is the chemist dependent on the ecologist for his analysis, or vice versa?

9. How would you distinguish between an organism's habitat and its ecological niche?
10. Man can be described as a two-footed mammal capable of adapting to a wide range of environments and of preying on many other species. Why is this an inadequate, though accurate, description of man's impact on his environment?

Glossary

abiotic not containing or related to living things

anaerobe an organism that does not require oxygen for respiration

biological oxygen demand (BOD) the amount of oxygen dissolved in a body of water required to break down organic wastes

biome one of the seven major climatic and geographical regions of the earth, each characterized by specific types of plant and animal life

carbohydrate a compound of carbon, hydrogen, and oxygen, such as sugar and starches; the ratio of hydrogen to oxygen is 2:1

community an integrated grouping of plants and animals in a specific location

consumers organisms that meet their energy needs by eating plants or other animals

decomposers organisms that meet their energy needs by feeding on the waste products and remains of other organisms

denitrifier a microorganism that converts ammonia into nitrogen gas in the nitrogen cycle

ecology the branch of biology which studies the relationships between living things and their biotic and abiotic environments

ecosystem the systematized relationships between the living and nonliving worlds

eutrophication excessive algal and bacterial growth causing oxygen depletion and eventual loss of animal life from a body of water

food chain the pattern of energy relationships between producers and consumers in an ecosystem *food web* the complex linkages between food chains that indicate the interrelationships in an ecosystem

greenhouse effect the heat-retaining ability of the earth's atmosphere, caused by carbon dioxide and water vapor

legume any one of a group of plants that possess nitrogen-fixing bacteria in their root systems

niche the occupational role of an organism in an ecosystem

nutrient cycle the pathway of conversion followed by nutrients between and within the living and nonliving worlds

organic compound compound containing carbon

photosynthesis the chemical process by which green plants use solar energy to form carbohydrates from carbon dioxide and water

producers organisms that meet their energy needs by fixing carbon dioxide and forming carbohydrates

respiration the universal process by which all cells release energy from carbohydrates

species a group of related organisms that normally mate with each other and produce fertile offspring

thermal pollution the release of dangerous and abnormal levels of heat energy into a body of water

4 Evolving

Man has always been curious about his origins. This curiosity is reflected in his religions and mythologies, all of which provide a point of origin for man and an explanation of his position in the universe. One of the scientific consequences of this human puzzlement was the formulation of the most unifying biological principle of all time, the theory of *evolution*.

In the mind of the ancient Hebrew there was no doubt that God created the world and all the living things in it by an act of His will. The book of Genesis explains the origins of the world and life in this way, as six successive days of divine creation of the earth and its living inhabitants. In the ancient Greek view of the world, the gods didn't create the universe—the universe created the gods. Heaven and earth were timeless entities that had always been. They were the first parents, and from them the gods arose. Likewise, all forms of life came originally from nonliving matter. While complex forms like men and horses reproduced themselves after their original creation, simple forms were still being created all the time. For example, rotting meat was the parent matter of maggots and warm mud that of frogs. This Greek idea, that all life came originally from nonliving matter, was called *spontaneous generation* (Figure 4–1).

The theory of spontaneous generation held sway in Europe throughout the Middle Ages and the Renaissance. It seemed very persuasive;

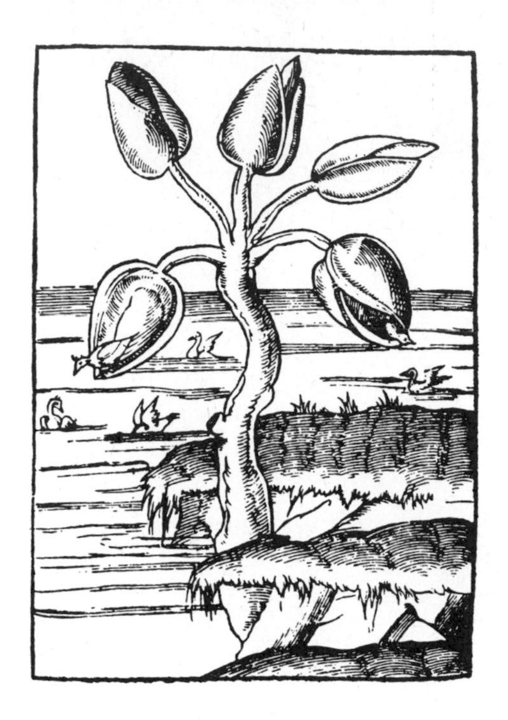

Figure 4–1 The ancient concept of spontaneous generation and lines of descent across species persisted well into the Middle Ages. This sixteenth century woodcut provides an explanation for the rise of water fowl, here shown emerging from the limbs of a seaside tree.

anyone could see that a piece of meat left to rot was soon covered with maggots and that the mud of a pond gave rise to frogs each spring. Philosophers and naturalists did wage arguments about where specific kinds of beings came from. Despite such controversies, however, practically all thinkers accepted without question one basic assumption: All creation had occurred at one time, and from that time the many varieties of organisms had remained unchanged. Until the beginning of the nineteenth century this assumption, called the *fixity of species*, was unchallenged.

Lamarck's heresy

The major challenge was presented by Jean Baptiste Lamarck, a French naturalist. Lamarck had spent most of his career studying *fossils* —remains of life forms left in rock—and he was particularly interested in simple life forms. What intrigued Lamarck was that such organisms seemed "simple" while life forms like the birds and mammals appeared "complex." Indeed, suggested Lamarck, was it not possible that the complex forms had developed from the simpler ones over the course of time? That is, all species had not been created at one time; instead, complex forms had developed from simple ones, and every species was undergoing change at all times.

What caused such change? If an organism encountered a new environment or a new stress, the changes brought about by adapting to the novel pressure would alter the species in some way. The changed organisms would then pass the new characteristic to their offspring. For example, most alpine, or mountain, plants grow only an inch or two in height and live out their life cycles hugging close to the ground. Lamarck thought that a sea-level plant transplanted to the alpine habitat would be stunted by the climate. The seeds of the newly stunted plant would produce stunted offspring, even if these seeds were returned to sea-level conditions. Figure 4–2 shows how Lamarck, using the same argument, accounted for the giraffe's long neck.

Like many scientific ideas, Lamarck's was a mixture of truth and error. His insight that not all species dated from one point in history and that species could change over time were important and accurate ideas. But Lamarck was mistaken about the mechanics of the process, the means by which changes are passed from parent to offspring. His assumption that physical changes produced in an individual organism during its lifetime appear in its offspring has been disproven by experiment. A lowland plant grown under alpine conditions will indeed be stunted, but the seeds of this plant returned to the lowlands will result in normal-sized offspring. Stunting is a response to the severity of the alpine climate, but the change, a merely temporary one, is not passed on to succeeding generations (Figure 4–3). Also according to Lamarck, it would be reasonable to predict that amputating the tails of several generations of mice would eventually produce mice without tails. But this doesn't happen. If the tail of every mouse in 50 successive generations is amputated, the fifty-first generation will still be born with tails.

Although Lamarck was seriously mistaken about heredity—as, indeed, was everyone of his time—he had developed the crucial idea that

Figure 4–2 Lamarck's explanation of the giraffe's long neck. The short-necked giraffe is able to feed from the tree (A). As the food supply along the lower portion of the tree is depleted (B), the giraffe must stretch to reach the taller branches (C). The long-necked giraffe which Lamarck thought would result (D) was able to pass this acquired trait on to future generations.

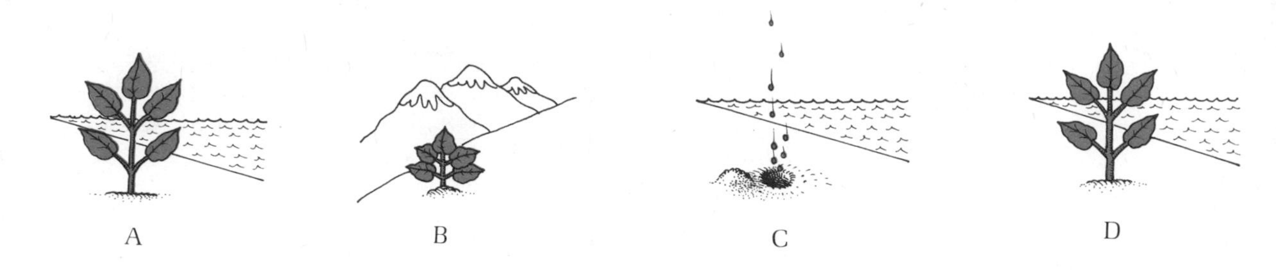

Figure 4–3 An example of non-Lamarckian evolution. The coastal plant grows to its full height at sea level (A). If transplanted to a harsh alpine climate, its growth will be stunted (B). However, if the seeds of the stunted plant are then raised at sea level (C), normal-sized plants will result (D).

species were not fixed and that one form could arise from another. With these ideas, Lamarck unlocked a new door for the science of biology. But it fell to a former theology student from England to open the door Lamarck had unlocked and to shine a light into the dark corners of the room it revealed.

Darwin: the origin of species

Reckoning on the basis of the generations recorded in the Bible, the date of the earth's creation was fixed at 4,004 years before the birth of Christ. If this date were accurate, then there was no doubt that Lamarck's theory was unworkable. Naturalists already recognized that species breed true. That is, if you mate dogs with dogs or poppies with poppies, the offspring are dogs and poppies, not cats and lilies. Any evolution of one species from another was most likely a slow process, an accumulation of changes. How could all the many, many species of plants and animals have evolved from simple predecessors in less than 6,000 years? The mere suggestion was mind-boggling. Evolution needed time.

Curiously, the evidence for that time came not from biologists but geologists. The most persuasive line of reasoning appeared in Charles Lyell's *Principles of Geology* published in the 1830s. Lyell argued that the strata of rock visible in the earth's crust and the fossils characteristic of specific layers proved that the earth had been around for a very long time. Although Lyell didn't say it explicitly, he made it obvious that the earth had been around long enough for some species to have evolved from others.

Figure 4–4 A. The tortoises of the Galápagos Islands. Among the extraordinary sights Darwin saw during the voyage of the *Beagle* were these mammoth beasts. B. The tarsier, a representative of the Prosimians (meaning "before the apes"). This tree-living creature, whose eyes are larger than its brain, is a very distant evolutionary relative of the apes, and by extension, man.

At about the same time as Lyell's book was published, Charles Darwin, a young student at Cambridge University in England, was bored by his studies and was thinking about dropping out. Darwin had been something less than a notable success at the university. He had intended to study medicine, but the spectacle of a child undergoing surgery without anesthetic so filled him with horror that he fled the operating room in panic. He then became a candidate for the clergy, but the prospect of a career in religion was less than enticing. A dedicated hunter, fisherman, and rock hound, Darwin was very interested in biology and geology, and was able to parlay his interest and knowledge into an unsalaried position as a naturalist on the H.M.S. *Beagle*, a survey ship bound for a five-year journey around the world (Figure 4–4). Not long before he left, Darwin bought a copy of the first volume of Lyell's *Principles*, planning to read it during some of those long and dull days at sea. The second volume reached him during the voyage.

The voyage of the *Beagle*

It was a wise choice of reading matter. With Lyell's argument in mind, Darwin turned fresh eyes on the life he saw along the South American coast. The number of species unknown in his native England amazed him, and the fact that as one moved along the coast different communities of species were arrayed one after another intrigued him. Where could all these forms of life have come from, he asked himself.

The question became particularly interesting when he visited the Galápagos, an isolated chain of islands some 600 miles off the coast of Ecuador. There an idle question became a disturbing puzzle. From geological evidence, Darwin could tell that the islands were much younger than the mainland. If species had been fixed at creation, one would have to assume that any plants and animals on the islands must be identical to those on the mainland. But not only were the species different from the mainland's, particular species—most notably tortoises and finches—even differed from one island to the next. It seemed impossible that each island had been the site of a separate creation. Was it not more likely that each of the species of, say, tortoise had developed from isolated groups of one original species? The problem plagued Darwin.

When Darwin returned to England, he read the comparatively old *Essay on the Principle of Population*, written by Thomas Malthus in the late 1790s. Malthus, an economist troubled by the migration of man from countryside to city and the constant growth of human population, warned that man would eventually outstrip his food supply and fall prey to severe famine. There was a natural balance, Malthus argued, between a population and its food supply. The food supply limited the size of the population. Darwin saw that this general rule applied not only to man but also to all species. Even though most organisms reproduce at rates that, if left unchecked, would double their populations in relatively short order, the number of organisms within a given population stays relatively constant. Consider an example Darwin himself drew. Imagine a population of eight pairs of birds. Each year four pairs raise a total of four young.

There was a natural balance, Malthus argued, between a population and its food supply . . . the food supply limited the size of the population.

At the end of seven years, the original 16 birds will have swelled to 2,048. However, any population in the natural environment would have remained about the same size, roughly 16 birds. How can this be? Obviously only a proportion of the offspring of a single generation survived long enough to reproduce.

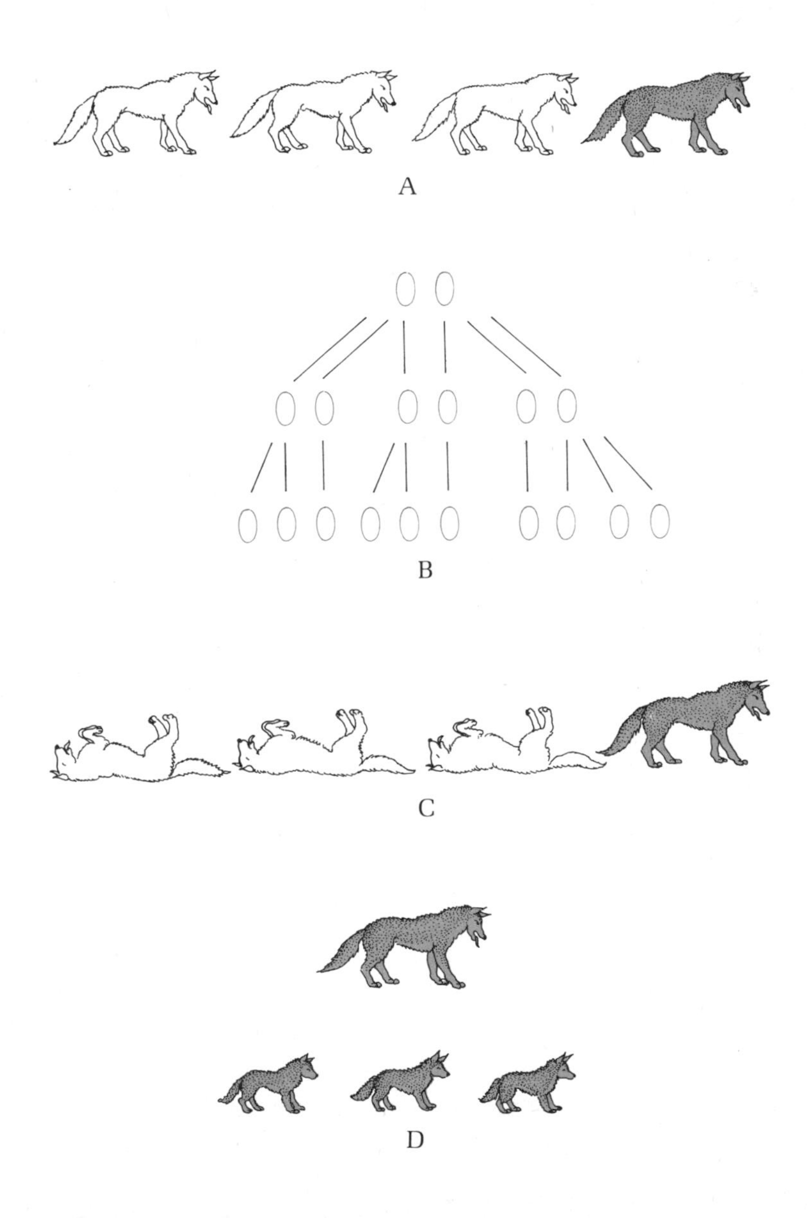

Figure 4–5 Darwin's four basic assumptions. First, all species display natural variation. In this case, one of the wolves has distinctive coat markings (A). Second, species are capable of producing more offspring than can survive into adulthood. Here a pair of wolves produces ten offspring within two generations (B). Third, some natural variations offer the individual a selective advantage for survival. Here the marked wolf has obviously survived some environmental threat that killed off the unmarked members of his species (C). Finally, the better-adapted organisms can transmit the favored trait to their offspring (D). Thus, the survival potential of the species increases over several generations.

The outline of a theory

It is said that the pieces of the puzzle—the diversity of species; the age of the earth; the effects of isolation; the balance between food, population, and the rate of reproduction—all fell into place at once and Darwin saw the finished picture. But he was a cautious man. Rather than rush into print, he spent the next 30 years making observations and gathering evidence to support his conclusions, and it was not until 1859 that Darwin published *On the Origin of Species by Means of Natural Selection*. The world hasn't been quite the same since.

In the book, Darwin made four basic assumptions (Figure 4–5). First, all species display natural variation. That is, each individual within a species is slightly different from the other individuals of the species. The next time you are in a good-sized crowd of people, compare them in terms of height, skin tone, musculature, and so forth, and you'll see what Darwin meant.

Second, the number of offspring produced by a species far exceeds the number that will survive into adulthood.

Third, some of the offspring are better able to adapt to their habitat than the others. Such adaptable offspring are more likely to survive. From this idea sprang the often misused phrase "survival of the fittest," which we will examine shortly.

Fourth, these better-adapted organisms, the ones that survive, transmit at least some part of the adaptation to their offspring. As a result, the species improves—that is, adapts better—over time. Darwin, though, couldn't explain how these characteristics were passed on. Genetics, the science of inheritance (which we will study in Chapter 8), was simply unknown at the time.

As Darwin saw evolution, the process resulted from the interaction between the natural variation of a population of organisms and the environment, an interaction called *natural selection*. By environment, Darwin meant not only the physical habitat and such factors as available water and shelter, but all other aspects of the world around the organism that affect its survival—the food supply, predators, competition both within the species and with other species, and so forth. Any generation of offspring contains more individuals than can survive into adulthood. The pressures of the environment select among the offspring. Only those that survive can reproduce, and the ones that reproduce pass their char-

We can think of the environment as a sieve, selecting for desirable traits and selecting against undesirable ones.

acteristics on to the next generation, which is likewise culled by natural selection. On and on the process goes, and over the long reach of time and by the cumulative effect of many small changes, species evolve.

A case in point

To make Darwin's idea clear, let's compare Lamarck's account of how the giraffe got its long neck with how Darwin would explain the same phenomenon. According to Lamarck, the giraffe was originally short-necked. Trying to feed on the upper parts of trees stretched the giraffe's neck a bit, and this additional neck length was passed on to the offspring. These giraffes fed on the same high trees, stretched their necks still more, and produced yet longer-necked offspring. After many generations and much stretching, an originally short-necked animal became long-necked. Darwin saw this quite differently. As he would explain it, all the giraffes in a population did not have necks of the same length. Some were relatively long-necked, others had much shorter necks, and the rest fell somewhere in between. Each generation of giraffes would include all these variations. However, some of the giraffes were better able to survive and reproduce than the others. The long-necked giraffes could reach more

On and on the process goes, and over the long reach of time and by the cumulative effect of many small changes, species evolve.

leaves than the short-necked ones, and they were more likely to live longer and produce more young, who would inherit the long-necked characteristic, than were the short-necked animals. With each generation, the long-necked giraffes would make up a greater percentage of the population, eventually displacing the short-necked animals altogether.

In Darwin's theory of evolution, characteristics passed on to the next generation were not acquired during the lifetime of the individual. A giraffe was born with a neck of a certain length, and it passed that characteristic of length onto its offspring, all stretching for the topmost leaves notwithstanding. In Lamarck's view, evolution affected the form of an organism during the organism's lifetime. But for Darwin, evolution worked on populations, not individuals. Natural selection affected the frequency with which specific characteristics appeared in a population—for example, the percentage of long-necked and short-necked animals.

Misnomers and misconceptions

A very common mix-up in understanding evolution involves the phrase "survival of the fittest." Those four words raise up visions of some dark and lush jungle where the inhabitants fight tooth and claw until from the struggle emerges a superbeast to claim his crown as king of the jungle. This is not what Darwin meant at all. In his theory, fitness refers only to an organism's ability to produce living offspring who are themselves able to reproduce. Take as an example a pointer judged the world's most perfect hunting dog. This dog could be the fastest, most cunning, and smartest representative of his breed ever to walk the face of the earth. But if he is unable to reproduce for any reason—kept away from any females or born sterile—his Darwinian fitness is zero. By contrast, the neigh-

The championship dog's many assets have no Darwinian significance if he cannot produce living offspring for any reason. The neighborhood mutt, on the other hand, has high Darwinian fitness.

borhood mutt, as mangy of coat, bad of breath, and crooked of leg as he may be, who regularly impregnates every female dog in the county and who leaves behind him a long, if undistinguished, line of fertile descendants, has high Darwinian fitness.

There is no way to determine the relative fitness of two organisms in the abstract. The organism that survives and produces fertile offspring is fit. The criterion for fitness is survival. In a way, "survival of the fittest" means "survival of the survivors."

In the late nineteenth century, twisted evolutionary ideas, called social Darwinism, were applied to human society with the purpose of showing that the upper classes had proven themselves more fit than the masses. Such ideas prevail even today, often with a racist tinge, but are quite erroneous. Any two humans equally capable of reproducing, no matter how different they might be in social status or physical beauty, are both fit in Darwinian terms.

A second misconception is somewhat more subtle and dates back to Lamarck. Lamarck felt that there was some purpose to evolution, an urge or force of the will pushing simple forms of life toward increasing complexity. Thus a simple, single-celled organism like an amoeba was "on its way" to becoming a man. It's an easy trap to fall into, to assume that evolution has a story line in the same way that a novel does. Darwin, though, said that evolution is the product of many subtle and seemingly arbitrary changes, a winnowing of the natural variation of any species. The only logic to it is survival.

A third mistake is to assume that modern species are descended from other modern species. In a debate over evolution, an opponent of the theory asked Thomas Huxley, a contemporary of Darwin and a champion of his ideas, whether he claimed descent from the apes on his mother's or his father's side. Whatever the worth of the remark as wit, it involved a misunderstanding of the theory. One modern species is not descended from another modern species—instead, they both share a common ancestor at some point in the past. In the case of man and the apes, this

In a debate over evolution, an opponent of the theory asked Thomas Huxley, a contemporary of Darwin and a champion of his ideas, whether he claimed descent from the apes on his mother's or his father's side.

ancestral species divided into two separate species. One eventually became the great apes, the gorilla, gibbon, orangutan, and chimpanzee. The other became man. Man is not the descendant of the apes—he is their distant cousin.

The case of the minority moth

It does seem nearly too incredible to believe that a creature as complex and varied as man could have come ultimately from some very simple organism as a result of the culling of natural variation over the long reach of time. But now and then nature provides an example of dramatic, easy-to-see evolutionary change.

A species of moth called the peppered moth was well known to British moth collectors of the last century. The peppered moth was active at night and spent its days resting on trees and rocks covered with the simple plant forms called lichens. Until the middle of the century, all known specimens of the moth were light colored, like the lichen. Then a black

Figure 4–6 An example of natural selection in the moth population of Manchester, England. The peppered moth and the alternate black form on a soot-darkened tree trunk (A). Note the selective advantage enjoyed by the darker form. The same moth forms in a soot-free environment (B). Here the peppered moth obviously is better camouflaged by the tree.

peppered moth was found near Manchester, England. A black phase, as this is called, is not necessarily uncommon. Black and yellow Labrador retrievers, for example, are two color phases of the same breed of dog; the black jaguar of South America is a black phase of the more common yellow-and-black-phase jaguar. But in the case of the peppered moth, the fact that only one black individual had been found indicated that the black phase was very rare (Figure 4–6).

But only for a while. The black-phase moth became more and more common. The overall population of peppered moths stayed the same size, so the black moths were actually displacing the light-colored ones. In time, the black-phase moth made up 99 percent of the population near Manchester, and the light-phase moth had become the rare form. What was going on?

It was a classic case of natural selection in response to a major change in the environment. Up through the first half of the nineteenth century, the light-phase moths were more successful at surviving and reproducing than were the black moths. When they rested on the lichen-covered trees and rocks during the day, the light moths were almost invisible to birds hunting for a meal. The black moths stuck out like sore thumbs and were picked off easily. But with the coming of industrialization, the tables were turned. Smoke and soot pouring out from Manchester's growing number of factories killed the lichen and turned the rocks and tree trunks black. Now the birds found the light-phase moths and passed over the black ones. Better able to survive, the black moths produced greater numbers of surviving black offspring, while the proportion of light-phase moths able to survive long enough to reproduce decreased steadily.

Just how crucial the environment is to evolution is shown by the increase in the number of light-phase moths in recent years. Measures aimed at controlling air pollution have reduced the amount of soot, returning some trees and lichens to their original lichen-covered color. A greater number of the light moths are surviving, and more of the black ones are falling prey to birds.

Even though the change in the peppered moth was quick enough to be dramatic and obvious enough to be visible, the change was actually quite small, involving only the color of the insect. On all counts other than color, the peppered moth remained exactly the same. This is the way evolution proceeds, by means of small changes, some of them so minute as to be imperceptible.

Environments and organisms

In a few instances we can see quite clearly how an environment has shaped and molded the material offered by a species to produce a variety of organisms that differ from each other on far more than just color. The animals that intrigued Darwin the most during his visit to the Galápagos were the finches. Despite the small size of the islands, Darwin spotted 14 different species of finch, all of them native to the Galápagos. On the mainland, the fewer species of finch lived on seeds. Some of the Galápagos finches lived on seeds, but others ate insects. In fact, the Galápagos species were so specialized that some fed only on ground insects or seeds while others restricted themselves to foraging in the trees. On the mainland, other birds, not found on the Galápagos, competed for these other sources of food. Also missing from the Galápagos were woodpeckers, which use their long beaks and tongues to capture grubs and insects that live inside tree bark. On the Galápagos, one of the species of finch performed the role of woodpecker. It used a cactus spine to probe for insects in crevices in tree bark—making it the only known tool-using bird. How had the finches become so diversified (Figure 4–7)?

Darwin reasoned correctly that the finches had all evolved from ancestors that came originally from the South American mainland, probably blown to the Galápagos in a severe storm. Since there were no nuthatches,

Figure 4–7 The evolution of the Galápagos finches. According to Darwin, all of the birds evolved from a single ancestor species to fill vacant ecological niches.

warblers, grosbeaks, or woodpeckers on the islands, a variety of ecological niches were unfilled. Any finch able to make use of these untapped sources of food and shelter—perhaps because of variations in beak shapes—was able to survive and reproduce successfully. The process continued until every ecological niche available to the finches was filled. What had been one species became 14. The phenomenon—development of a group of organisms into a variety of species to fill the available niches—is known as *adaptive radiation*.

Environments, though, can make organisms similar as well as different. Life forms that occupy similar ecological niches often come to look much like each other. Figure 4–8 shows photographs of a porpoise and a shark. Each animal descended from very different ancestors. The shark developed from some of the earliest kinds of fish, while the porpoise, like all members of the whale family, developed from a swamp-dwelling mammal that forsook the land and returned to the sea. But both animals are fast swimmers that prey on fish, and they have evolved similar body shapes that enable them to swim at high speed. When two or more organisms of different ancestry develop similar characteristics, the phenomenon is known as *convergent evolution*.

When the first European explorers and settlers reached Australia, many of the animals they saw looked quite similar to species common back home. But closer examination showed that all the Australian mammals were *marsupials*. Marsupials carry their young for only a short time inside their bodies before giving birth; the immature animal finishes its development in a special pouch on the mother's stomach. Marsupials

Figure 4–8 An example of convergent evolution. Despite their outward similarities, the porpoise (A) and the shark (B) have completely different ancestry. Adaptations to common environmental factors have resulted in their close resemblance.

are found all over the world, but there are relatively few species. For example, the only North American marsupial is the opossum. The other mammal species are all *placentals*, which carry their young internally far longer than the marsupials and have no stomach pouch. Australia had no placental mammals to compete with the marsupials, and the marsupials filled ecological niches occupied by placentals on the other continents. In some cases, adaptive radiation of marsupials on the one hand and of placentals on the other produced animals that occupied similar niches and that bore a striking resemblance to each other (Figure 4–9).

Figure 4–9 The hare-wallaby and the hare resemble each other closely but differ in one important way. The hare is a placental; the hare-wallaby, a marsupial. They have evolved to fill the same ecological niche but in different geographical areas.

The unity of the cell

At the same time that Darwin was writing his theory of evolution, other biologists and naturalists had come to realize that living things were composed of microscopically small units called *cells* (Figure 4–10). Cells are the building blocks of life, and all living things—from a one-celled bacterium to a trillion-celled elephant—comprise at least one cell. (However, partially upsetting this convenient organization of the living world into cells are the *viruses*. Having almost none of the structural features of cells, viruses appear nonliving when examined in their dormant state. However, when they invade a host cell, they are able to reproduce quickly.

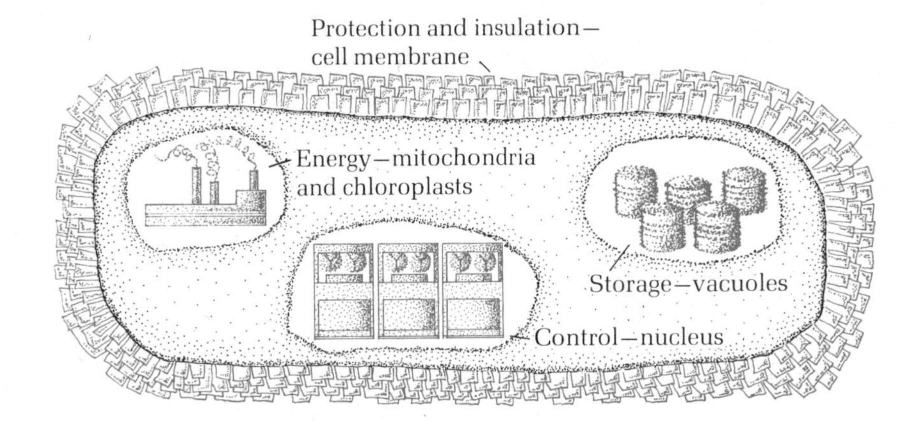

Figure 4–10 Despite its small size, the cell is subdivided into organelles, some of whose functions are shown here. In plant cells, chloroplasts serve as powerplants, using the sun's rays to produce carbohydrate molecules. In all cells, both plant and animal, the mitochondria are the site of respiration, in which energy is released for cellular processes. The vacuoles store food and waste products. The nucleus is the control center of the cell, regulating its chemical processes and containing its hereditary information. Surrounding the entire cell is the cell membrane, which insulates the cell from its environment and controls the flow of materials in and out of the cell.

As your last virus-induced cold probably indicated to you, viral growth within a host organism can cause disease.) There is considerable variety among cells. For example, a plant cell from the root tip of an onion looks different from a cell in the human brain, but cells share common characteristics.

Surrounding the cell is the *cell membrane*. The membrane keeps the cell's chemical composition different from that of the environment. The nature of the outside environment varies for different cells; for example, fresh water in the case of a one-celled amoeba, blood for many animal cells, and sap for plant cells. The living matter within the cell comprises the *cytoplasm* and the *nucleus*. The cytoplasm contains a number of small structures known as *organelles* (literally, "little organs"), which carry out specific chemical tasks. Except in bacteria and blue-green algae, the nucleus is separated from the cytoplasm by the *nuclear envelope*, a membrane similar to the cell membrane.

Cells are more than just structural units in living things, like bricks in a wall. They are the site of all life processes. In fact, all the characteristics of life listed in Chapter 1 hold true for cells: They are organized in a definite pattern; they maintain a chemical composition different from the environment; they take in and use energy; they respond to stimuli; they reproduce; and each structure within the cell and the cell itself fills a life-serving role. The realization that all living things, both the simplest and the most complex, comprise cells lent a special order to the living world. All life was shown to be of the same units. Although specific cells varied somewhat from each other in structure and function, all cells were variations on the same basic theme. The ubiquity of cells underscored Darwin's contention that all life shares a common origin.

But there was a problem. After much debate, controversy, and experimentation, it was proved that cells could come only from preexisting cells. If all modern species descended from an original group of ancestor cells, where did these original cells come from? If there was no unique creation, if all cells—and therefore all life forms—came from preexisting cells, how did the first cells come into being? What was the starting point of life?

Cell theory was still so new at the time of the publication of *On the Origin of Species* that Darwin did not deal with this problem. But for his followers the question of the first cells became a very tricky and seemingly insoluble issue. By working with fossils, they were able to reconstruct much of the evolution of life on the earth, but only to a certain point. Always the problem remained: How did life first begin?

Organisms through time: the length of life

Only in the last 20 years have scientists working on this question been able to draft a reasonable explanation based on a long and ingenious series of experiments. It appears that in the early conditions of the earth, simple molecules bonded into ever more complex compounds that, over many millions of years, aggregated into cells. Let's go back some time before life actually arose and see how the stage was set (Figure 4–11).

The early eons

According to the most widely accepted hypothesis, appropriately called the big-bang theory, the universe originated some 7 billion years ago in an explosion of unimaginable force. This explosion, centered in some huge mass of concentrated matter, sent smaller bits of that matter

If all modern species descended from an original group of ancestor cells, where did these original cells come from?

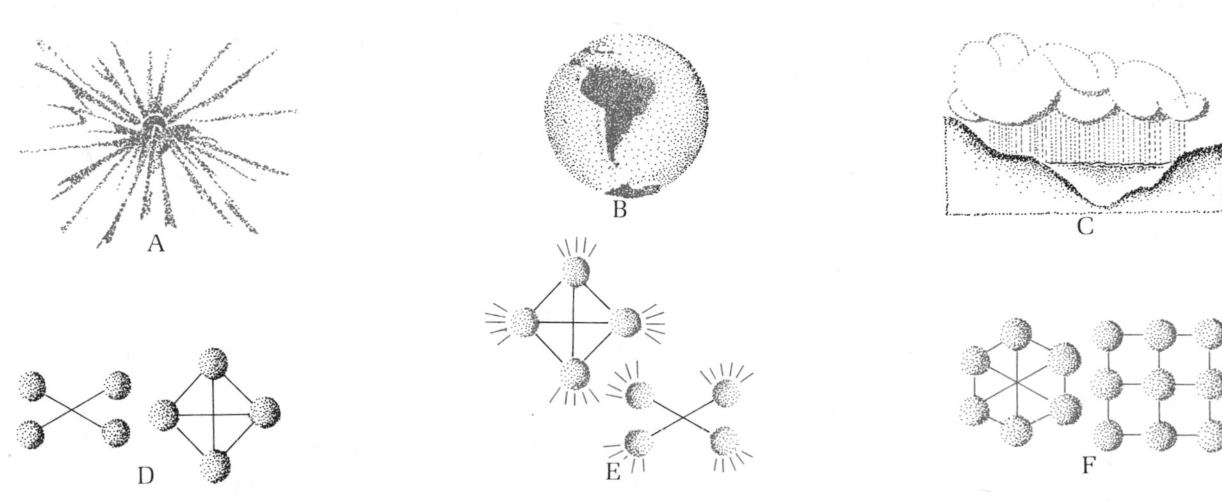

Figure 4–11 A possible explanation of the formation of the first organic chemicals. After a huge explosion of cosmic materials (A), the earth became a separate planet (B). As rainfall collected in the hollows of the earth's surface (C), simple compounds were dissolved in water (D). As ultraviolet light energized these simple compounds (E), they reacted to form more complex chemical substances (F).

flying in all directions into the trackless and empty space. As these bits traveled, they gradually decelerated and cooled until they became our present galaxies, stars, and planets.

The earth probably became an independent planet sometime between 5 and 6 billion years ago. At the time, the earth was far too hot to support life, and most of the water necessary to life was locked away in the interior of the planet. As the planet continued to cool, a primordial crust formed, folded, and became pocked with uplands, basins, plateaus, and mountainlike volcanoes. Steam was pushed out from the earth's interior through fissures into the atmosphere, where the steam cooled into liquid water and fell back onto the barren rocks of the surface. Slowly the basins and hollows filled with water and became ponds and lakes. As rainwater ran from the uplands into these low-lying reservoirs, it dissolved many materials from the rocks and carried them off, even as rainwater does today, to accumulate in the ponds, lakes, and infant oceans. The seas became a hot, diluted soup containing all the simple nutrients essential for life.

Ultraviolet light falling on the earth's surface and lightning in the primitive atmosphere provided the energy needed for chemical reactions between these simple nutrients. In time, the earth's waters contained a relatively great assortment of complex molecules. It is somehow ironic that ultraviolet light, which as we noted in Chapter 3 is lethal to life as we now know it, was a major source of energy for the reactions leading to the original store of organic molecules. This isn't just a theoretical idea; scientists have been able to "reconstruct" this chemical Genesis by mixing ammonia, water, and methane (the probable constituents of the basic soup) and energizing them with ultraviolet light and electric sparks. These simple compounds then react through a series of stages to form more complex organic molecules, identical to many that we find in living systems today.

However, a supply of organic molecules is by no means the same thing as life. A storeroom in a chemistry lab may contain every item essential to life, but that storeroom is not alive. How did life spring from the soup? No one knows exactly. Apparently, organic molecules collected into ever larger and more complex groups. At some point these collections or systems became capable of reproducing themselves accurately. Also, these same collections became able to take in energy, store it, and use it for their own purposes. These molecular systems eventually became the first organisms—single cells capable of taking in energy and storing it and of reproducing.

The first cells were *heterotrophs*, that is, organisms that need a supply of complex organic molecules as a source of energy. We humans are heterotrophs, as are all other animals, fungi, and some one-celled organisms. We take in complex food molecules and break them down by respiration within our cells to release the energy they contain. These original heterotrophic cells needed a steady supply of organic molecules to maintain life. But, as the number of cells grew, the quantity of organic molecules in the sea decreased, resulting in competition between the cells for food. Cells that needed the larger complex molecules were less successful at surviving and reproducing than were cells able to use simpler molecules. Because of natural selection, there evolved cells, called *autotrophs*, capable of producing their complex compounds from simple components. The

A storeroom in a chemistry lab may contain every item essential to life, but that storeroom is not alive.

most successful autotrophs were organisms capable of photosynthesis, which used the energy of sunlight to build carbohydrate molecules from carbon dioxide and water.

When these first photosynthetic cells arose, the earth's atmosphere was considerably different from what it is today. It contained carbon monoxide, carbon dioxide, methane, nitrogen, hydrogen, ammonia, and hydrogen sulfide. Only with the first photosynthetic organisms did free oxygen, which is a product of photosynthesis, appear in the atmosphere. The oxygen reacted with the other gases, reducing the amounts of methane, carbon monoxide, and hydrogen sulfide. Four billion years ago there was no or almost no oxygen in the atmosphere. It took almost 3.5 billion years for the level to hit just 1 percent of what it is today. Then it increased sharply. In only another 300 million years—a short while in the overall scheme of things—the amount of free oxygen increased to the present level, where it has remained until now. A balance had been struck between oxygen production and oxygen use.

A beginning in simplicity

So far the picture we have drawn is largely an educated guess. When we turn to the question of when life began, then we can make use of the fossil record laid down in the rocks of the earth's crust.

Life probably began about 4.5 billion years ago. The oldest fossils of early life were found in exposed rocks near the gold-mining town of Barberton, on the border between South Africa and Swaziland. A little more than 3 billion years ago, this land was covered with a warm, shallow sea or bay. Simple organisms lived in thin sheets at the bottom of the silica-rich seas. Some of these organisms were trapped in the bottom sediments, and geological pressures transformed the sediments into chert, a hard and incompressible rock that preserved the specimens even better than plastic (Figure 4–12).

Similar collections, even richer in fossils, have been found along the Ontario shore of Lake Superior and in the heart of Australia. The Australian deposit, dating back 1 billion years, contains bacteria, blue-green algae, fungi, and green algae. In this deposit details are so well preserved that the cells can be seen dividing.

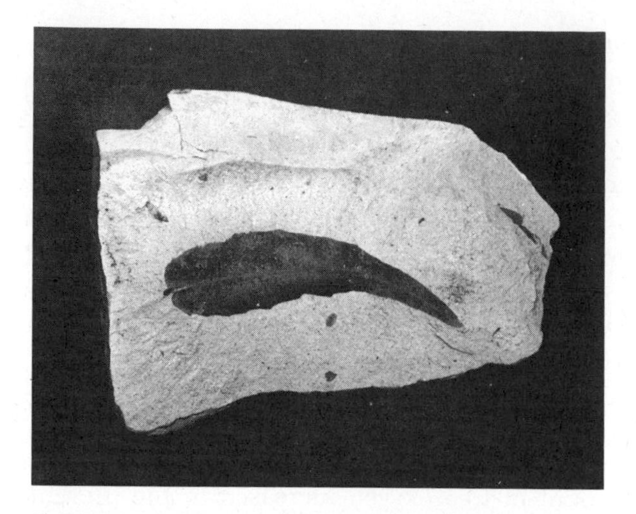

Figure 4–12 Fossil evidence of a long history of life on earth: a fossilized leaf (A); a fossilized fern (B); a fossilized trilobite (C), ancestor of the modern crab and lobster. The discovery of fossils in the nineteenth century led geologists to suspect that life had existed on earth for millions, rather than thousands, of years.

In the 400 million years after the Australian fossil deposit was laid down, living things became increasingly complex and diversified into many forms. The seas teemed with life: trilobites (ancestors of crabs and lobsters), algae and other simple plants, sponges, jellyfish, and a wide assortment of shellfish. As oxygen became more abundant in the atmosphere, the ultraviolet light was filtered out, and the land became a possible site for life.

From the sea to the land

Six hundred million years ago, the land was low-lying and flat, and great inland seas moderated the climate. As a result, the major seasonal changes and the northsouth differences in climate true today were not true then. Instead, most of the planet was graced with a warm and humid climate much like that found now in the tropics.

About 400 million years ago, the first land plants appeared. They were small and leafless things, but in the short span of 25 million years, the first trees evolved from them. Fish, the first animals with backbones, or *vertebrates*, spread throughout the seas. Some of these fish were developing a novelty called lungs that enabled them to explore the coastal fringes of the land for brief periods of time. From the lunged fish evolved the *amphibians*, who live in both land and water; frogs and salamanders are modern examples of amphibians. At the same time, insects flourished. Even today insects are the most complex *invertebrates*, or animals without

Figure 4–13 Luxuriant swamp forests such as this one provided the plant and animal materials that eventually were the basis of our supply of fossil fuels.

backbones. But 350 million years ago, they enjoyed a true heyday. Some insect forms very similar to the cockroach grew to a length of nearly 6 feet!

These huge insects and early amphibians lived in a landscape of unbroken and very lush swamp forests (Figure 4–13). Since the land was low, even a minor increase in the level of the oceans flooded a portion of the land and drowned the plant life. When the sea level fell again, another forest rose on the land and the remains of the previous forest. This pattern of alternating flood and retreat laid down thick deposits of plant remains, and from these deposits came the fossil fuels—coal, gas, and oil—we burn so freely today. Fossil fuel deposits are still being laid down; peat bogs, such as the ones in Ireland, are a first step to the formation of coal and oil. But the amount of fossil fuel being formed now is insignificant compared to deposits laid down 300 million years ago during the Coal Age.

The coming of reptiles and mammals

From the amphibians, who spend at least part of their life cycle in water, evolved the *reptiles*. Lizards and snakes are modern reptiles, and they share the same characteristics as the first members of the clan: They are cold-blooded, egg-laying vertebrates totally adapted to life on the land and independent of the water for their life cycles. About 250 million years ago, the reptiles replaced the giant insects as the dominant creatures, and from these reptiles came the largest land animals the world has known: the dinosaurs (Figure 4–14). Not all the dinosaurs were large or land

Figure 4–14 The dinosaurs dominated the earth's surface about 250 million years ago. Despite their strength and massive size, they were unable to survive environmental changes, and so died out before the advent of the warm-blooded species.

bound, however. Some were small, others took to the air and looked much like birds, and others returned to the sea and resembled dolphins and porpoises.

About 200 million years ago, there appeared the first warm-blooded organisms, animals that maintained a constant body temperature independent of the environment. These were the *birds* and *mammals*. Birds, distinguished by their feathers, lay eggs similar to those of reptiles, while mammals bear their young alive (except for two primitive egg-laying forms) and nurse them from milk-producing glands. Man is a mammal, as are all the fur-bearing animals. Today birds and our fellow mammals seem very prominent, but when the birds and mammals first appeared on the land, they were small and inconspicuous, as well as few and far between.

At the same time that the mammals and birds appeared, the land was changing from swamp forests to a forest dominated by tree-sized ferns and coniferous, or evergreen, trees similar to the contemporary firs and pines. Then the dinosaurs disappeared. No one knows what caused their sudden disappearance; it may have been a sudden and drastic cooling of the climate. Into the place left empty by the vanished dinosaurs moved an ever-increasing number of mammals. Birds likewise became abundant, as did insects resembling modern forms. Flowering plants appeared in the forests and meadows, as well as insects and birds that fed on them. The climate and major geographic features of the land and sea became essentially what they are today.

Mammals became ever more diversified in form. From early and small insect-eaters came larger predators. Primitive horses and camels, much smaller than their modern descendants, appeared. As grasslands replaced forests, large, browsing animals evolved. And in the trees, a group of mammals called primates made their homes, and from those trees and the primates in them would come man.

The tale of man

In the previous pages, we have drawn the very barest of bare outlines of the evolutionary process. We will now turn to man in more detail, but keep in mind that we are giving him far more coverage than his evolu-

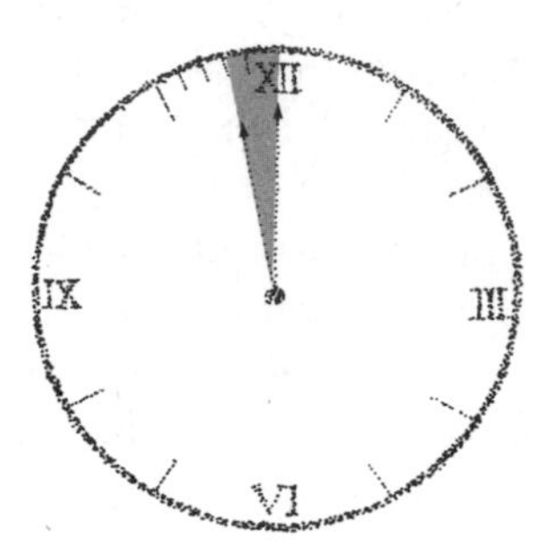

Man's relatively short history on earth. On a 24-hour clock, our lifetime as a species would take up only the last 2 minutes.

tionary seniority allows. If the history of the world from the beginning 5 or 6 billion years ago to the present were depicted by a 24-hour day with midnight representing the present, then man appeared at about 11:58 P.M. We've looked briefly at the first 23 hours and 58 minutes of the day. Now we'll take a more detailed look at those last couple of minutes before midnight.

The rest of the family

About 50 million years ago, the first members of a family of mammals known as the *primates* appeared. As a group the primates share a number of important features. Primates' hands and feet are enlarged, and they contain five fingers or toes. The thumb of the foreleg, or arm, is set opposed to the remaining four fingers, allowing the primate to grasp objects firmly and with good control. All primates have eyes that are set close together and in the front of the head, allowing three-dimensional vision. Most of the modern primates live in trees, and it seems reasonable to assume that the original primates were tree dwellers in the tropical forests of central Africa.

About 10 to 20 million years ago, some groups of primates left the forest and began to live in the neighboring East African grasslands. With the change in habitat came gradual changes in life style and physical form. Originally four-legged, these plains primates began to spend more time on their two hind legs. They began adding meat to their vegetarian fare. With the mixed diet came a mixed set of teeth: knifelike cutting teeth in the front of the mouth for meat eating and flat grinding teeth in the back for vegetable matter.

The exact fossil record of man's evolution from one form to the next is very sketchy, like a few photographs taken during a long and adventurous childhood (Figure 4–15). The oldest known fossil ape-man, *Austra-*

If the history of the world from the beginning 5 or 6 billion years ago to the present were depicted by a 24-hour day with midnight representing the present, then man appeared at about 11:58 P.M.

lopithecus africanus ("the southern ape of Africa"), dates from 2 million years ago. Though an ape by our standards, *Australopithecus* walked upright and made and used tools. But when did such early forms as *Australopithecus* evolve enough to cross the divide between ape and man? For that matter, what does that divide consist of anyway?

An environmental IQ test

Man is one of the more successful species to have existed, able to live in and dominate a great portion of the world. To what is this success due? Man can run relatively fast, but many other species can easily outdo him in both speed and stamina. His sense of smell is weak. His eyesight is good, but not nearly so highly developed as that of an eagle or a hawk. He possesses little body armor or protection, and in terms of strength and fighting ability he can be easily bested by a fierce dog weighing only two-thirds as much as he does. On only one scale does man surpass all other animals, and that scale is size of the brain relative to the body. To get a sense of just how big man's brain is, consider this comparison: A dog, a gorilla, and a man of equal weight would have brain weights of $\frac{1}{2}$, 1, and 3 pounds, respectively. Most of this difference is accounted for by the

Figure 4–15 Despite the gaps and question marks in the fossil record, evidence does connect man with the other primates in an evolutionary pattern.

relative sizes of the cerebrum, the brain's locus of conscious thought.

Twenty million years ago, the ancestors of man had brains no bigger than those of present-day chimpanzees or gorillas. This brain size held true for early primates until 1.3 million years ago. Then, in only 1 million years, brain size tripled—an increase in brain weight of roughly 1 ounce every 30,000 years. Given the glacial pace of evolution, such a change was somewhat akin to running a 2-minute mile.

How could such a change occur so quickly? The species of man was shaped during the period of the four Ice Ages. During each of these Ice Ages, the planet cooled, and glaciers, vast "seas" of ice and rock, crept out from the poles and covered much of the land. Periodically the glaciers receded. The earth warmed, and those areas previously covered with ice were swampy and tropical. These severe, alternating climatic changes had drastic effects on plant and animal life. If you imagine evolution as a way of trying alternate models, the Ice Ages were a very harsh proving ground.

Lacking the strength, speed, or teeth and claws of solitary predators, man became a group hunter. Social or group skills became important; speech developed as a mode of communication during the hunt. To kill big game, man needed something more than just force of numbers, and weapons came into being. And through it all the thing that guided man's success was his ability to think—whether for outwitting fleet-footed game, for figuring out how to fashion a knife from a piece of broken flint, or for communicating with other members of his band. The more intelligent a man, the better his chances of survival. In perhaps the same way that the giraffe's neck grew longer, man's brain grew larger.

Has brain size continued to climb at a steady rate? No. In fact, there has been no appreciable change for the past 150,000 years. Has some physical limit been reached? Probably not. Most likely a balance has been reached between brain size and the demands of the environment.

The spread of man

The first species of what is definitely man has been given the name *Homo erectus*, the name honoring, of course, the creature's ability to stand on his own two feet. *Homo erectus*, who first appeared about 1 million years ago, quickly spread from eastern Africa to appear in a num-

Figure 4–16 Cro-Magnon man. Famed for his paintings on the walls of caves in southern France, he probably bore a close resemblance to his modern-day descendants.

ber of geographically separate places, which have lent their names to the varieties: Java man (southeast Asia), Peking man (north China), and Heidelberg man (western Europe). *Homo erectus* was relatively similar to modern man, but his brain was not quite so large.

By about 200,000 B.C., *Homo erectus* had given rise to a more advanced form with a brain at least as large as ours, Neanderthal man. Neanderthal has been found in Africa and Asia, but the largest population was in Europe. Neanderthal did not survive. About 50,000 years ago, he was replaced by Cro-Magnon man, who is definitely *Homo sapiens*, the species we belong to. Properly shaved and attired, a Cro-Magnon man would be indistinguishable from one of us (Figure 4–16).

We don't know precisely where each type of early man arose. It's possible that *Homo erectus*, Neanderthal, and Cro-Magnon each developed in Africa and then migrated north into Europe and Asia, displacing the species that had gone before. Unfortunately, since much of our direct fossil evidence comes from Europe, we don't know what was happening in Asia. In addition, we don't know when North and South America were populated. According to one theory, Asiatic peoples crossed the Bering Strait between Russia and Alaska and populated the New World. But that seems an exceptionally short time for these people to have filled up two continents stretching all the way from Alaska to the tip of Tierra del Fuego, and to develop into cultures as different as Eskimos and Incas.

We do know that about 10,000 years ago, at least some groups of humans, most likely in the rolling hill country of northern Iran, began growing food crops and living in permanent dwellings. This was the first page of human civilization, a relatively advanced level of social and political organization. By that time, humans had taken up residence in nearly every corner of the world, and from that time the world has been increasingly subject to their whim and fancy.

The significance of the theory

The theory of evolution is easily the single most important idea in biology. The cell theory provided the unifying idea that all living things were composed of the same units. Darwin advanced an even more basic idea—that the form, structure, and behavior of organisms were the result of natural selection, and that the environment "judged" how well every aspect of an organism contributed to its ability to survive. Before Darwin, biology was just a jumble of facts, a chaos of information that had no order and therefore no meaning. The fact that man looks something like the apes or sharks like porpoises, that dogs turn in circles before they lie down, or that flowers have sweet smells and pretty colors were simple curiosities, facts without significance. Evolution gave them that significance, lent them meaning. Similarities between organisms could be traced to common ancestors or to the influence of a shared environment. As to the sleeping habits of dogs and the attractiveness of flowers, one could now ask how these things better enabled the organism to survive. There is no phenomenon in biology, from the structure of the molecules that make up cells to the complicated behaviors of social animals, that has not been given order and significance by the ideas that Darwin first drafted.

Summary

Until the time of Lamarck, it was believed that the varieties of living things had been fixed and unchanged since the moment of creation. Lamarck proposed that complex organisms had evolved from simple ones over time. He cast the inheritance of acquired characteristics as the mechanism of evolution.

Darwin formulated an alternative theory that made four main points.

(1) All populations of organisms show variation.
(2) Each generation of offspring contains more individuals than will survive to maturity.
(3) Only the offspring best adapted to the environment survive.
(4) The surviving organisms transmit at least some aspects of their ability to survive to their offspring. The culling of the population to yield the fittest individuals is called natural selection.

In Darwin's theory, the only determinant of fitness is the ability to survive and to produce fertile offspring; two organisms equally capable of surviving and reproducing are equally fit no matter what else may be true of them. Also, evolution has no purpose or direction, except for the effects of natural selection. And, finally, modern species have not evolved from other modern species; they are all derived from common ancestors at some more or less distant time in the past.

Adaptive radiation describes the phenomenon where a single ancestor evolves into a number of different species, each of which occupies a previously unfilled ecological niche. The Galápagos finches are an example.

Convergent evolution appears when two organisms derived from different ancestors but occupying similar niches resemble each other, for example, the North American placentals compared with the Australian marsupials.

Darwin's ideas were buttressed by the realization that all living things are composed of cells. But it is only in recent years that scientists have explained how conditions on the early earth could have given rise to the first cells. After the earth cooled, water vapor condensing into liquid filled low spots in the crust, resulting in the first seas. These seas contained dissolved nutrients that bonded into ever more complex molecules. These molecules aggregated into the first heterotrophic cells about 4.5 billion years ago. Subsequently, the pressures of natural selection favored the development of autotrophs capable of photosynthesis.

Evolution of life forms continued in the sea with the development of many kinds of algae and invertebrate and vertebrate animals. The first land plants appeared about 400 million years ago, and with them the giant insects. Amphibians, the first vertebrate land dwellers, developed from lungfish. Next came the reptiles who quickly dominated the land, replacing the giant insects. Subsequent cooling of the climate then favored the warm-blooded mammals and birds.

Man developed from the order of mammals known as the primates. The rapid increase in the size of man's brain has been the major factor contributing to his success. Modern man first arose as a recognizable species about 1 million years ago, and became essentially what he is today about 50,000 years ago.

Evolution is crucial to modern biology as the major unifying principle of the science. It has given meaning and significance to a welter of data.

Questions to consider

1. What is the doctrine of fixity of species?
2. In Lamarck's theory of evolution, what was the mechanism of change?

3. What are the basic assumptions of Darwin's theory?
4. What precisely did Darwin mean by natural selection?
5. In a speech given to a Sunday school class, John D. Rockefeller, the founder of Standard Oil and one of the most powerful nineteenth-century American businessmen, said: "The growth of a large business is merely survival of the fittest. . . . The American Beauty rose can be produced in the splendor and fragrance which bring cheer to its beholder only by sacrificing the early buds which grow up around it. This is not an evil tendency in business. It is merely the working-out of a law of nature and a law of God." (Quoted in Richard Hofstadter, *Social Darwinism in American Thought*, Boston: Beacon Press, 1965, p. 45.)
   a. Was Rockefeller interpreting "survival of the fittest" as Darwin would?
   b. On what points would they agree and disagree?
6. a. In evolutionary terms, how can it be explained that insects, birds, and bats, although descended from greatly different ancestors, have very similar wings?
   b. What is the name given to such a phenomenon?
7. a. What is a cell?
   b. What are the typical parts of a cell?
   c. How do we know that the cell is a living thing?
8. How does the universality of cell theory support Darwin's evolutionary ideas?
9. What were the major events in the formation of life between the time of the earth's cooling and the first fossils?
10. The movement from the sea to the land provided life forms with a new environment in which to adapt and evolve. What were the major routes of evolution among the land animals?
11. How can the relatively rapid evolution of man's brain be explained?

Glossary

adaptive radiation the process by which an ancestor species evolves into several species able to fill unoccupied ecological niches in a new habitat *amphibians* a class of cold-blooded animals that lay their eggs in the water and complete their adult years as land animals

autotroph an organism that manufactures its food directly from inorganic substances

birds a class of warm-blooded animals characterized by external feathers and shelled eggs

cell the basic structural unit of living things

cell membrane the outermost boundary of the cell

convergent evolution the process by which species of different ancestry evolve to share similar characteristics

cytoplasm the various substances that compose the nonnuclear portion of the cell

evolution the theory that all life forms are descended from earlier life forms and that the environment imposes a selective force over successive generations

fixity of species the outdated theory that all species were created at one time in the past and that they are unchanging

fossil the remains of a life form or the impression of that life form

heterotroph an organism that is dependent upon autotrophs for its food needs

invertebrates animals that do not have backbones

mammals a class of warm-blooded animals that are covered with fur or hair and which provide milk for their young

marsupials mammals that give birth to live, underdeveloped young which then continue their growth in an external pouch on the mother's body

natural selection the process by which the environment favors specific traits of certain variant forms, thus permitting the more successful reproduction of that form

nuclear envelope the membrane enclosing the cell's nucleus

nucleus (of the cell) the portion of the cell that contains the hereditary information, separated from the cytoplasm by the nuclear membrane

organelle one of the various structural and functional units of the cell *placentals* mammals that carry their young internally for an extended period of time before birth

primates the order of mammals that includes man, characterized by five digits on the fore and hind limbs and three-dimensional vision

reptiles a class of cold-blooded animals characterized by lungs and shelled eggs

spontaneous generation the theory that living things originally arose from nonliving substances

vertebrates animals that have backbones

viruses extremely simple, noncellular life forms that require a "host cell" in which to reproduce

5 Energy

If you took the two previous chapters to be a complete and accurate description of life, you'd have a rather wooden and static idea of what life is. Thus far we've talked about the physical setting and chemical composition of living things, and we've outlined the history of life on the earth. But we have avoided talking about living things in the way we most commonly think of them—that is, not so much as what they are, but as what they do.

If someone told you to describe each of a series of organisms with only one word, you might very well choose to describe them by their typical actions or activities. Life is dynamic—it involves movement and change. To talk about this aspect of life, we have to first get some idea of the concept of energy. We've looked at the chemistry and history of life; now we'll see something of its physics. In addition, we'll take a look at the flow of energy through living systems.

Basic ideas about energy

Energy is simply the capacity to do work, while *work* is force exerted through a distance. If you bend over and pick up a book that has fallen from a desk to the floor and put it back on the desk, you've done work, and performing that work involved an expenditure of energy. In this case the energy was *kinetic*, involving motion. *Electrical energy* involves the invisible movement of electrons. *Sound energy* results from vibration, a movement of matter. *Heat energy* likewise involves movement of the molecules making up some sample of matter; molecules in hot water move faster than they do in cold water. The rays of the sun contain *light energy*, which we feel as heat. *Chemical energy* involves the making and breaking of chemical bonds, a topic we shall look into more closely in a short while.

Energy can be stored as well as used, and such stored energy is called *potential energy*. Potential energy is energy stored against a force, as shown in Figure 5–1, a diagram of a typical spring scale, the kind fisher-

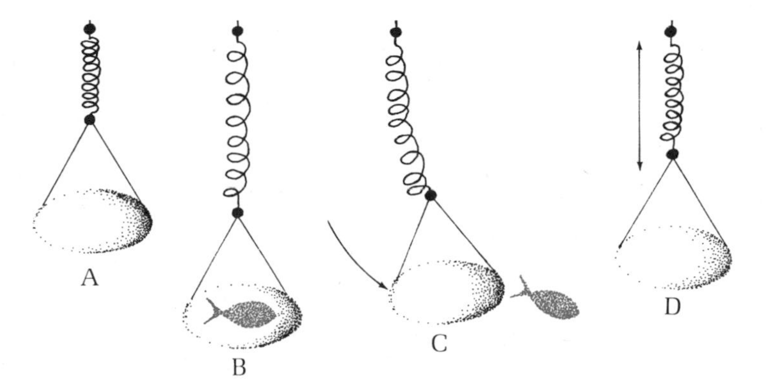

Figure 5–1 A simple demonstration of the conversion of potential energy to kinetic energy. The coiled spring has potential energy by virtue of its position (A). The spring is lengthened by the weight of the fish (B). If the fish is bumped off (C), the potential energy is converted into the kinetic energy used in keeping the scale vibrating up and down (D).

men use to weigh their catch. As long as there is no weight hanging from the spring, the scale is at rest and contains no potential energy. When a fish is hung on the scale and the spring uncoils, it contains potential energy, as does the fish. The fish is hanging against the force of gravity, which would otherwise pull it down, and the spring is extended against the force of its own contraction. If the fish should slip off the scale, say someone bumps it, then the potential energy stored in both the fish and the scale would be converted to kinetic energy as the fish falls to the ground and the scale springs back to its normal shape.

Since we are used to thinking about energy only in terms of machines, it is easy to forget that living things make use of every one of the types of

Figure 5–2 Living things use energy in many different forms. They can convert the energy stored in chemical bonds into electrical energy (electric eel), sound energy (the call of a bird), mechanical energy (athlete), light energy (flask of luminescent bacteria), and heat energy (hibernating bear). With the exception of nuclear energy, living systems can use and convert all energy forms.

energy mentioned. Consider a few examples. A cricket's chirp involves sound energy; a firefly's flare is light energy; a fish's rhythmic swimming is an example of kinetic energy. Electrical energy, except in the case of the electric eel, might seem hard to isolate in living things, but the nerve impulses that direct the movement of all animal life are a form of electrical energy (Figure 5–2).

A resting animal that wakes from a nap and then trots away to find its dinner has obviously tapped some store of potential energy and used it to power the mechanical activity of walking. But potential energy is not only a power source for activities such as walking or chirping. Potential energy might also be called the energy of organization, and in this role it is crucial to life.

The first law

In Chapter 2, we explained a law called the conservation of matter, which said that the earth has a given and unchanging store of matter. A similar law applies to energy, and it is appropriately and similarly named the *law of the conservation of energy*. This law states that energy can be neither created nor destroyed, only changed from one form to another. The branch of physics and chemistry that studies changes in the form of energy is called *thermodynamics* (from the Greek, meaning the flow or movement of heat), and the conservation of energy is often referred to as the *first law* of thermodynamics.

The first law is really an accounting principle, a sort of double-entry ledger of energy, as is the conservation of matter. If you took any living system and measured the energy entering and leaving, you would find that the two amounts would equal each other exactly. What we popularly call an energy-producing process—say, burning wood to heat and light a room—is, from the scientific point of view, an energy-converting process—changing potential chemical energy into heat energy and light energy.

The first law is really an accounting principle, a sort of double-entry ledger of energy.

The first law is not the whole story of energy, though. In fact, if the first law alone described the energy transactions in both the living and the nonliving worlds, earth would be a very strange place indeed. For example, assume that the first law is the whole truth about energy. Now build that fire in the cold and dark room, and predict the results. If you say that the room will grow warmer and lighter, you are using knowledge not contained in the first law. On the basis of the first law alone, all that could be accurately predicted is that all the transformed energy would remain in the fireplace. The fireplace would be extremely bright and extremely hot, while the rest of the room would remain just as dark and as cold as it had been before the fire (Figure 5–3).

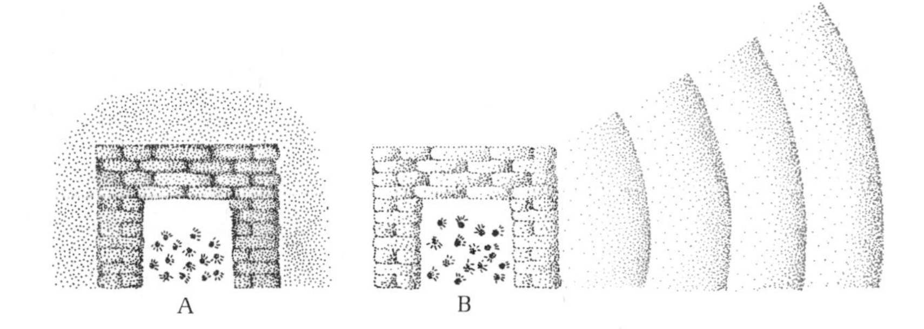

Figure 5–3 The first and second laws of thermodynamics. The first law of thermodynamics tells us that the heat and light from the fireplace would stay in one corner of the otherwise dark room (A), being neither wasted nor lost. However, we know from experience that the heat and light will dissipate throughout the room (B), reaching a more random, less concentrated state.

But we know from common sense that such things don't happen. We need another law of thermodynamics to explain what we already know from everyday experience.

The second law

The law we need is the *second law* of thermodynamics, which can be stated in two ways. The first is that all conversions of energy from one form to another involve an increase in heat, which is dissipated into the surroundings. In the case of our fire in the room, it is impossible for all the chemical energy of the wood to be converted to light energy. A certain proportion of that chemical energy is converted into heat. Some of this heat can be put to work, say, in cooking a pot of stew over the fire. But some of it escapes into the surroundings where it is not readily available to do work.

Efficiency is the term used to indicate the ratio of useful energy—energy available to do work—to total energy. Because of the second law, efficiency is always less than 100 percent. If the efficiency of a certain process is 70 percent, then 30 percent of the energy has dissipated as heat. Humans, for example, are relatively inefficient energy converters. To get the energy the average adult needs each day, he has to eat food that would theoretically yield 10 times that much. Thus humans have an efficiency of about 10 percent. What happens to the remaining 90 percent? Some energy is simply passed through the body without being transformed and is stored as potential energy in the feces. Most of it, though, is given off as heat. Some of this heat serves the useful purpose of maintaining the constant human body temperature of 98.6°F, but most of it simply dissipates into the surroundings.

The second statement of the second law says that in any closed system (that is, any system to which no new energy is added), the energy of that system moves toward the least concentrated and least orderly areas. This is the reason that the room doesn't remain cold and dark once the fire is burning. If all the heat and the light energy of the burning wood remained within the confines of the fireplace, then the system of the room would contain an area of high energy concentration—the fireplace—and an even larger area of low concentration—the rest of the room. Instead, the energy moves from the area of high concentration to the area of low concentration. This movement entails an increase in *entropy*, which is a measure of disorder or randomness. The more ordered a system is, the less its entropy.

The two statements of the second law add up to the same thing, for an increase in the heat of the surroundings involves an increase in entropy. An ice cube, which is water in the solid state, contains little heat and little entropy. Add heat so that the ice melts into liquid, and the entropy increases. Add still more heat so that the hydrogen bonds break and the water molecules escape into the air, and the entropy has increased still more.

The second law provides the most basic explanation of a common phenomenon and irritation: the fact that things fall apart (Figure 5–4).

The second law provides the most basic explanation of a common phenomenon and irritation: the fact that things fall apart.

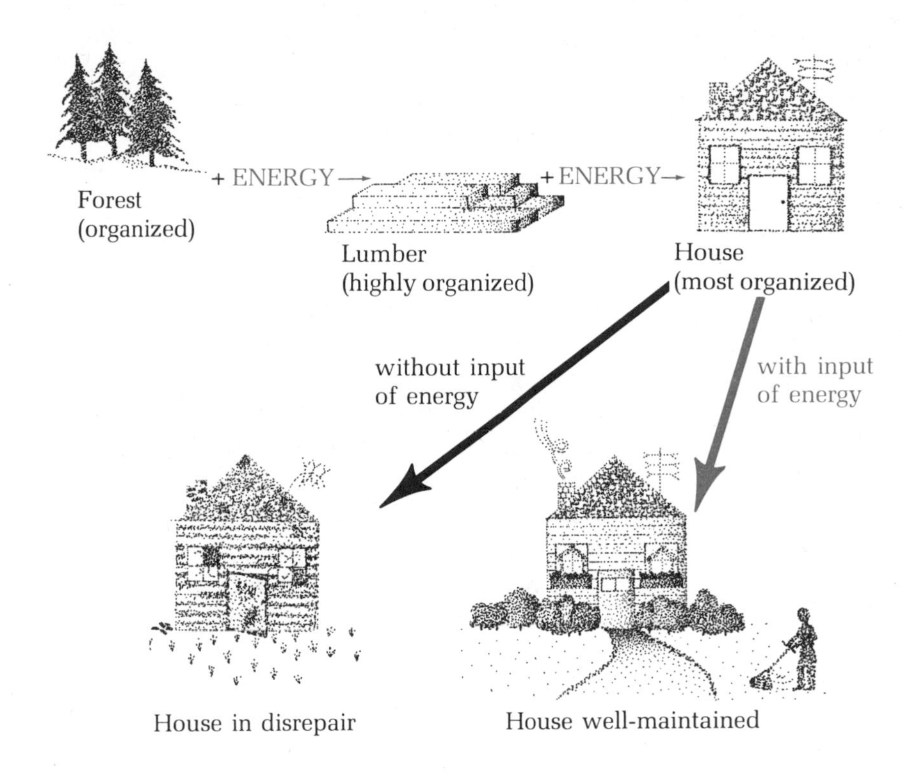

Figure 5–4 The natural tendency for things to fall apart. Energy is required to convert the forest into lumber and to organize the lumber into a house. Once the house is built, it begins to fall apart and will continue to do so unless its owner uses further energy to maintain it. Despite even his best efforts, however, the house will never be restored to its most "organized," least random new condition.

Any system with potential energy has less entropy than one that contains no potential energy at all. According to the second law, all natural processes tend to increase entropy. Recall the example of the spring scale and the suspended fish. This system is highly ordered, containing a great amount of potential energy. When the fish slips off and the spring jumps back to its normal shape, potential energy is converted to kinetic energy. At the end of the process of falling and springing, the system that was orderly and concentrated in terms of energy has become relatively disorderly and randomized. This is a simple example of a constant phenomenon, the movement from lesser entropy (higher potential energy) to greater entropy (lower potential energy). Thus mountains wear away, rivers flow from high ground to sea level, wind blows from a high pressure center to a low one.

Thermodynamics and life

The laws of thermodynamics affect living things just as they do nonliving, and they provide us with some basic and crucial insights into the processes of life. The first law requires that there be a constant source of energy for life. Go again to the example of the fire in the room. If wood isn't added to the fire at regular intervals, the blaze will burn down to ashes, and the room will again become cold and dark. The same thing holds true of energy for living things; living things need a steady source of energy.

The second law adds to our understanding of the nature of life's energy needs. In addition to any organism's requirements for the energy needed to power its daily activities, it requires a constant source of potential energy. Any living thing, even the simplest one, is a highly ordered arrangement of matter, and like all such orderly and high-energy systems it tends to disintegrate, to move toward a state of lowest order, least energy, and greatest entropy. Living things need energy not only for performing mechanical actions. They also need it, very literally, just to stay together.

Two important questions

So far we have avoided two important questions. The first is how any living organism gets energy from the food it consumes. Food is obviously a different fuel from wood, gas, or coal, and an organism is different from a furnace or a fireplace. How is energy stored in a hamburger or a candy bar, and how is it released? The second question deals with where the energy comes from in the first place. Somewhere, somehow, energy must enter living systems. How does this happen?

First we'll investigate the form of biological energy. Then we'll look for the ultimate source of that energy.

Living things need energy not only for performing mechanical actions. They also need it, very literally, just to stay together.

The energy in bonds

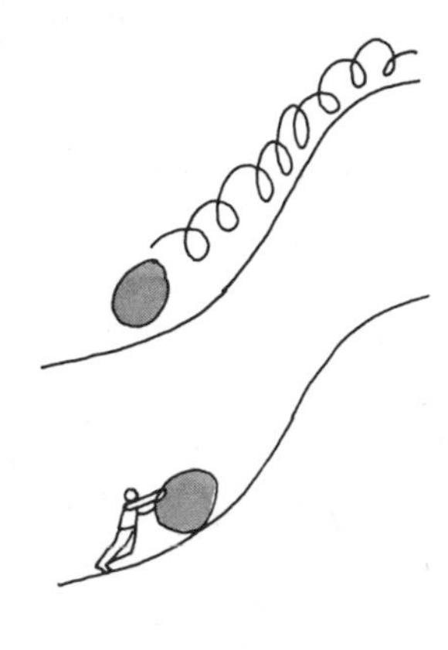

Biological energy is chemical energy. Accustomed as we are to thinking of energy as somehow related to heat and burning, it may be somewhat difficult to see how energy can be stored and released from chemicals.

Chemical energy resides in the chemical bond. All chemical reactions—which are essentially the making and breaking of bonds—have an energetic factor. If no outside energy is added to a group of atoms, the only reactions that will occur are those in which the bonds of the product molecules contain less energy than the bonds of the original reactant molecules. Spontaneous reactions or series of reactions are sometimes called *downhill reactions* because, like a boulder tumbling down the side of a mountain, they entail a release of potential energy. Reactions that require energy are called *uphill* because they involve an input of energy, as does pushing the boulder back up the mountain.

Series of reactions arranged uphill and downhill serve as the energy conduits in all organisms. Uphill processes are used to store energy, while downhill reactions release that energy for an organism's activities. In all organisms, the making and breaking of chemical bonds is the mechanism for energy flow.

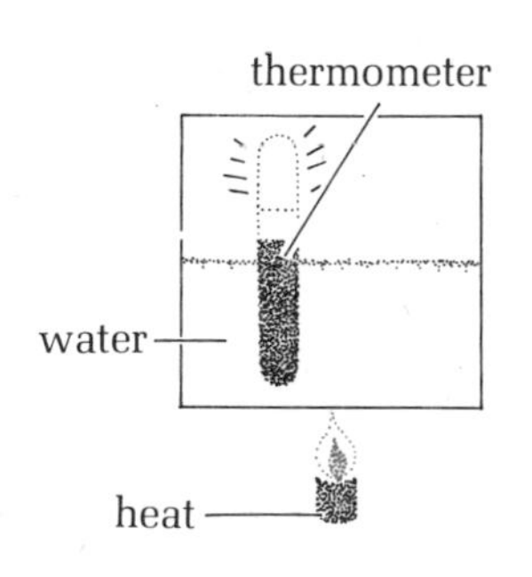

Figure 5–5 A calorie is the unit of heat energy required to raise one gram of water one degree Celsius. Multiplied by 1,000, it would equal a Calorie.

For convenience sake, bond energy is generally measured in terms of the *calorie,* which is the amount of heat needed to raise the temperature of 1 gram of water 1° Celsius (C). The energy value of food, though, is measured not in these relatively small units, but in the bigger unit known as the *kilocalorie,* which is equal to 1,000 calories. The kilocalorie is also designated as the "*big calorie,*" or simply Calorie with a capital C (Figure 5–5). Food values are stated in Calories, the bane of every dieter worried about the bulge on his or her middle.

Though we have answered that first question—the one about the form of biological energy—we still have to deal with the second one. Where does the energy stored in the chemical bonds of food come from to begin with?

Again, the green plant

In our discussion of nutrient cycles in Chapter 3, we found that green plants play an important role in taking essential elements from the nonliving world and transforming them into compounds that other organisms can use. Plants play an equally crucial role in the flow of energy through the living world. Through the process of photosynthesis, green plants capture the energy of the light of the sun and fix that energy into the bonds of carbon compounds.

If the world is again imagined as the cold and dark room, the fire that warms the cold and lights the dark is the sun. Within the sun occurs precisely the same reaction that provides the enormous energy of a hydrogen bomb. However, of all the light produced by the sun, only a very small proportion strikes the earth. About 30 percent of the light is reflected back into space by the upper atmosphere, and 47 percent is absorbed by the oceans, the crust, and the atmosphere and converted to heat. Most of the remaining 23 percent provides the power for the water cycle—evaporating water, carrying it into the atmosphere, and dropping it back to the earth and the oceans. Only a meager 1 percent of the light energy falling on the earth is captured by green plants, and this small fraction of a small fraction of the energy produced in the hydrogen bomb of the sun is the power source for the countless life forms on the earth (Figure 5–6).

The green pigment

Light energy comes in a number of different wavelengths, and each has a characteristic energy. What we see as color is a reflection of this fact. The white light of the sun is actually a mixture of all the colors of light, as shown in Figure 5–7. An object has color because it absorbs certain wavelengths and reflects others; the color you see is the reflected one. When a color is absorbed by a pigment, the pigment takes up some of the

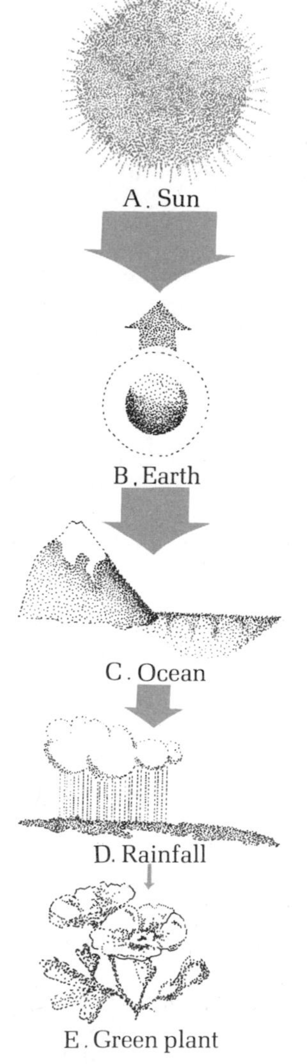

Figure 5–6 Accounting for the earth's share of the sun's energy. Of the solar energy that reaches the upper atmosphere of the earth (A), about 30 percent is reflected back into space (B). Approximately 47 percent is absorbed by the crust and the oceans (C), and about 23 percent provides the power for the water cycle (D). Only 1 percent of the total original energy is absorbed by the earth's green plants to serve as the power source for photosynthesis.

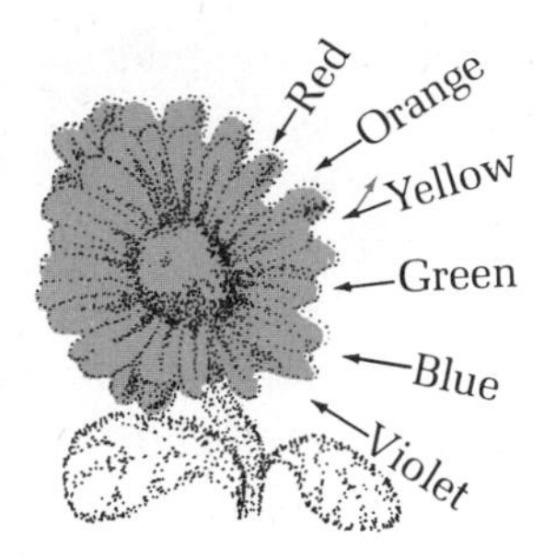

Figure 5-7 White light is actually made up of six basic colors. An object is said to have a certain color if it reflects light of that particular color, while absorbing all other colors. Plant leaves are green because their pigments reflect green light.

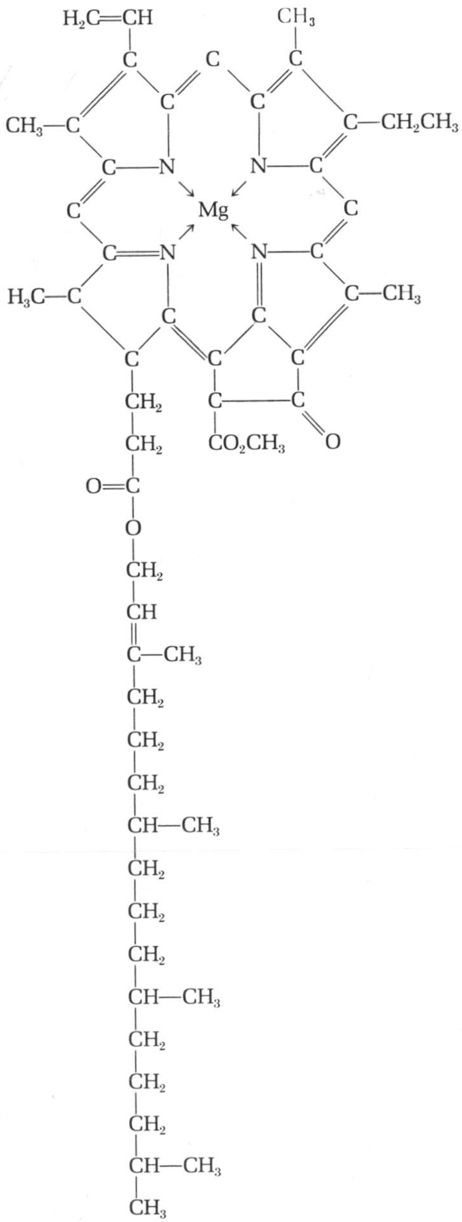

$H_2C{=}CH$ CH_3
C C C
CH_3—C C C C—CH_2CH_3
C—N N—C
C Mg C
C=N N—C
H_3C—C C—CH_3
C C
C C C
CH_2 C—C
CH_2 CO_2CH_3 O
O=C
O
CH_2
CH
C—CH_3
CH_2
CH_2
CH_2
CH—CH_3
CH_2
CH_2
CH_2
CH—CH_3
CH_2
CH_2
CH_2
CH—CH_3
CH_3

Figure 5-8 Chlorophyll *a*, one of the common forms of the chlorophyll molecule. This complex chemical substance is the basic energy trap for green plants, serving as the portal of entry for the solar energy needed in photosynthesis.

energy of the light. Photosynthesis is built around the ability of pigments to absorb energy.

The most important photosynthetic pigment is *chlorophyll*, a complex carbon-nitrogen-magnesium-hydrogen compound found in all land plants, all sea-borne algae, and a few kinds of bacteria (Figure 5–8). Chlorophyll absorbs blue and red light and reflects green, thus giving plants their characteristic green cast. There are several different types of chlorophyll, each a little different chemically from the others and each characteristic of certain species, but they all serve roughly the same function. Plants do contain pigments other than chlorophyll; such pigments produce the characteristic orange of carrots and provide the blaze of color in autumn's leaves. In terms of photosynthesis, though, the task of these pigments is simply to absorb the energy in other wavelengths of light and to pass the energy on to the chlorophyll (Figure 5–9).

Chlorophyll and light

When light strikes a molecule of chlorophyll, an electron of the chlorophyll molecule becomes excited. The chemical use of the term excited is not unlike the popular meaning of the word. We think of an excited person as being more energetic and active than a sedate one, and the comparison holds true of electrons as well. In Chapter 2, we noted that the electrons of a molecule are arranged in shells about the nucleus and that these shells contain characteristic numbers of electrons. When light falls on a chlorophyll molecule, the molecule absorbs the energy of the light, and this absorbed energy boosts the electron from one shell to the next. In this boost, the light energy of sunlight is transformed into chemical energy. So transformed, sunshine becomes the energy of life.

From chlorophyll to carbohydrate

The excited electron is unstable, tending to return to its original, unexcited state and to release the absorbed energy in the process. This released energy is transferred to other molecules. Such an energy transfer between molecules is actually a transfer of electrons: A molecule rich in

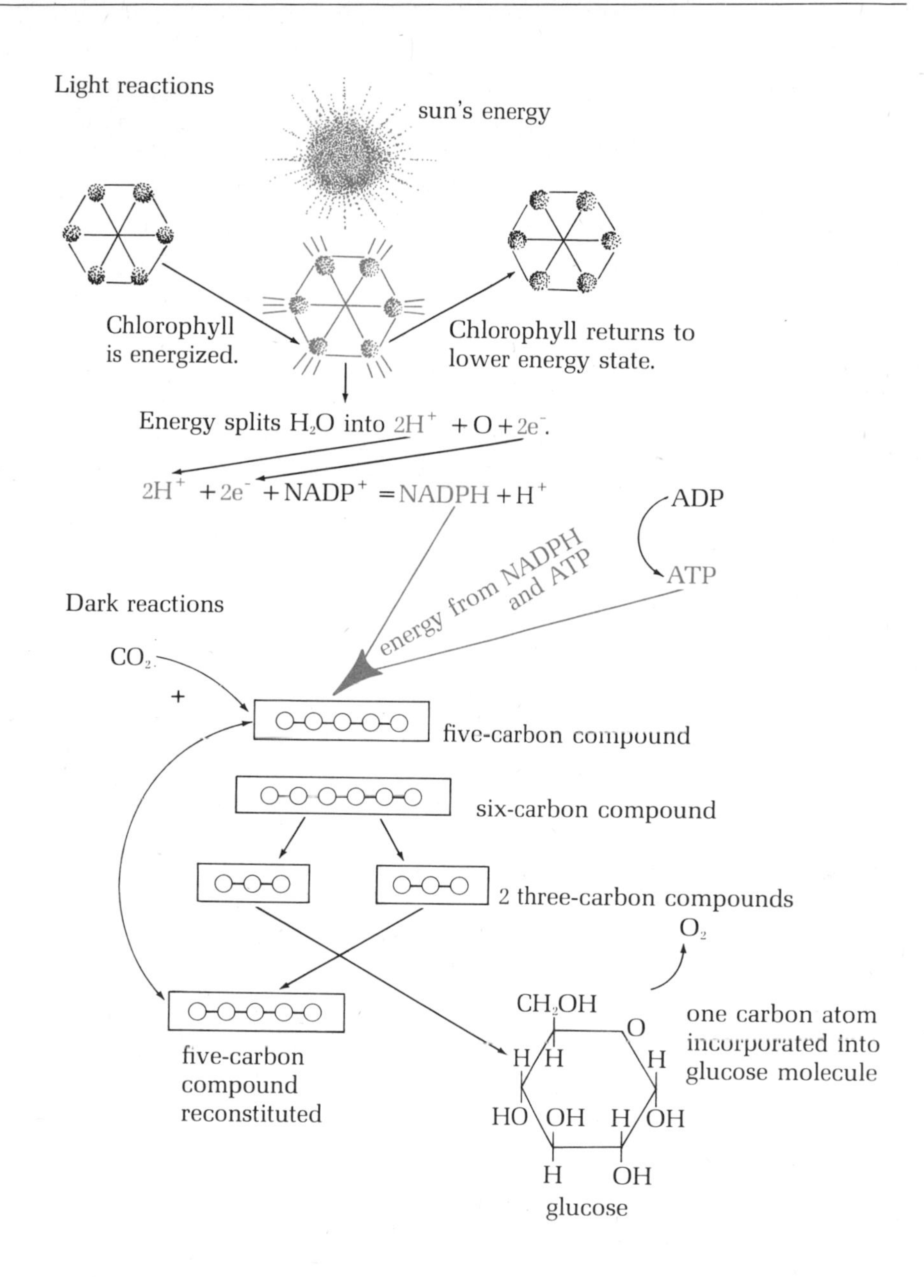

Figure 5-9 A summary of photosynthesis. In the light reactions, energy is stored in the bonds of ATP and NADPH. This same energy is then used to fuel the reactions of the dark phase of the process—the fixing of atmospheric carbon dioxide into the six-carbon glucose molecule.

electrons, and therefore in energy, passes electrons along to a molecule poor in electrons. With the added electrons, the second molecule moves from a state of low energy to one of high energy.

As the chlorophyll electron drops from its excited state, the released energy is picked up by two chemical compounds, NADP and ADP. Having gained electrons, the NADP becomes the higher-energy form NADPH. The ADP becomes ATP, which is also a higher-energy form. At the same time, water molecules absorbed by the plant are broken apart, yielding two hydrogen ions, one oxygen atom, and two free electrons for each molecule of water. And, as the second law predicts, some of the energy is lost as heat.

These reactions, activated by the energy of sunlight, are known as the *light reactions*. At the end of this series, the energy of the sunlight has been stored in the NADPH and ATP molecules. These molecules serve as the energy source for the next series of reactions, called the *dark reactions* because they do not require the presence of light.

The dark reactions

The leaf of a green plant contains carbon dioxide that has entered from the air through microscopic pores in the leaf's surface. This carbon dioxide bonds with a particular carbohydrate containing five carbon atoms. The resulting six-carbon molecule quickly splits into 2 three-carbon molecules. Now the energy stored in ATP and NADPH enters the picture and powers the remaining steps. The three-carbon molecules are first taken apart, then reassembled. At the end of the process, the original five-carbon compound has been regenerated, and the carbon dioxide has been fixed into carbohydrate, representing a net gain in the energy of the organism. Also, the component atoms of the water broken apart in the first stages of photosynthesis enter into the process: The hydrogen is incorporated into the carbohydrate molecules, while the oxygen is released as a gas into the atmosphere.

Once the simple carbohydrate is formed, it may follow any one of a number of routes. It may be used as fuel in the plant's own cellular respiration, which powers its growth and reproduction. It may serve as a simple building block, in addition to other elements, for complex molecules found in the plant. Some of the simple sugars are combined into long, chainlike molecules of repeating units known as *polysaccharides*. Much of the plant's carbohydrate is stored in the form of the polysaccharide called starch.

The sum of the process

Most of the nutrients and all of the energy available to living things begin in green plants, which are quite successful at their energetic task. About 30 to 35 percent of the solar energy absorbed by plants is finally trapped as chemical energy, which makes plants some three times as efficient as humans in extracting useful energy from their energy source.

But trapping energy in the first place is only part of the story and half of our concern. How does an animal that eats a plant transform the potential energy stored in the grass into a form that it can use? To answer that question, we must look in more detail at respiration.

An outline of cellular respiration

While photosynthesis is the special province of green plants and a few other organisms, cellular respiration is common to virtually every kind of living organism, from the simplest to the most complex, and to each and every cell within those organisms. It is the most basic and universal process of life (Figure 5–10).

Light energy is stored in the plant in the form of bonds in carbohydrate molecules. To make use of that stored energy, the cell must convert the energy in the sugar into ATP. ATP is the power source for cells. Why is it that this molecule rather than any other performs such a crucial function?

ATP, whose full name is adenosine triphosphate, comprises three units: a nitrogen-containing molecule known as adenine; the five-carbon sugar ribose; and three phosphate groups, each group composed of one atom of phosphorus bonded with four atoms of oxygen. The phosphates are linked together like a chain with one end attached to the ribose. The

About 30 to 35 percent of the solar energy absorbed by plants is trapped as chemical energy, which makes plants some three times as efficient as humans in extracting useful energy from their energy source.

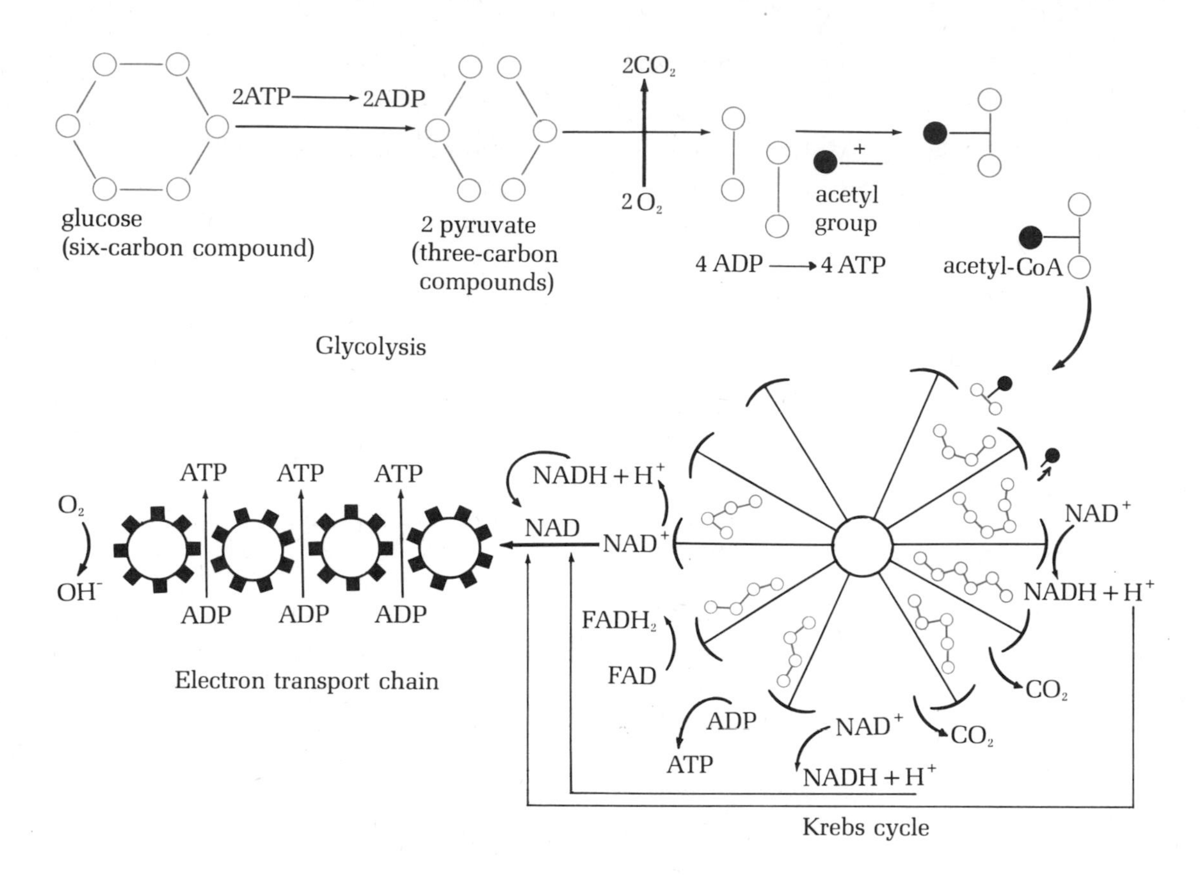

Figure 5–10 A summary of cellular respiration. In glycolysis, ATP is used to convert the six-carbon glucose molecule into two three-carbon pyruvate molecules. In the presence of oxygen, the pyruvate reacts to yield carbon dioxide and acetyl CoA. Fed into the Krebs cycle, the acetyl CoA activates a series of chemical reactions, the ultimate purpose of which is to store energy in the compounds NADH and $FADH_2$. This energy is passed into the electron transport system, in which ADP is converted to ATP. The energy in the high-energy bond of ATP can then be used to serve the energy needs of the organism.

key to the energetic role of ATP is the phosphate-phosphate bond, which is a type of *high-energy bond* (Figure 5–11). The term high-energy bond refers not to a strong bond but to one that yields its energy readily. Breaking one phosphate-phosphate bond of ATP—that is, making ATP into ADP (adenosine diphosphate)—releases energy useful to the cell. Thus the basic role respiration plays is to use the energy provided by carbohydrate to combine ADP with phosphate to yield ATP.

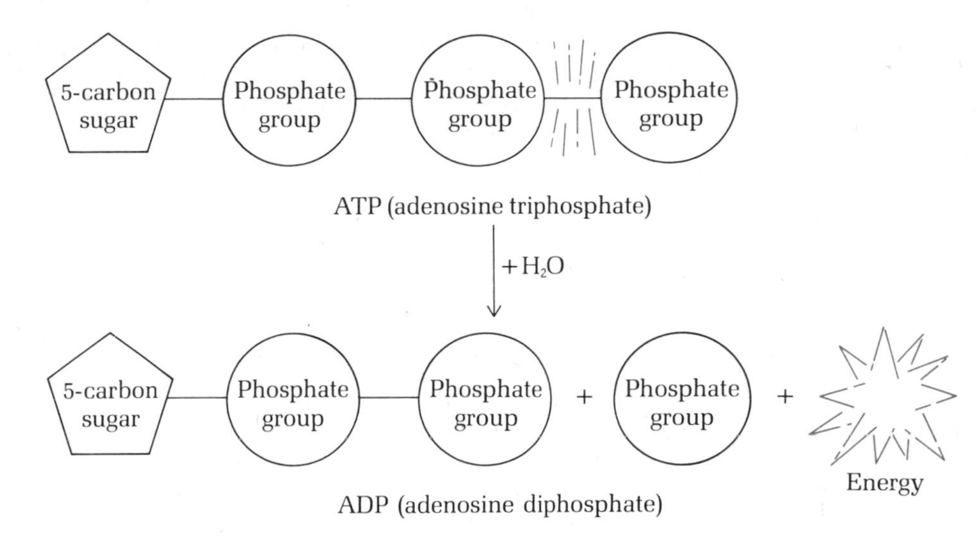

Figure 5–11 The bond between the second and third phosphate groups in ATP is a high-energy bond. When it is chemically broken, it yields a great deal of energy, as well as forming ADP and a free phosphate group.

Chemically, respiration is a highly complex process, involving many series of reactions. In outline, though, the process is simple. In most animal cells, the carbohydrates produced in photosynthesis arrive in the form of glucose, a sugar molecule. The glucose molecule is then split, the hydrogen atoms are stripped from the carbon atoms and combined with oxygen to yield water, and the energy released in these reactions is used to make ADP into ATP. Let's look at these three steps in more detail.

Glycolysis

The word *glycolysis* means the unbinding of glucose, and that is precisely what happens in these first reactions. The glucose molecule is broken into 2 three-carbon molecules called pyruvate. The process requires the expenditure of two ATP molecules, but glycolysis results in a net gain in energy for the cell. The pyruvate molecules still contain a relatively great amount of energy, and the next two steps in cellular respiration break them down.

Glycolysis does not require oxygen, and it therefore has the alternative name of *anaerobic respiration*. It is very likely that glycolysis evolved very early in the history of life, before the atmosphere contained much oxygen. The ability to extract energy from carbohydrate when no oxygen is present is still important to many organisms. Yeast cells, living under oxygenless conditions and provided with a supply of sugar, make alcohol from the pyruvate remaining after glycolysis. The yeast can stand only so much alcohol. When the medium is about 12 percent alcohol, the yeast dies and the process stops. Cut off from the air in special vats and fed on the sugars found in grain or fruit, yeasts produce all our drinking alcohol. We humans are also capable of anaerobic respiration. When we exercise strenuously, the muscle cells quickly run out of oxygen. Under these conditions, the muscles convert the pyruvate into lactic acid. And the soreness of overexerted muscles is due to accumulated lactic acid.

The Krebs cycle

When oxygen is present, the pyruvate molecule reacts so that it yields a molecule of carbon dioxide and a two-carbon fragment. This fragment bonds with a special carrier molecule to make a compound called acetyl CoA, which enters the *Krebs cycle*. In the course of this chain of reactions, the acetyl group is broken off from the acetyl CoA, and then reacts in such a way that all its energy is released. Some of this energy escapes as heat, but most of it is transferred to the electrons of hydrogen atoms released during the breakdown of the acetyl group. These energy-rich hydrogen atoms are passed to the receptor molecules NAD and FAD.

The electron transport chain

In this set of reactions, the energy-rich NAD and FAD molecules are passed "downhill" in such a way that their energy is released to convert ADP into ATP. At the end of the process, the released electrons combine with protons to yield hydrogen atoms, which in turn bond with oxygen and form water.

Respiration and energy

At the end of the whole cycle of cellular respiration, the energy of each glucose molecule has been packaged into 40 molecules of ATP.

Since two ATP molecules are expended to start glycolysis, the cell nets an energy gain of 38 molecules of ATP. Respiration achieves an efficiency of about 60 percent, a rather remarkable figure when you consider that an automobile engine runs at an efficiency of only 25 percent.

And what is the fate of the ATP? The energy released by the breaking of the high-energy bond powers the vast array of the cell's biochemical machinery, which is the basis of all life. Chemically, ATP is the form in which solar energy is eventually used by the cell.

Energy and ecosystems

When we talked about the carbon cycle in Chapter 3, we noted that all organisms can be classified as producers, consumers, and decomposers. Roughly the same set of labels apply to the pattern in which organisms obtain food and the energy it contains. Solar energy is converted to food by photosynthetic producers. *Primary consumers* get their energy by feeding on plants. They are also called *herbivores*, meaning plant eaters. *Secondary consumers* eat the primary consumers, which earns them the alternate name *carnivores*, or meat eaters. *Tertiary consumers* are the carnivores who eat other carnivores. The waste matter and dead bodies of all these organisms form the diet of the decomposers.

To a certain extent, any such classification scheme is flexible, and its categories may overlap. A mountain lion feeding on a jackrabbit is a secondary consumer, while a meal of coyote would qualify him as a tertiary consumer. Man feeds on a wide variety of other organisms. We eat great quantities of vegetables and grains, and feed regularly on such primary consumers as cattle and pigs, while most of the fish we prefer—tuna, salmon, and catfish, for example—are carnivores. Man belongs to a class known as *omnivores*, organisms that eat both plant and animal matter. Probably the only organism more omnivorous than man is the black bear, whose varied fare includes fresh salmon, marsh grass, rotten venison, honey, and ants.

Assigning these labels to the various organisms in an ecosystem provides one with a good idea of how energy flows through that system. Each class of organism in an ecosystem constitutes a *trophic*, or feeding, *level* in the flow of energy:

I producers → II primary consumers → III secondary consumers → IV tertiary consumers

decomposers

As the second law states, any transfer of energy involves some loss of useful energy to the surroundings in the form of heat. For the trophic levels of an ecosystem, this means that each level must contain less energy than the one below it and more than the one above it. A series of trophic levels diagramed in terms of size looks not like a set of steps but like the pyramid shown in Figure 5–12.

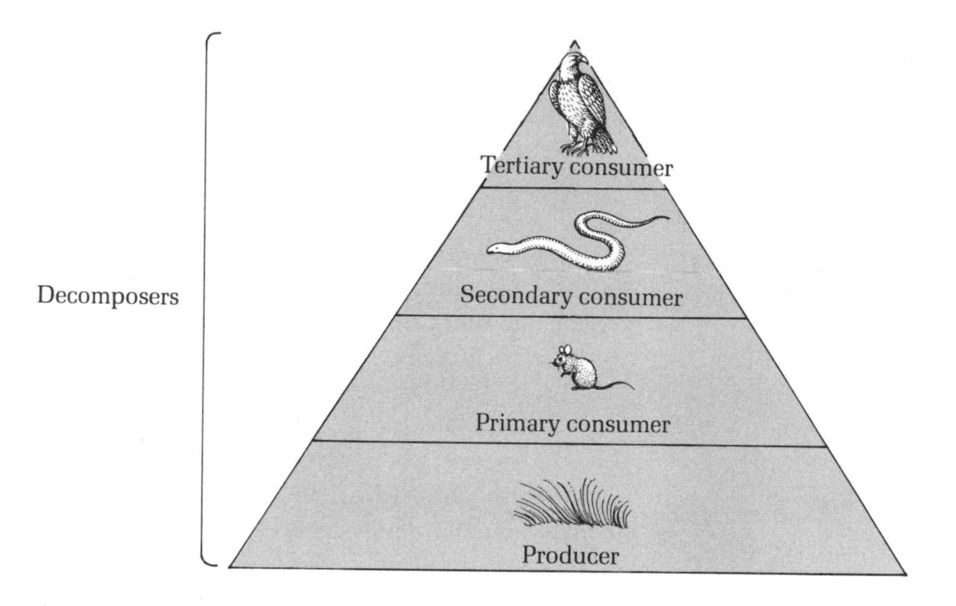

Figure 5–12 Energy relationships shown by trophic levels. Each level fulfills its energy needs by drawing on the level beneath it.

While it is possible to measure the precise amounts of energy transferred and lost from level to level in a complex ecosystem, the picture becomes very confusing very quickly. The main points about energy trans-

Each trophic level must contain less energy than the one below it and more than the one above it.

It takes considerably more energy to make a pound of hamburger than it does to make an ear of corn.

fer can best be seen in a simpler ecosystem consisting of only three levels: corn plants, cattle that fed on the corn, and humans who subsist on the meat and milk produced by the cows. Let's assume that in the course of a year, 1,000 units of energy fall on the cornfield. The corn plants trap and retain only 10 units of the solar energy, roughly 1 percent. The cattle graze on the corn, but they don't make use of the whole plant or even of all the energy available in what they do eat. They retain only 1 of the original 1,000 units. Man is quite inefficient in his utilization of the Calories in the cattle's milk and meat. We get only 0.1 unit of the 1,000 that fell on the corn.

Obviously, it takes considerably more energy to make a pound of hamburger than it does to make an ear of corn, and this is part of the reason that diets high in meat cost more than vegetarian fare. The amount of energy lost from one level to the next is comparatively staggering. As a general rule, available energy is reduced a hundredfold in the step from sun to carbohydrate, and about tenfold in each successive step. In our simple food chain of three levels, the energy used by secondary consumers is approximately:

% energy available to secondary consumer
= % energy passed from sun to plants
× % energy passed from plants to primary consumer
× % energy passed from primary consumer to secondary consumer

$$= \frac{1}{100} \times \frac{1}{10} \times \frac{1}{10}$$

$$= \frac{1}{10{,}000} \text{ of solar energy}$$

Great amounts of plant matter are required to support relatively few herbivores and even fewer carnivores. This tremendous loss of energy

Available energy is reduced a hundredfold in the step from sun to carbohydrate, and about tenfold in each successive step.

from one level to the next explains why it is that most food chains have only three or four trophic levels and never more than five.

Some energetic facts of life

All of life is organized about exchanges of energy. The key to life is photosynthesis, which converts the energy of sunlight into a form cells can use. All cells share cellular respiration, in its aerobic or anaerobic versions. At all levels of organization of living things, from the molecules that compose the fuel cells to the trophic levels of an ecosystem, the rules governing energy exchanges shape life and impose limits on it.

The inevitability of waste

Because of the second law, living things exist at the expense of their surroundings. Any living thing, even the smallest and simplest bacterium, can maintain its order only by increasing the entropy of the environment. Waste is an inevitable by-product of life.

The increase of entropy can be seen in a number of specific ways. To begin with, all the energy transformations of life entail a dissipation of waste heat to the environment. Also, the products entering any living system—whether that system be a single cell or a complex ecosystem like a human city—are considerably more ordered than the products that leave it (Figure 5–13).

Eutrophication as an energetic phenomenon

Eutrophication is the process where conditions in fresh water change so that animal life is displaced by algae and bacteria. In Chapter 3, we

The products entering any living system are considerably more ordered than the products that leave it.

Figure 5–13 The modern city provides a useful example of the relationship between order and randomness. Ordered, containerized goods enter the city, are processed within the systems of the city, and their remains leave the city in a highly disordered, randomized form—garbage and pollution.

discussed it as a disruption of the normal nutrient cycle. It can also be seen in terms of energetics as a massive shift in trophic levels. When fresh water is the dumping ground for human wastes and fertilizers, the water becomes filled with molecules that algae, but not animal life, can use as an energy source. The number of algae increases. As the increasing numbers of algae die, decomposer bacteria are presented with an increasing source of energy, and their numbers also grow. The cellular respiration of the algae-consuming bacteria uses up oxygen dissolved in the water. Animals at a higher trophic level can no longer get the oxygen they need for the aerobic phases of their respiration and can no longer meet their energy needs. As a result, they die.

Biological magnification

The animals of the second trophic level feeding on the plants of the first are able to make use of only part of the energy stored there. As we saw previously, it takes several acres of corn to support one steer, and several steers to support one human. Generally, each trophic level is smaller in size than the one below it. Certain poisons, generally ones not found in nature but introduced by man's activities, are picked up by feeding plants or animals. These poisons are not broken down by the feeder, nor are they excreted. Instead they gather in the tissues, and, as the organism eats more, the concentration of the poison increases. This load of pollutants is passed from one trophic level to the next, and what was trace poisoning in level I rapidly increases to a toxic concentration in levels II and III. The phenomenon is known as *biological magnification.*

A good example of biological magnification occurred at Clear Lake, located in northern California near San Francisco. In many respects, it was an ideal summer resort. Tucked into the low mountains of Lake County, its setting was scenic, and fishing for crappie and catfish was excellent. The one drawback was the clouds of annoying gnats.

The immature gnats lived on the bottom of the lake, where they were particularly vulnerable to insecticides. In 1949, 1954, and 1957, the insecticide DDD, a close chemical relative of DDT, was mixed into the water of Clear Lake. Little poison was used, only 0.02 molecule of DDD to a million molecules of water, or 0.02 ppm (parts per million). But 99 percent of the immature gnats were killed, and Clear Lake's bug problem seemed to be solved.

However, only a few months after the 1954 application of DDD, more than 100 dead western grebes, a species of large, fish-eating birds, were found around the lake. None of the birds showed any sign of disease or violent death. A large number of dead grebes was again found after the 1957 dose of insecticide. Some curious researcher analyzed fatty tissue from the dead grebes for DDD and made an astounding discovery: The concentration of DDD in the grebes' fat was 1,600 ppm, a concentration some 80,000 times greater than that of the lake water!

Closer analysis of the Clear Lake food chain, shown in Figure 5–14, revealed that the DDD became more and more concentrated at each trophic level. The water contained 0.02 ppm DDD, while the phytoplankton (microscopic water plants) contained 5 ppm, and small organisms that fed on the phytoplankton hit 15 ppm. Fish that ate these small organisms and that were in turn food for the grebes averaged 1,000 ppm. In low concentration, DDD was lethal only to the tiny, immature gnats. But, at such great accumulations, the DDD became highly toxic to larger animal forms. In 1949, 1,000 pairs of grebes rested at Clear Lake and produced many young. In 1961, there were only 32 nesting pairs and no young.

Clear Lake is no isolated case. Many species of carnivorous birds living at the edges of lakes and oceans have been affected by accumulations of DDT. Besides killing some birds outright, DDT alters the egg-laying chemistry of the survivors so that the shells of their eggs are too thin to be incubated by a sitting bird. As a result, several species, among them the osprey and the pelican, are in imminent danger of extinction.

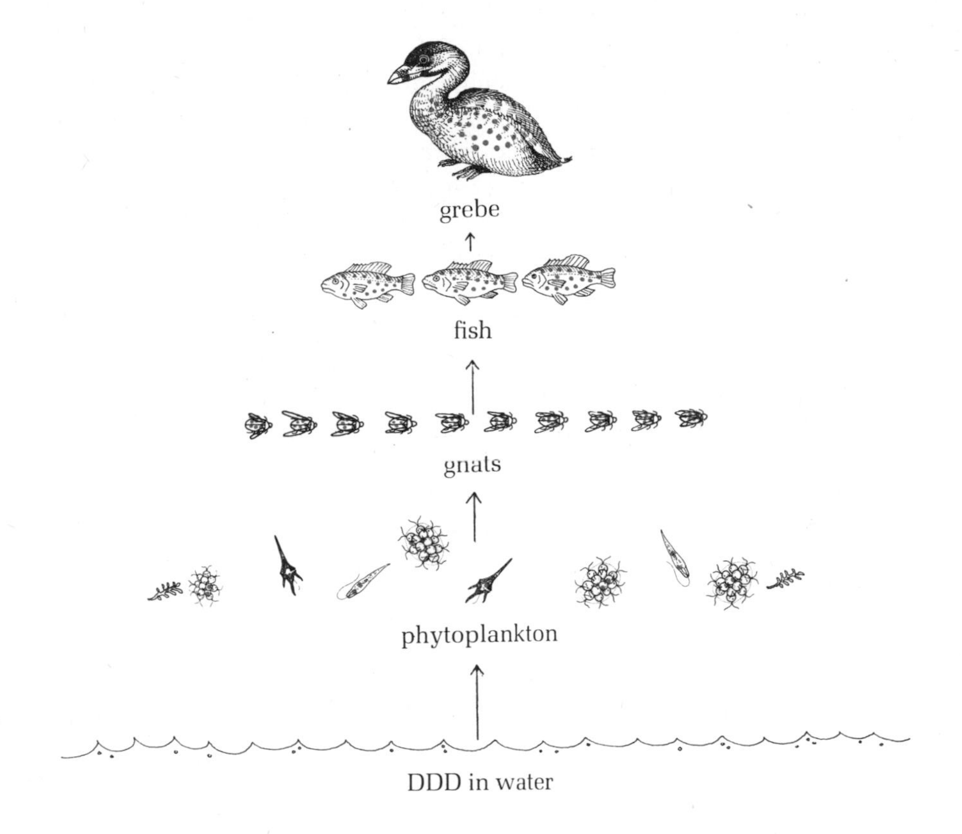

Figure 5–14 The fatal food chain of Clear Lake. As the DDD molecules were transferred from one trophic level to the next, they became increasingly concentrated, finally reaching a fatal level in the grebe.

Biological magnification not only affects birds. The tissues of the average American contain 11 ppm DDT. DDT also concentrates in human milk, where the average concentration of about 0.2 ppm is several times higher than the amount of DDT allowed by federal law in cow's milk transported across state lines.

Biological magnification also occurs with other poisons, such as strontium 90, a radioactive waste produced in atomic tests, and methyl mercury, a by-product of certain industrial processes.

Summary

Energy exists in a variety of forms: kinetic, electrical, sound, heat, chemical, potential. All energy conversions are governed by the laws of thermodynamics. According to the first law, energy is neither created nor destroyed, only converted from one form to another. The second law states that in all energy conversions some energy dissipates to the surroundings as heat and that with each energy conversion the entropy of the universe increases.

The energy of living systems is chemical energy. Downhill reactions release energy, and uphill reactions require it.

All the energy available for life begins in photosynthesis, by which solar energy is converted to chemical energy. In the light reactions, sunlight falling on the chlorophyll molecule excites an electron. When the electron returns to its normal state, the energy of excitement is picked up by the molecules NADP and ADP, becoming NADPH and ATP. These molecules power the dark reactions, in which carbon dioxide is fixed into usable carbohydrate.

Through cellular respiration, the energy of the carbohydrate is converted into the form of ATP, which the cell can use for its functions. The key to ATP's role is the high-energy phosphate-phosphate bond. There are three parts of the process: glycolysis, the Krebs cycle, and the electron transport chain.

Ecosystems can be divided into trophic levels based on the energetic roles of the organisms: producers, consumers (primary, secondary, tertiary), and decomposers. Each level is smaller than the one below it. Characteristically, the first level captures only 1 percent of the energy potentially available to it. Each level thereafter gets only 10 percent of the energy of the preceding level.

The laws of thermodynamics shape the form of life in special ways. Among them are waste, eutrophication, and biological magnification.

Questions to consider

1. What is potential energy?
2. What is the first law of thermodynamics?
3. What is the second law? What does it mean to say that one system has more entropy than another?

4. What is chemical energy? How can chemical reactions store and release energy?
5. What are the crucial events of both the light and dark reactions of photosynthesis?
6. What are the major events of the three stages of cellular respiration?
7. Why would the second law lead one to predict that each trophic level is smaller than the preceding one?
8. What is biological magnification?
9. How is biological magnification an energetic phenomenon?

Glossary

anaerobic respiration phases of cellular respiration that do not require oxygen

ATP adenosine triphosphate; the energy-storage molecule of living systems

biological magnification the phenomenon in which substances are concentrated as they pass up through the trophic levels

calorie a measure of heat energy, the amount of heat needed to raise 1 gram of water 1 degree Centigrade

Calorie a measure of heat equal to 1000 calories; a convenient unit for measuring food energy

carnivore an animal that meets its energy needs by eating other animals

chemical energy energy stored in chemical bonds

chlorophyll the complex chemical pigment which absorbs solar energy, thus providing energy for photosynthesis

conservation of energy the principle that the amount of energy in the universe is fixed and that energy can be neither created nor destroyed

dark reactions the fixing of carbon from carbon dioxide into carbohydrate, a process that does not require light energy

downhill reaction a reaction that releases energy

efficiency the ratio of useful energy gained to total energy used in a reaction

electrical energy energy carried in the movement of electrons

electron transport chain the third phase of respiration, in which much of the energy released in the Krebs cycle is used to form ATP from ADP

energy the capacity to do work

entropy the disorder randomness of a system

first law of thermodynamics the law of conservation of energy

glycolysis the anaerobic phase of respiration in which glucose is split into pyruvate

heat energy energy carried in the movement of molecules

herbivore an animal that meets its energy needs by eating plants

high-energy bond a chemical bond that readily releases energy

kinetic energy the energy of motion

Krebs cycle the second phase of respiration, which completes the breakdown of glucose

light energy energy carried on light waves

light reactions the photosynthetic reactions that require light energy captured by chlorophyll

omnivore an animal that meets its energy needs by eating both plant and animal matter

polysaccharide a large molecule formed of repeating units of simple sugars

potential energy stored energy

primary consumer an organism that meets its energy needs by eating a producer organism

second law of thermodynamics the principle that the randomness of a system increases with each reaction or conversion in that system

secondary consumer an organism that meets its energy needs by eating primary consumers

sound energy energy carried on sound waves

tertiary consumer an organism that meets its energy needs by eating secondary consumers *thermodynamics* the study of energy conversions

trophic level a feeding—or energetic—level in an ecosystem

uphill reaction a reaction that requires an input of energy to proceed

work the exertion of a force through a distance

Part Two

The Processes of Life

In Part One, we looked at the most basic facts of life, energy and matter, and saw how these two ingredients affect the composition of life. In Part Two, we will examine some of the most basic activities of organisms, particularly those that have to do with maintaining and continuing life. Chapter 6, *Breathing and Circulating*, examines the mechanisms by which the gases of cellular respiration are transported and exchanged. In Chapter 7, *Eating and Excreting*, we'll look at what food is, how it is digested, and how wastes are disposed of. Chapter 8, *Inheriting*, will explore the transmission of traits from one generation to another. In Chapter 9, *Reproducing*, we will examine how species propagate, and we will pay particular attention to human sexuality. Chapter 10, *Developing*, will chart the course of a growing organism from its beginning as a single cell. Chapter 11, *Moving*, will be a study of how organisms get around in their search for food and the other needs of life. Chapter 12, *Communicating: Internal*, looks at the self-coordination and regulation of organisms. Chapter 13, *Communicating: External*, describes how organisms learn about their environments through their senses.

6 Breathing and Circulating

Except for the few anaerobic bacteria and the yeasts, which can live without oxygen, all cells must take in oxygen and get rid of carbon dioxide. The oxygen is needed in the steps of cellular respiration that follow glycolysis, and the carbon dioxide is a waste product of the Krebs cycle. This process of gas exchange is what we commonly call breathing.

Breathing alone, the taking in of oxygen and the expelling of carbon dioxide, is only part of the picture. A simple organism can exchange gases with its environment across the membrane of its single cell. But the respiratory activities of each cell in a complex being comprising billions of cells, many of which are a long way from any regular supply of oxygen, poses a problem. How can oxygen be brought to the cells and carbon dioxide taken out? Among the animals, the answer consists of two sets of specialized structures. The first, known as the *respiratory system*, performs the task of removing carbon dioxide from a special liquid medium and of adding oxygen to it. The liquid is part of the *circulatory system*, which also includes the pathways the liquid follows to and from the cells and any specialized structures that pump the liquid.

In this chapter, we'll look at breathing and circulation in a number of organisms, particularly man. But there's more to air than the oxygen we need and the carbon dioxide we give off, so we'll also delve into how air pollution affects life and health.

The variety of breathing

The many kinds of organisms employ a great variety of ways of obtaining oxygen and getting rid of carbon dioxide. This variety is not simply a curiosity, but a list of alternative "answers" posed by evolution to the "problem" of gas exchange. Each one results from an interplay of natural selection, the history of the organism, and the type of environment it lives in. Beneath this variety lies the logic of evolution.

The land plants

We rarely associate breathing with land plants because the structures of gas exchange are too small to be seen readily by the naked eye. How-

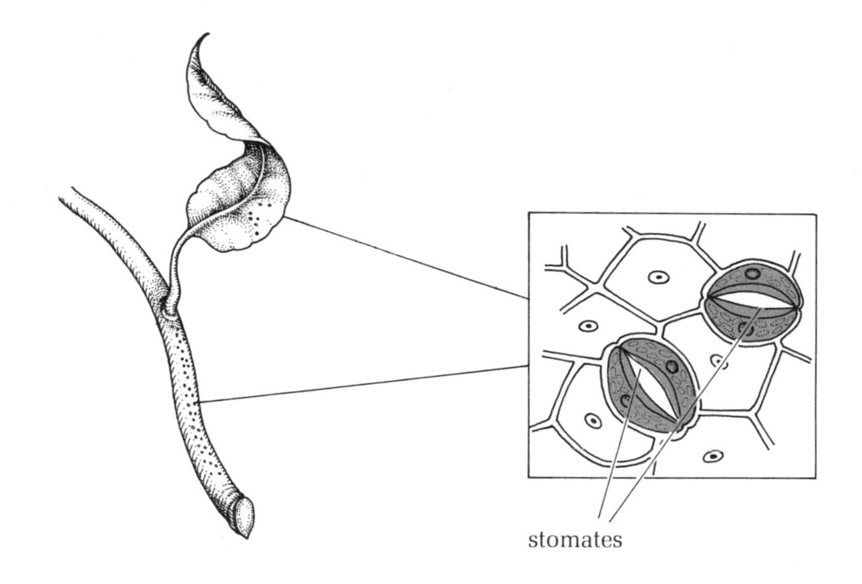

Figure 6–1 The stomates of a typical land plant. These small openings make possible the exchange of gases with the atmosphere.

We rarely associate breathing with land plants because the structures of gas exchange are too small to be seen readily by the naked eye.

Figure 6–2 The sea anemone is an example of an organism whose interior–exterior ratio makes possible a simplified breathing system based on the passage of gases directly through its exterior surface.

ever, gases do pass in and out of plants through openings in the stems and leaves. Figure 6–1 points out the small white flecks typical of twigs. These flecks are clusters of loose-fitting cells through which oxygen and carbon dioxide can pass. The leaves contain smaller holes, visible only under a microscope, called *stomates*. Although the stomates appear small to the eye, they are cavernous in comparison to a molecule of oxygen or carbon dioxide. If such a molecule were your size, a stomate would look like a door a third of a mile high and equally wide. Once oxygen has passed through the stomate, the gas is dissolved in the watery film that coats all the plant's cells and is thus carried to and from each cell.

The surface breathers

A number of simple plants and animals that live either in water or in very moist environments simply exchange gases through their surfaces, even without the aid of structures like stomates. The cell membrane is not solid, as is a brick wall. It contains small openings that small molecules like carbon dioxide and oxygen can pass through. The simplest surface breathers are one-celled organisms like bacteria, blue-green algae, and protozoa. Others are multicelled organisms like sea anemones, worms, snails, strands of algae, tufts of moss, sponges, and mushrooms (Figure 6–2).

Diffusion through a cell membrane. As the cell carries out its metabolic functions, carbon dioxide accumulates within the cell (A). The second law of thermodynamics requires that oxygen will diffuse into the cell across the cell membrane and that carbon dioxide will diffuse out of the cell, thus reducing the initially high concentrations on both sides of the membrane (B).

How is it that the carbon dioxide and the oxygen flow the "right" way? The answer lies in the second law. Water surrounding a cell contains dissolved oxygen. As the cell respires, carbon dioxide builds up within it. Thus there is a concentration of oxygen outside the cell and of carbon dioxide inside. Because all systems tend toward the lowest order, the gases flow from the place of greatest concentration to that of lowest concentration until the concentration is equal throughout. The phenomenon is known as *diffusion*. This is exactly the same thing that happens when a bottle of strong-smelling ammonia is opened on one side of a room, and the highly concentrated ammonia molecules move quickly to disperse throughout the room's air space.

The drawback to diffusion as a mechanism of gas exchange is that it works well only over a short distance. All of the surface-breathing organisms are small, and the ratio of surface area to volume is large. Organisms larger than a sea anemone have many body cells too far from their surfaces for their oxygen needs to be met by gas exchange through the skin alone (Figure 6–3). But skin breathing acts as a supplement to the more complicated respiratory mechanisms in frogs, salamanders, and

How is it that the carbon dioxide and the oxygen flow the "right" way?

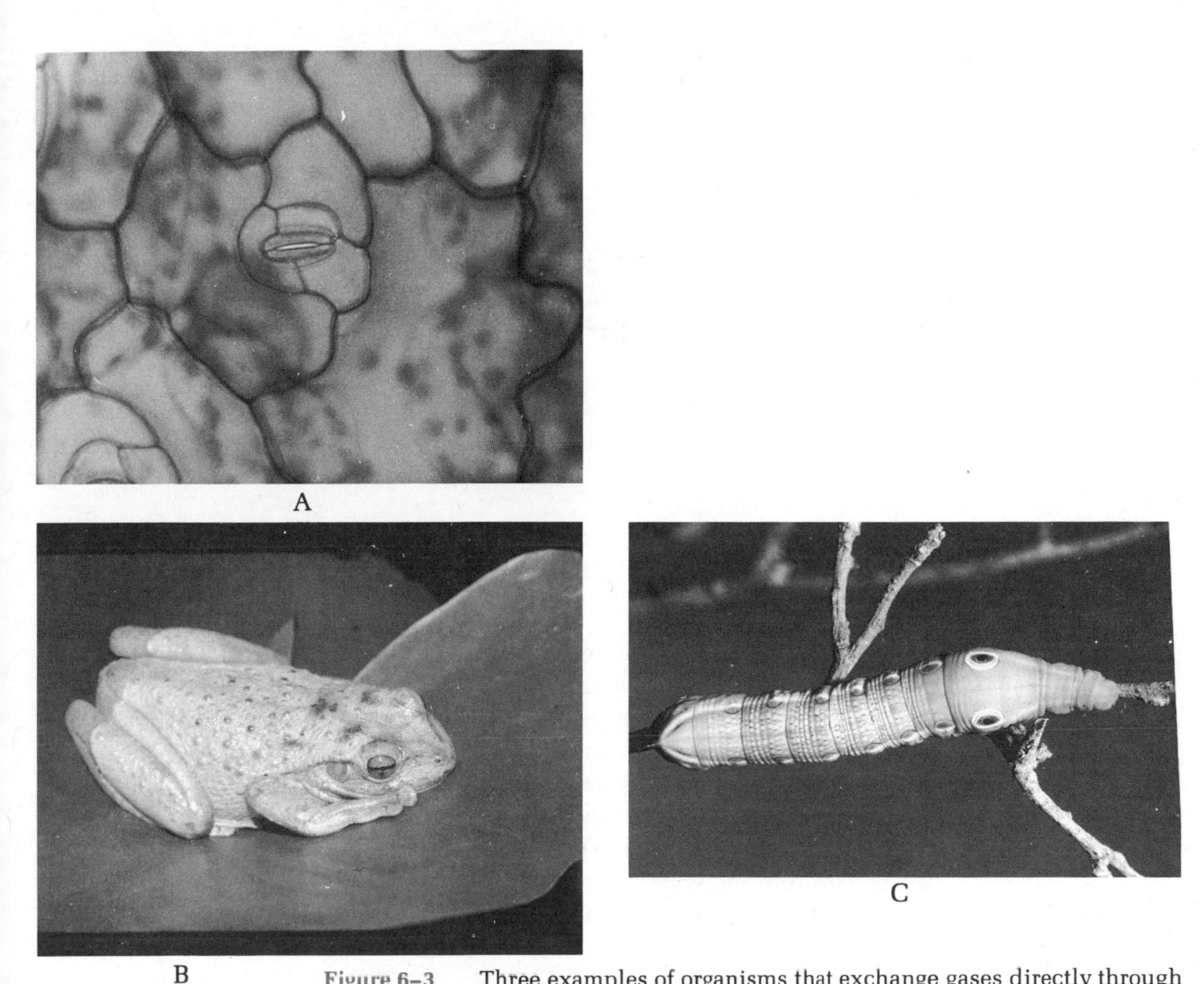

Figure 6–3 Three examples of organisms that exchange gases directly through their outer surfaces: (A) a photomicrograph of a stomate of a green plant leaf, (B) a frog, and (C) a tomato cutworm.

fish. In fact, frogs rely entirely on skin breathing when they hibernate for the winter in the mud at the bottom of their pond.

Breathing through gills

Fish breathe through *gills*, which in many species are concealed and protected by *gill covers*. The gills are sets of folded tufts or loops filled

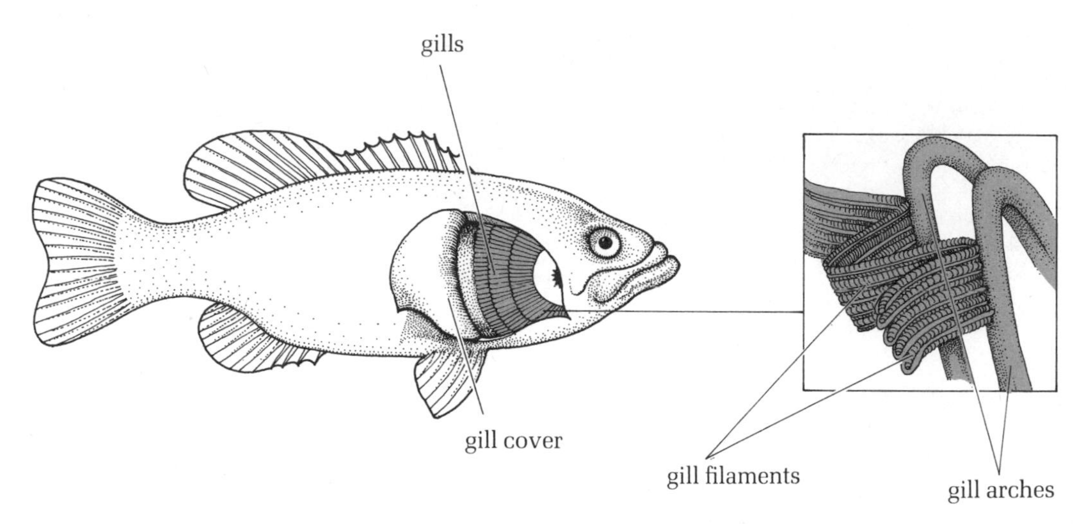

Figure 6–4 The breathing apparatus of fish. The blood circulating in the gill arches picks up oxygen dissolved in the water flowing over the gill filaments.

with tiny blood vessels close to the surface of the skin. The blood acts as the special medium for transporting gases through the body. Figure 6–4 shows the component parts of a typical set of gills, particularly the *gill filaments,* the workhorses of gas exchange. By means of the gulping that we commonly associate with fish, the fish keeps water flowing over, around, and through the network of gill filaments. Oxygen dissolved in the water passes into the fish's blood by diffusion, and carbon dioxide moves into the water.

The amount of dissolved oxygen varies from one body of water to another, but even under the best conditions, water contains only about one part of oxygen in 250 parts of water—relatively little compared to the one part of oxygen in five parts of air. Breathing water is hard work. Even though the gill filaments provide a large surface for gas exchange, a fish expends up to 20 percent of its total energy doing the work of keeping water flowing across the gill filaments. If water is such a poor oxygen source, why do fish die in the richer environment of the air?

In water the gill filaments are opened out and all their surface area exposed, but out of the water they stick together like the pages of a waterlogged book, much reducing their surface area. More importantly, the water on the surface of the gill filaments evaporates. Without this coating of moisture, diffusion cannot occur, and the fish suffocates.

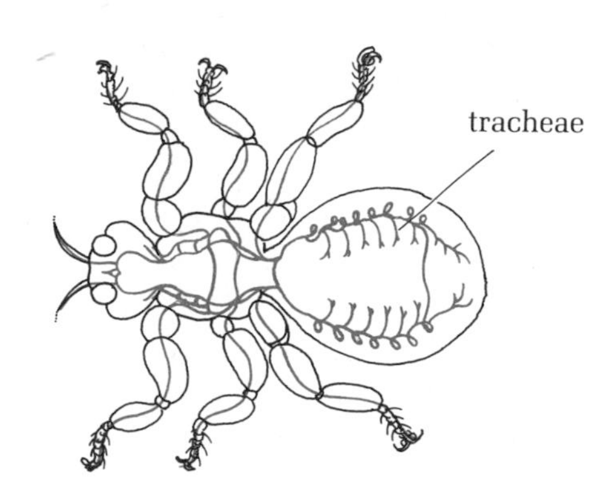

Figure 6–5 The breathing apparatus of insects. A branched series of tubes, the tracheae, supplies oxygen from openings on the body surface of an insect to the interior tissues.

Breathing among the insects

Insects have their own special form of respiratory organ called *tracheae*, a series of tubes opening to the air on both sides of the body. The size of the opening is not fixed; it can be made wider or smaller depending on the insect's needs. Inside the body, the tracheae branch into smaller and smaller tubes that supply oxygen dissolved in fluid directly to the cells. The tracheae system is very lightweight and thus well adapted to flying creatures such as insects. However, tracheae are suitable only for small organisms. Air moves fairly slowly through the tracheae, and in any organism larger than an insect a great proportion of the cells would be too far down the tubes for their respiratory needs to be filled (Figure 6–5).

Breathing in man

Birds, reptiles, mammals, and some amphibians all use the *lung* as their mechanism of breathing. Essentially, a lung is a cavity within the body where gas exchange can take place. Each group of organisms has lungs particularly suited to its needs and to its place on the evolutionary tree. The lungs found in man and other mammals are particularly adapted to the respiratory needs of large, land-dwelling animals. Figure 6–6 shows the major parts of the human respiratory system.

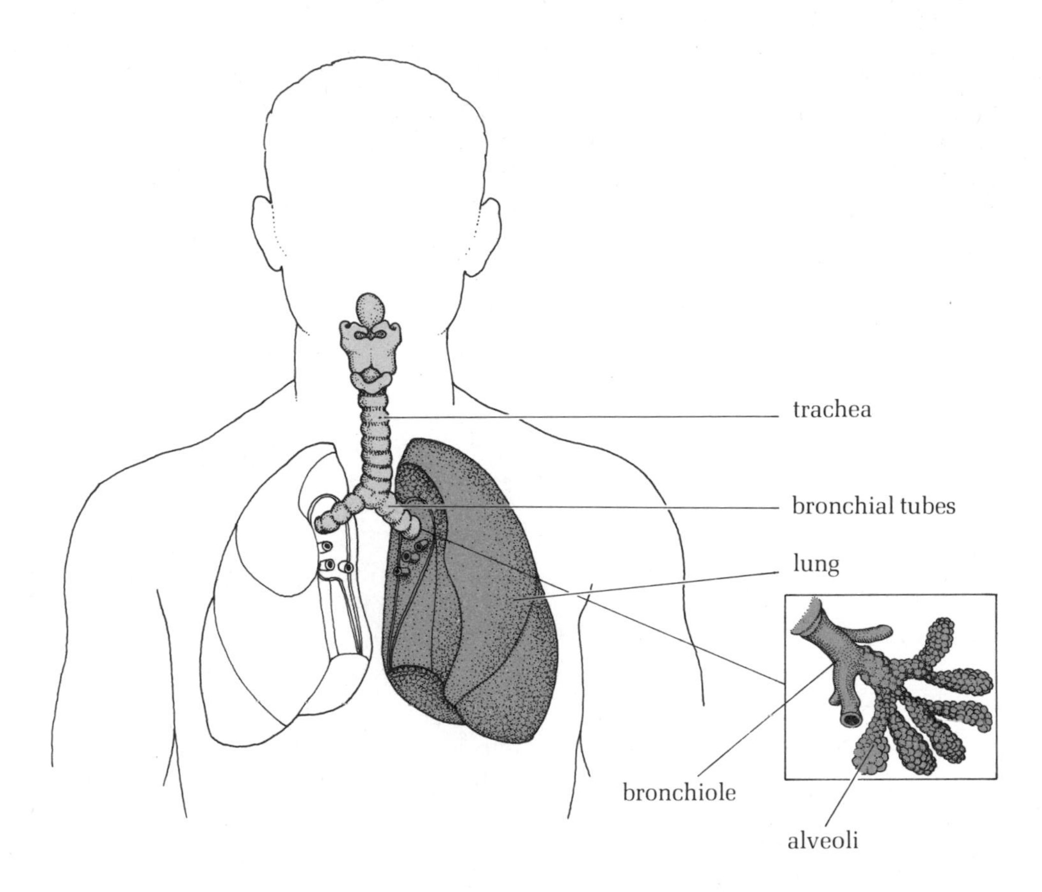

Figure 6–6 The human respiratory tract. After the air is brought into the lungs from the trachea, it comes into contact with the bloodstream in the alveoli.

The parts of the system

Before the air that enters the nose reaches the lung, it passes through a series of passageways and chambers where it is warmed or cooled, as needed, moistened, and cleaned. The nose contains a great many blood vessels that either warm or cool the air to make it roughly equal to the temperature of the body. The hairs in the nose serve to trap foreign matter like dust or soot coming in with the air. The air is warmed further in the *pharynx*, the cavity that extends from the back of the nose, before passing into the *trachea*, the hard, ringed tube that you can feel with your fingers

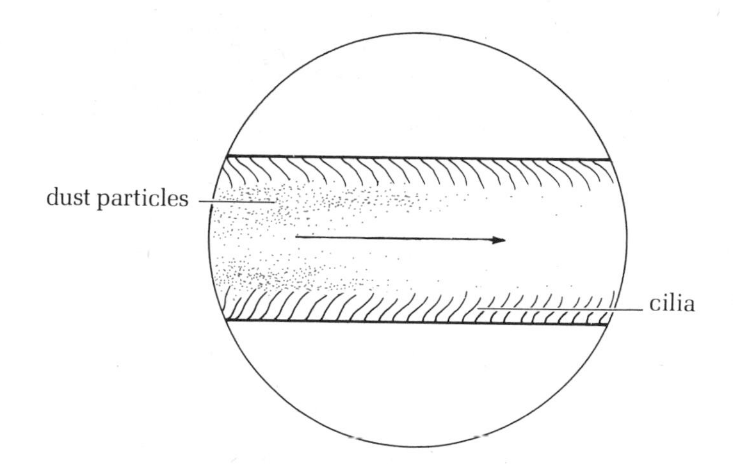

Figure 6–7 The cleansing action of the cilia of the respiratory tract. Incoming air is cleaned of dust and other particles as it proceeds farther into the lungs.

just under your voice box. The trachea is lined with millions of tiny, cellular appendages called *cilia* (Figure 6–7). The cilia wave constantly, and push dirt and foreign matter trapped in *mucus*, the thick, white secretion of the nose and throat, back up the trachea toward the pharynx. Coughing, sneezing, and clearing the throat keep these passages clear and free of excess mucus and the particles of dirt and dust it contains. One of the first noticeable effects of cigarette smoking is the destruction of the ciliated cells in the trachea. As a result, irritating foreign matter that would normally have been trapped can get into the lungs and damage them.

Warmed and cleaned, the air passes from the trachea into the two *bronchial tubes*, each of which leads to one of the lungs. Just as a tree branches into smaller and smaller twigs, so the bronchial tubes divide into ever tinier passageways called *bronchioles*. The smallest bronchioles, visible only in a microscope, end in an air sac shaped something like a bunch of grapes and called an *alveolus*. The alveolus is the actual site of gas exchange.

Why is the lung such a complex and convoluted structure? Why doesn't man get along with a simple air sac? Man is far too large to survive on surface breathing; the amount of oxygen that could diffuse through the surface of the body, were that surface thin enough to allow diffusion, would not be enough to supply all of man's body cells. In the fish, the many fine gill filaments increase the surface area available for gas exchange. The lungs accomplish the same purpose. The bunchings and

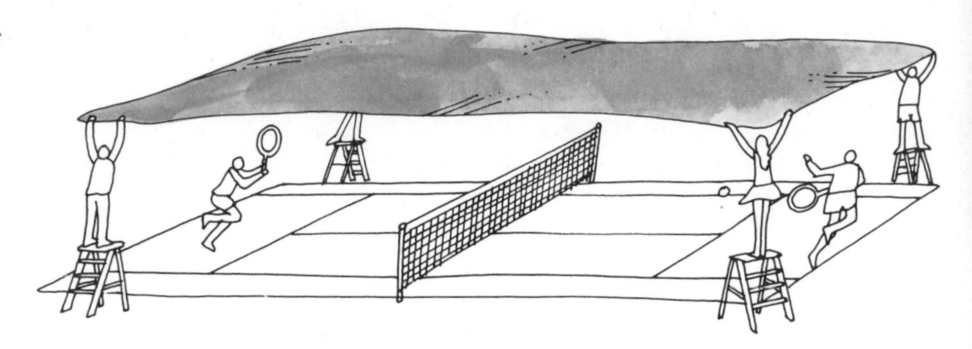

The folded structure of the lining of the lungs makes possible a surface area approximately the size of a tennis court.

foldings of the lungs' 300 million alveoli provide 750 square feet of gas exchange surface; that's an area roughly the size of a tennis court!

Now we know the path the air follows, the road it travels from here to there. But precisely how do we breathe?

The mechanics of taking a breath

The answer lies in the common fireplace bellows. Opening a bellows creates a partial vacuum; air from the outside rushes in to fill up the space. When the bellows is squeezed, the air rushes out. The lung, though, contains no muscles and can't do its own squeezing. For that it has to rely on the structures surrounding it.

The lungs are enclosed in what amounts to a box, as shown in Figure 6–8. The sides of the box are the ribs and the muscles attached to them, while the bottom is the *diaphragm*, a sheet of muscle that separates the chest cavity from the abdomen. When we inhale, the diaphragm contacts and moves downward, and the rib muscles move the ribs outward. This action enlarges the chest's volume, creating a partial vacuum that is immediately filled by the rushing in of outside air. Exhaling consists of the relaxation of the diaphragm and the rib muscles, returning the chest cavity to its original size and pushing out the air.

By no means do the lungs fill and empty completely with each breath. Usual breathing involves only about 10 percent of the lung's capacity, but very deep or labored breathing of the sort a distance runner or swimmer experiences in the course of a race can raise the amount to 80 per-

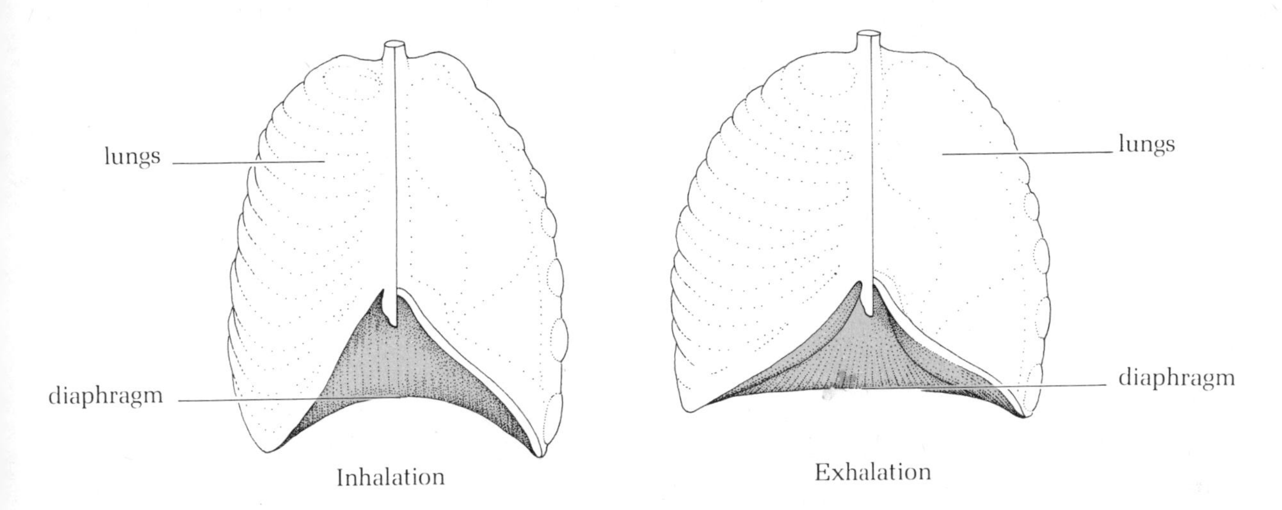

Figure 6–8 The action of the diaphragm. Enclosed within the "box" formed by the rib cage and the diaphragm, the lungs expand to inhale and collapse to exhale.

cent. The remaining 20 percent of the lung's air is replaced only very slowly, so that the carbon dioxide–oxygen transfer continues between breaths.

The fine workings of the alveolus

By the time the inhaled air has made its way into the alveolus, it has been dissolved in the watery fluid that coats the interior surfaces of the lung. In man, as in all other organisms, gas must be in solution before it can be carried by the blood or taken in by a cell to fuel its respiration. Each alveolus is surrounded by microscopic blood vessels called *capillaries*. The walls of the alveolus and the capillaries are so thin, often only one cell layer thick, that dissolved gases diffuse from one into the other. Carbon dioxide dissolved in the blood passes from the capillary into the alveolus, while oxygen travels the other way. Because the total surface area provided by all the alveoli of both lungs is so great, this exchange can take place very rapidly.

But how does an oxygen molecule in solution in the blood get to where it is needed—say, fueling cellular respiration in a muscle cell of the foot? It travels there by means of the circulatory system.

From lung to cell and back again

The circulatory system is made up of two components. The first is *blood,* the liquid that acts as a medium for the materials transported by the circulatory system. The second is the *heart*—the pump that keeps the blood moving—and the vessels the blood passes through.

The blood

Blood is not a homogeneous solution, like salt and water, but a mixture of cells and large molecules in a watery base. The base is known as *plasma,* and it is about nine parts water to one part proteins, salts, gases, and other molecules and compounds. Although blood contains a number of kinds of cells, we need to be concerned with only one—the *red blood cell* shown in Figure 6–9.

The red blood cell derives its characteristic color from the nearly 265 million *hemoglobin* molecules it contains. A single hemoglobin molecule can bond with four oxygen atoms. If blood contained no hemoglobin and simply transported the oxygen dissolved in the plasma, it could carry only one-fourth as much oxygen. Hemoglobin can also bond with carbon dioxide, but most of the carbon dioxide is converted to sodium bicarbonate in the blood and is transported in this form back to the lung.

The double pump

The simplest heart is nothing more than a muscular bulb in the wall of a blood vessel, such as the ones found in the common earthworm. The

If blood contained no hemoglobin and simply transported the oxygen dissolved in the plasma, it could carry only one-fourth as much oxygen.

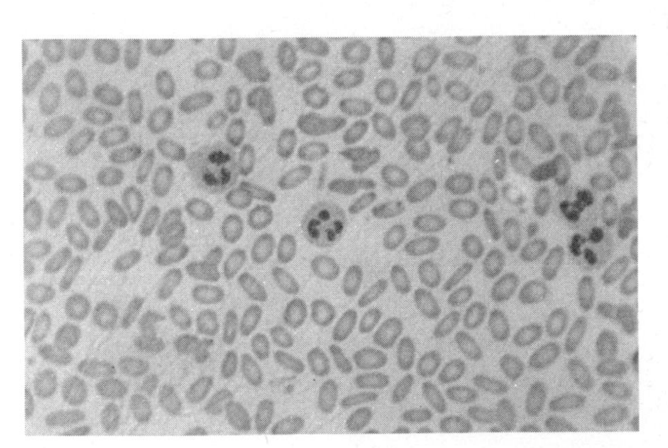

Figure 6–9 A. Red blood cells. These small, concave disks have no nuclei and in the mature form live for about 120 days before being removed from the blood by the liver. B. A model of the structure of the hemoglobin molecule. This complex molecule forms a reversible bond with oxygen.

contractions of such a heart help keep blood moving through the vessels. Among the fish, the heart contains two separate chambers, one that receives blood returning from the body and one that pumps it to the gills to pick up oxygen and dispose of carbon dioxide (Figure 6–10). In the heart characteristic of amphibians and reptiles, the receiving chamber is divided in two—one for the blood coming from the lungs and the other for the blood coming from the body. In all of these circulatory arrangements, the blood moves relatively slowly. This poses no problem for these animals, for they live slow-paced lives. But mammals and birds need a high-pressure, high-flow circulatory system to allow for their constant body temperatures and high rates of activity.

The human heart serves as a good example of such a system. Man's heart is not one pump but two—one on each side of the heart. Used, or oxygen-poor, blood returning from the body enters the heart at the *right atrium*, then flows into the *right ventricle*, which pumps it into the lungs. In the lungs, the used blood gives up its carbon dioxide, picks up a load of oxygen, and returns to the *left atrium* of the heart. The blood passes on into the *left ventricle*. A contraction of the thick and powerful wall of the ventricle sends the oxygen-rich blood out to the body (Figure 6–11).

The various sections of the heart are separated from each other by valves that open only in one direction. The structure of these valves allows blood to flow, for example, from the atrium into the ventricle, but

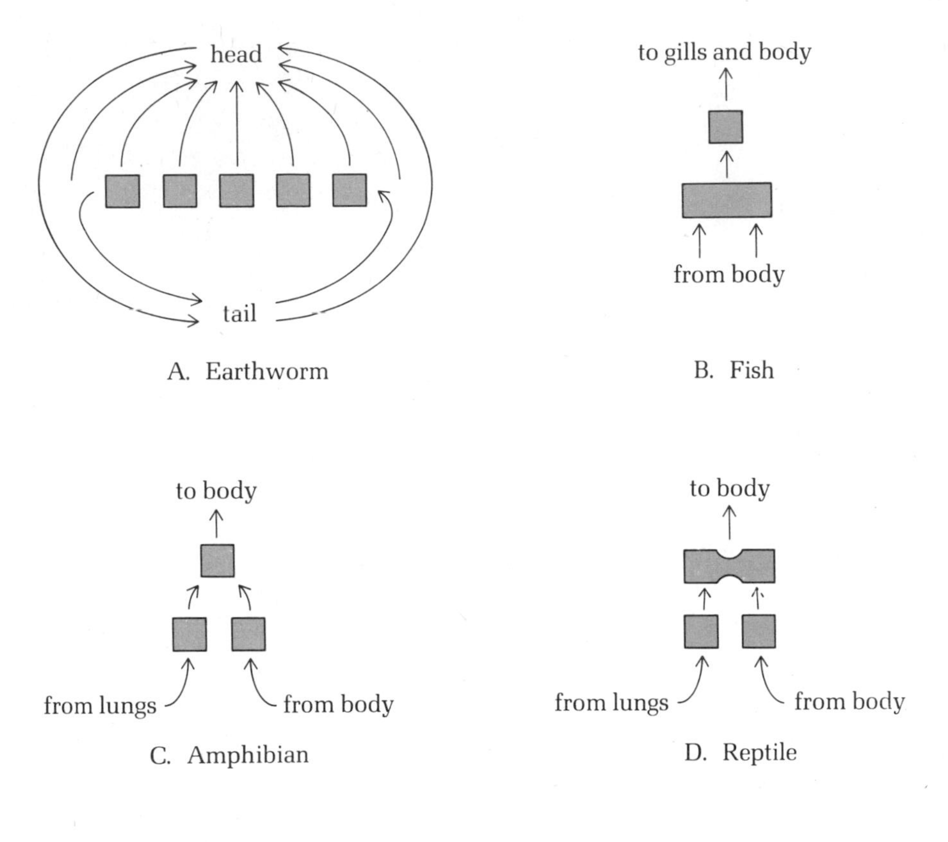

Figure 6–10 A comparative view of four animal circulatory systems. A. The earthworm has a simple system of five pumping chambers. B. In fish all blood from the body is collected in one chamber and then passed through to the gills. C. In amphibia one chamber receives blood from the lungs, another from the body; then one pumping chamber circulates it to the body. D. The more complex three-chambered system found in reptiles appears to be a precursor of the four-chambered system found in mammals.

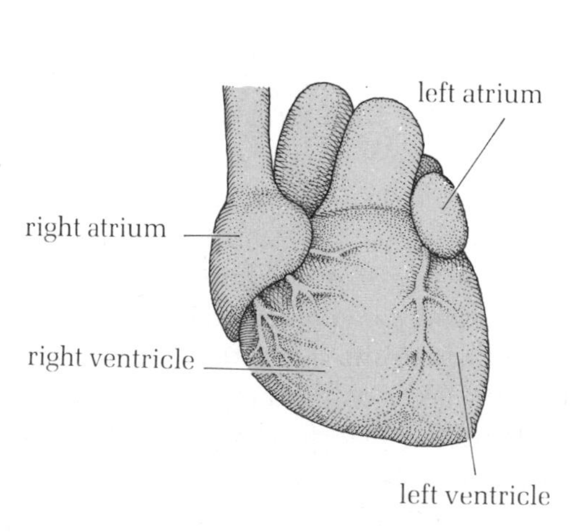

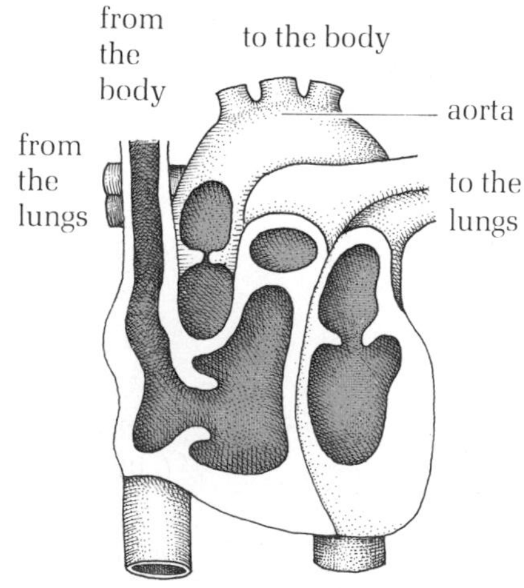

Figure 6–11 The human heart. The four chambers of the heart receive oxygen-poor blood from the body and circulate it to the lungs where it is oxygenated. The blood then returns to the heart, which then pumps it to the body.

when the ventricle pulses and blood starts to flow back into the atrium, the valve shuts tightly, stopping backflow. Such a valve is located between both atria and ventricles and in each of the blood vessels leading to the lungs and to the body. The lub-dub we associate with the heartbeat is the sound of these valves opening and closing.

Except for the brief period between beats, the heart never really rests. It goes on beating from well before birth until the very moment of death. Usually the heart beats 70 to 80 times a minute, but the speed can vary considerably depending on the activity of the body. In sleep, the heart slows down to around 50 or 60 beats, while the strain of vigorous exercise can push it toward 200. Even at the relatively slow, average pace, the heart beats about 100,000 times a day, sending 13,000 quarts of blood around the body. In other words, the body's volume of blood circulates through the arteries and veins about 2,200 times each day.

The double-pump arrangement of the human heart, which is also found among the other mammals and the birds, is what allows such a steady, copious flow. Blood returning from the body has traveled a relatively great distance and is under little pressure. The double-pump system segregates this low-pressure blood from the oxygenated blood being pumped into the body. As a result, the pressure generated by the pumping of the left ventricle affects only the oxygenated blood, sending it on its way under high pressure. None of the pressure is dissipated by mixing with the returned blood, as happens among fish, amphibians, and reptiles. And this high pressure allows for a steady, fast flow of blood and the oxygen it carries.

The body's pipeline

Any blood vessel that carries blood away from the heart is called an *artery*. Blood coming from the left ventricle courses into an artery named the *aorta*, the largest blood vessel in the body and certainly the most important. Blood fresh from the heart is under terrific pressure, and the aorta has thick, elastic, muscular walls equipped to handle such pressure. When blood is pumped into the aorta, its walls expand to receive it. Then they contract, sending a rhythmic wave down the aorta and into all the arteries that branch off it, a wave that can be felt close to the skin as the *pulse*. The wave keeps the blood flowing at a fast pace.

Smaller arteries branch off the aorta and travel toward specific regions of the body. The carotid arteries, for example, carry blood up through the neck to the brain, while the renal arteries go to the kidneys. Once they reach their destination, these arteries divide and divide again and divide still again until the blood is flowing in capillaries so thin that only one red blood cell at a time can pass through. In the capillaries, the blood releases its oxygen to the cells and receives carbon dioxide from them. The capillaries are so arranged that there is a capillary within rapid diffusion distance of every cell of the body. In fact, the average adult human contains more than 60,000 miles of capillaries!

After passing through the capillaries, the blood, turned purplish blue by the loss of its oxygen, makes the return trip to the heart through the system of *veins*. The blood moves much more slowly through the veins than it did through the arteries. It has lost most of its pressure, and the veins have no thick and muscular walls to give the blood a helping squeeze. That squeeze has to come from the movement of muscles surrounding the veins. To keep the blood from flowing downhill toward the feet and legs, the veins are fitted with one-way valves (Figure 6–12).

The respiratory accelerator

Our description of the mechanics of oxygen–carbon dioxide exchange has omitted a very important fact: that the body's needs for oxygen may change from one minute to the next, even in a matter of seconds. A person reading a book or watching TV is using a relatively small amount of energy. Cellular respiration is occurring at a low rate, and little oxygen is being used. Compare that resting body with one engaged in strenuous exercise like running. Energy is being used in great amounts, and cellular respiration is taking place at a very high rate. In the change from resting state to running, the body's oxygen needs may increase by as much as 20 times. How does the body know to shift gears?

The average human contains more than 60,000 miles of capillaries.

In the change from resting state to running, the body's oxygen needs may increase by as much as 20 times.

Notice first what happens to your respiratory and circulatory systems when you exercise vigorously. Let's take running a mile as an example. When you first begin running, your rate of breathing and heartbeat remain the same for the first 100 or 200 yards. Then they begin to speed up. Breathing becomes deeper and faster, increasing from a usual 18 breaths per minute to perhaps 50. A larger volume of air is taken in with each breath. At the same time, the heart beats stronger and faster, possibly tripling its normal rate. Even after the distance has been covered, the body doesn't return to normal immediately. It takes several minutes for breathing and heartbeat to settle back to the usual pace. How has this change occurred?

Since carbon dioxide is the waste product of cellular respiration, the level of carbon dioxide contained in the blood in the form of sodium bicarbonate indicates how much oxygen the body's cells need. A respiratory center in the brain monitors the blood for its sodium bicarbonate

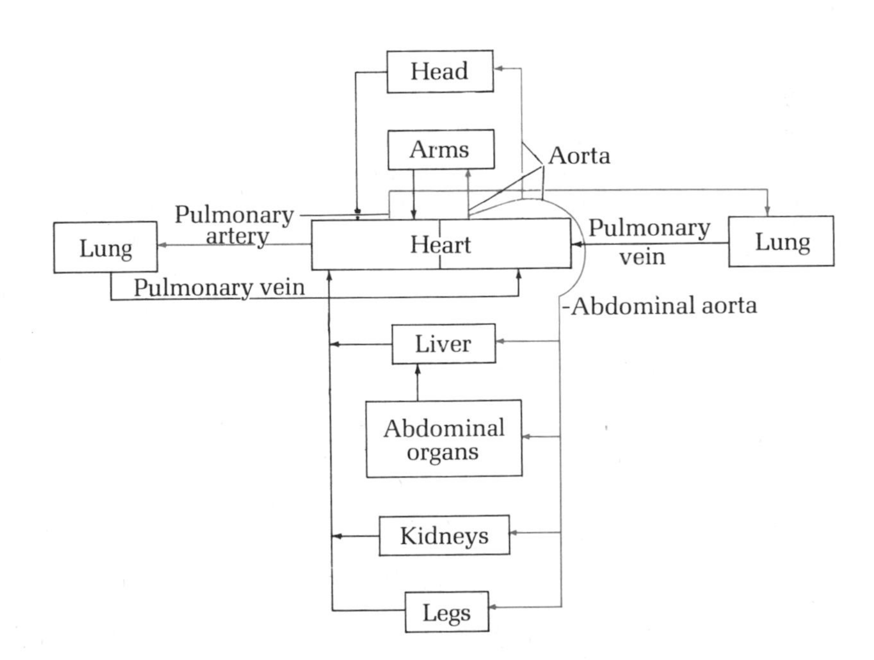

Figure 6–12 A flowchart of the human circulatory system.

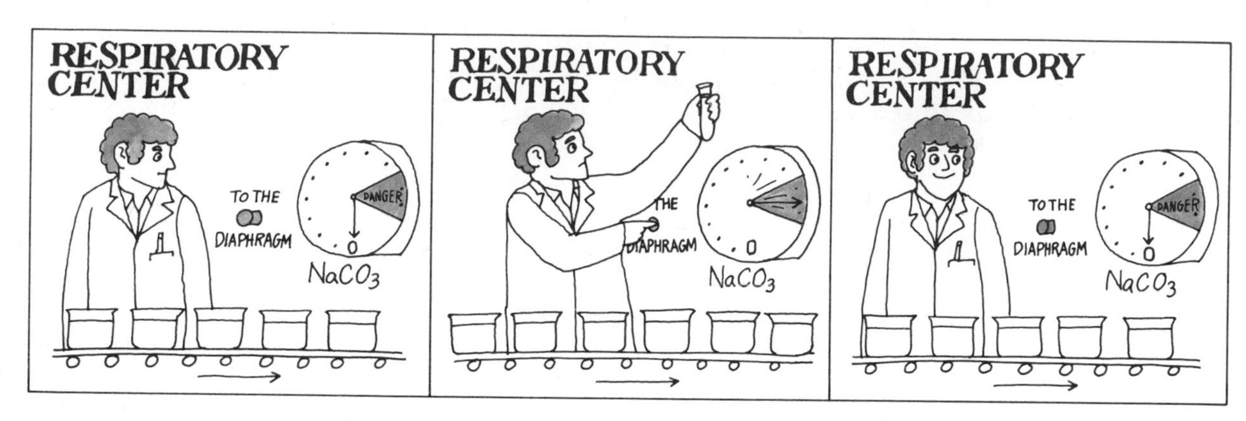

The body's clue to the need for increased breathing activity is the amount of sodium bicarbonate in the blood. The brain continually tests the bloodstream and will stimulate the diaphragm and the heart to increase the circulation of oxygen to the cells if needed.

content, and on the basis of that information sends impulses along nerves leading to the muscles of the chest and the diaphragm, which control the rate of breathing. (In Chapter 12, we'll take a closer look at how nerves pick up information and transmit directions.) If the blood's sodium bicarbonate level increases, the respiratory rate increases with it. At the same time, the heart picks up its beat to send blood more rapidly into the lungs and then into the body. Both sets of organs continue to work as fast as necessary until the sodium bicarbonate level in the blood returns to normal.

In the first part of the mile run, increased cellular respiration, particularly in the muscle cells of the legs, puts more sodium bicarbonate into the plasma of the blood. As soon as the respiratory center detects the change, impulses sent along the nerves into the chest speed the muscles of breathing and spur the heart. Even after the legs have ceased the hard work of running, the blood still contains a high level of sodium bicarbonate. As the sodium bicarbonate is expelled from the lungs as carbon dioxide, breathing settles back to its routine pace, and the heart again beats normally.

More than meets the eye

Throughout this discussion, we've been talking as if breathing involved only oxygen and carbon dioxide. These are only two gases that the respiratory and circulatory systems were required to handle. Today's air contains considerably more than oxygen, and some of these pollutants can damage and even kill the organisms that breathe them in. It has been estimated that each year the air in the United States is the dumping ground for 139 million tons of pollutants. In terms of sheer volume, the automobile is the worst offender, contributing a massive 70 percent of the total. The remainder comes from industrial sources. However, weight is not the only relevant statistic because some pollutants are highly poisonous even in very small amounts.

Keeping in mind what we know about the basic mechanics of respiration and circulation, we'll look at a selected few of these pollutants to see how they affect life.

Carbon monoxide: the oxygen thief

One of the pollutants dangerous even in small amounts is *carbon monoxide*, CO, a particularly insidious poison for all organisms whose blood contains hemoglobin. Carbon monoxide's affinity for hemoglobin is 200 times that of oxygen, and the bond made is tighter as well. Thus any organism inhaling air heavy with carbon monoxide is starved for oxygen. Typically, the brain cells are the first affected. If there is enough carbon monoxide, the gas is lethal—and that "enough" is a relatively small amount. Normal, clean air contains about 0.1 ppm of carbon monoxide. Raise that to 60 ppm, and the first symptoms of oxygen deprivation appear—headache, distorted vision, feelings of drowsiness. Continued exposure to 100 ppm causes serious oxygen starvation. And carbon monoxide levels over 100 ppm can be tolerated only for short periods of time. However, most of the effects are counteracted if the person can breathe in fresh air afterward and "wash out" his lungs.

These figures are only guidelines, and they apply to healthy adults. As is the case with all pollutants, people with respiratory or circulatory troubles are affected even more drastically. So are children and old people.

Carbon monoxide's affinity for hemoglobin is 200 times that of oxygen, and the bond is tighter as well.

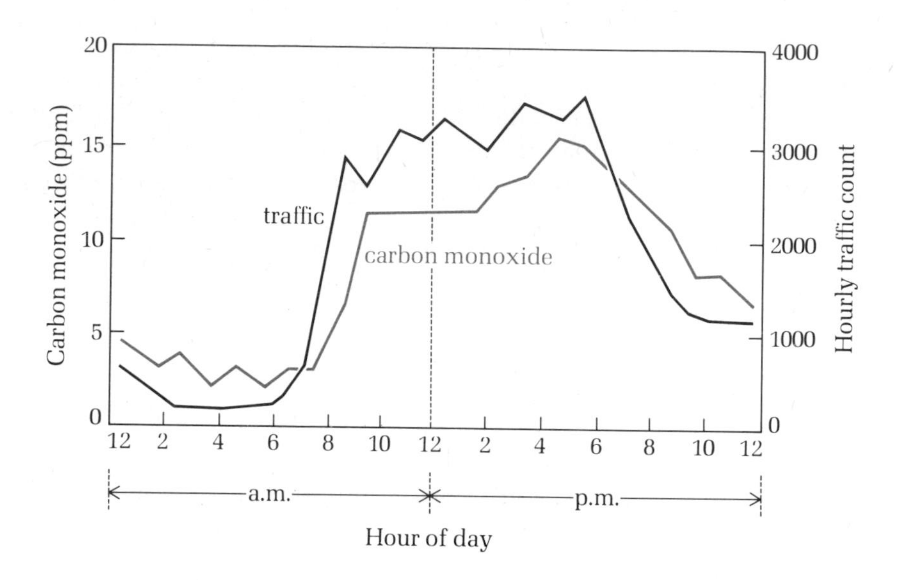

Figure 6–13 The correlation between traffic congestion and carbon monoxide levels in Manhattan.

The principal source of carbon monoxide is automobile exhaust, and the level of carbon monoxide in city air depends on how heavy the traffic is. On a typical day in Manhattan, the average concentration of carbon monoxide varies from 3 ppm in the wee hours of the morning, when only a few cabs are about, to 15 ppm at 5 o'clock, when every roadway in the borough is clogged with cars (Figure 6–13). And the closer you are to those cars, the more carbon monoxide you inhale. People driving during the rush hour are subjected to a steady concentration of 50 ppm, and for brief periods they may hit enclosed or trapped air that contains a whopping 300 ppm.

Automobile exhausts are not the only source. Cigarettes also give off quite a bit of the gas. One study showed that smoking half a pack a day triples the carbon monoxide in the blood; two packs or more per day bumps the figure up to five times the normal level. One of the subtle effects of smoking is gradual oxygen starvation of many of the body's cells because of carbon monoxide poisoning.

Lead: it only takes a little

Since 1923, lead has been added to gasoline to increase octane ratings, allowing lower-quality gasoline to be used in high-compression

Figure 6–14 Air pollution is a serious and steadily worsening threat in many American communities.

engines. The lead is not burned in the engine, and it comes out of the exhaust pipe in particles so small that they are not trapped by the protective hairs and mucus of the respiratory system. Once in the lungs, the lead enters the bloodstream and is carried to various parts of the body.

Lead is a powerful poison, which, like carbon monoxide, affects hemoglobin. But instead of simply bonding with the hemoglobin, lead prevents its formation in the first place. At high concentrations, lead actually destroys red blood cells. It can also do serious damage to the kidneys, the brain, various nerve centers, and the reproductive organs. Adults can tolerate more lead than children, who may suffer irreversible mental retardation after exposure to relatively small amounts of the metal. Lead is poisonous to plants as well as to animals, but for some unknown reason plants are more adept at filtering lead out of the air. Even plants growing along the edges of a highway show little lead in their interior tissues.

How much lead can people tolerate safely? Very little. Symptoms of lead poisoning begin when the body's 5 quarts of blood contain a mere

Symptoms of lead poisoning begin when the body's 5 quarts of blood contain a mere 0.013 ounce of lead.

0.013 ounce of lead. Apparently as little as 0.001 ounce causes hemoglobin production to fall off. The peculiar sensitivity of children to lead means that for safety their blood should contain no more than 0.00002 ounce. According to these criteria, lead poisoning of a minor sort affects many Americans. The average American adult has 0.003 ounce of lead in his blood, three times more than the amount known to affect the production of hemoglobin. Also, city dwellers carry more lead in their bodies than do their country cousins (Figure 6–14).

Sulfur oxides: the acid producers

When coal and oil are burned, the sulfur molecules they contain combine with oxygen to form sulfur dioxide, SO_2. The sulfur dioxide is released into the atmosphere as a gas. By itself, sulfur dioxide can damage plant and animal tissues, but in the air it recombines with oxygen and other molecules to produce sulfur trioxide, SO_3, also a gas and an even more potent pollutant. When sulfur trioxide combines with water, it becomes sulfuric acid, H_2SO_4, a very corrosive and damaging substance. This reaction occurs when sulfur trioxide comes into contact with the watery insides of a leaf or a lung. Long-term exposure to sulfuric acid can severely damage a hard material such as marble. Imagine what it does to the soft tissues of your lungs!

Plants, though, are far more sensitive to low concentrations of *sulfur oxides* than is man. Plants exposed continually to air with as little as 0.02 ppm sulfur oxides show the effects; some of the plants' cells die, and the rate of photosynthesis declines. Crop losses resulting from such pollution amount to $30 million each year. By comparison, human beings don't experience pronounced lung and throat irritation until the air contains 5 ppm sulfur oxides, and the level must go higher yet before breathing is impaired.

The automobile is the source of so many air pollutants that it is almost surprising to find it innocent in the case of sulfur oxides. Although sulfur is a natural component of all fossil fuels, sulfur is removed from gasoline in the refining process. Airborne sulfur oxides come principally from diesel fuel and from the industrial burning of coal and oil for power and light.

Synergism: 1 + 1 = 3

Each one of the principal pollutants is a poison in its own right. But what makes air pollution a particularly serious problem is that the air is polluted not merely by one poison, but by a whole bunch. The entire array of pollutants magnifies the effects of each one. Such a phenomenon is known as *synergism*, the general name given to any phenomenon where two or more things, be they poisons or organisms, have an effect greater than the sum of their individual effects.

A prime example of the synergistic effects of air pollution is the killer fog that beset London, England, in 1952. During a 5-day period in December of that year, a cold, foggy calm fell on the city. No winds blew, and all the city's pollution accumulated until the air was thick and ghostly. By the time the weather changed and winds came to sweep away the pall of poison, 4,000 people had died because of the severe air pollution. Which specific pollutant was responsible for the deaths? The levels of carbon monoxide, sulfur oxides, and such particles as soot and ash were all high, but not one was high enough to cause death directly. Sulfur oxides, for example, were at only 1.3 ppm, well under the level of troublesome irritation. Those 4,000 people died because of the combined effect of the various pollutants, not because of any one of them. And the pollution killed mainly the most vulnerable in the population: the very young, the very old, and those suffering from respiratory or circulatory disease.

A word to the wise

Recall the main point of Chapter 5: Life must have energy to survive. Translate that general statement into practical terms, and we come up with two equally crucial statements: All organisms require a source of food, and all organisms, apart from a small percentage of anaerobes, need oxygen to turn that food into usable energy. Being able to take a deep breath of clean air is not just pleasant—it's necessary. Air pollution is more than a matter of irritated eyes and throats or scummy air. It affects the basic gas exchange process of the energy-releasing mechanism of the cell. To tamper needlessly with the atmosphere is like playing Russian roulette for the very highest of stakes—the continued existence of all forms of life.

Summary

Practically all organisms are involved in gas exchange with their surroundings. In the land plants, this takes place through openings on the stems and through stomates on the leaves. Some organisms breathe through their surfaces by means of diffusion, but this works well only over short distances. Fish use gills made of a network of fine filaments; blood flowing inside the filaments exchanges gases with water passing outside. The breathing system of the insects is the tracheae.

Man, like the reptiles, birds, mammals, and some amphibians uses lungs. Air passes through the nose, the pharynx, and trachea and enters the lung through the bronchial tube and the bronchioles branching off it. The alveoli are the sites of gas exchange. The structure of the lungs serves to increase the available diffusion surface.

The circulatory system comprises the blood and the heart and blood vessels. Blood plasma contains several kinds of large molecules and cells, among them the red blood cells. Red blood cells contain hemoglobin, which increases the blood's ability to carry oxygen by bonding readily with it. All hearts are muscular pumps that keep blood flowing through the body. The mammalian heart contains two pumps. The right side receives oxygen-poor blood from the body and pumps it into the lungs. The left side receives the newly oxygenated blood and pumps it into the body. This separation allows for a high-speed, high-pressure system. The blood travels away from the heart in arteries, which subdivide into fine capillaries. The blood then returns to the heart through the network of veins.

The body is able to adjust heartbeat and respiratory rate to its needs of the moment because of monitors in the brain. A rise in the blood's sodium bicarbonate level, indicating increased cellular respiration, triggers a rise in the heartbeat and the rate of breathing, which continues until the blood's chemistry returns to normal.

Besides oxygen and carbon dioxide, air contains various pollutants. Carbon monoxide bonds with hemoglobin and starves the body of oxygen. Lead inhibits red blood cell formation and attacks certain organs. Sulfur oxides can become the highly corrosive sulfuric acid. The synergism of air pollution multiplies the overall effect of these atmospheric poisons.

Questions to consider

1. Why do all organisms apart from the anaerobes need to take in oxygen and dispose of carbon dioxide? What would happen if their supply of oxygen ran out?
2. Why is surface breathing suitable only for relatively small organisms?
3. What path does air follow from its entrance into the body to its ultimate destination in the lungs?
4. What purpose does hemoglobin serve in the blood? Why isn't the oxygen dissolved directly in a watery medium?
5. What are the advantages of the double-pump plan of the human heart?
6. How is the body able to adjust the heartbeat and the breathing rate to its needs?
7. What are the principal physiological effects of the following air pollutants: carbon monoxide, lead, sulfur oxides?
8. Daily reports of air pollutant levels often show that the air contains less than a damaging dose of each poison. Does this mean that no harm is being done? Why?

Glossary

alveolus a small outpocketing on the interior surface of the lung; the site of gas exchange between the respiratory system and the circulatory system

aorta the large artery through which oxygenated blood flows from the heart into the body

artery any blood vessel through which blood flows away from the heart

atrium a heart chamber that receives blood from the body or the lungs *bronchial tube* the narrow pathway leading from the trachea to the lung

bronchiole a fine branching of the bronchial tube

capillary a small, thin-walled blood vessel joining an artery and a vein; the site of gas exchange between the bloodstream and the body cells

carbon monoxide a compound of carbon and oxygen which has poisonous effects because of its high bonding affinity for hemoglobin

cilium a fine, hairlike appendage on certain cells

circulatory system the liquid medium of food and gas exchange, its pumping mechanism, and its vessel structure

diaphragm the sheet of muscle separating the chest cavity from the abdominal cavity; its position regulates the expansion of the lungs

diffusion the movement of a dissolved substance from an area of greater concentration to an area of lesser concentration

gill the respiratory organ of fish and some other aquatic organisms; a projection of the body surface exposed to water which is filled with blood vessels

gill cover a flap of bone that protects the gills in bony fishes

gill filament the site of gas exchange in the gill

hemoglobin the complex specialized molecule in the blood that transports oxygen to body cells

lung an expandable sac-like cavity which is the main breathing organ of most land animals

mucus a thick, semifluid substance which coats the interior surface of most of the respiratory system

pharynx the cavity between the back of the nose and the voice box

plasma the liquid, noncellular portion of the blood

pulse the rhythmic wave of pressure through the arterial system which follows each heartbeat

red blood cell the simple, hemoglobin-containing cells in the blood

respiratory system the organs of gas exchange

stomate the small openings on plant leaves that are the site of gas exchange *synergism* a combination of effects greater than the individual effects of the components of a system

trachea the hard, ringed tube that leads from the pharynx to the bronchial tube

tracheae the system of breathing tubes in insects

vein any blood vessel through which blood flows toward the heart

ventricle a heart chamber that pumps blood to the lungs or the body

7 Eating and Excreting

Unlike the land plants and the other autotrophs, which can manufacture their food from simple ingredients through photosynthesis, animals depend on plants or other animals for food. Once inside the organism, food serves two purposes: as fuel for cellular respiration and as raw material for maintenance and growth. But putting food to work poses problems for the animal. Most food comes in the form of molecules far too large to enter cells. As a result, the animal must have some way of breaking food into appropriately small units. Any such breakdown inevitably produces wastes. For example, remember that water and carbon dioxide are by-products of cellular respiration and that an important task of the respiratory system is disposing of the carbon dioxide. By-products are likewise produced in the breakdown of food. The animal has to have some way of getting rid of them, lest it find itself awash in its own wastes.

In this chapter, we'll be concerned with food and with the way it is handled by animals, particularly man. We'll look at the kinds of molecules that make up food, at how food is broken down and absorbed, and at how wastes are disposed of.

The variety and similarity of food

We pointed out in Chapter 5 that man is an omnivore, a class of animals characterized by the ability to eat both plant and animal matter. In fact, man is a particularly adept omnivore. Although individual people or groups of people may shun certain foods as inedible, the range of animal and plant products consumed by humans is extraordinary. To the Greenland Eskimos, raw seal liver is a delicacy. The Masai of Kenya and Tanzania subsist on a mixture of cow's milk and blood curdled with urine. The Vietnamese consider chicken breast tasteless, much preferring the feet and head, and they often add savor to their rice with a sauce made by pouring oil over fish that has been sitting in the sun for a day or two. The milk we Americans consume in such great quantities is completely undrinkable to many Asians, while the British relegate our highly touted corn on the cob to the status of cattle feed.

However different each of these foods may look on the plate or to the palate, a chemist's analysis would reveal that they are all made up of just a few types of compounds. Since our food comprises the bodies of plants and animals, these classes of molecules are not only the essential units of food but also the important molecular building blocks of cells.

We'll examine each group of compounds in turn.

Proteins: the first team

The name *protein*, which means "of first rank," gives some idea of the importance of these complex molecules to living systems. They are the most abundant of the large organic molecules, representing as much as 50 percent of the dry weight of a typical animal cell. They serve as important structural components in the cell and play an indispensable role in its chemical activities.

Molecules of three dimensions

Although proteins are large molecules, they are made up of linked chains of simple units called *amino acids*, a class of organic compounds

The range of animal and plant products consumed by humans is extraordinary.

The commonly occurring amino acids

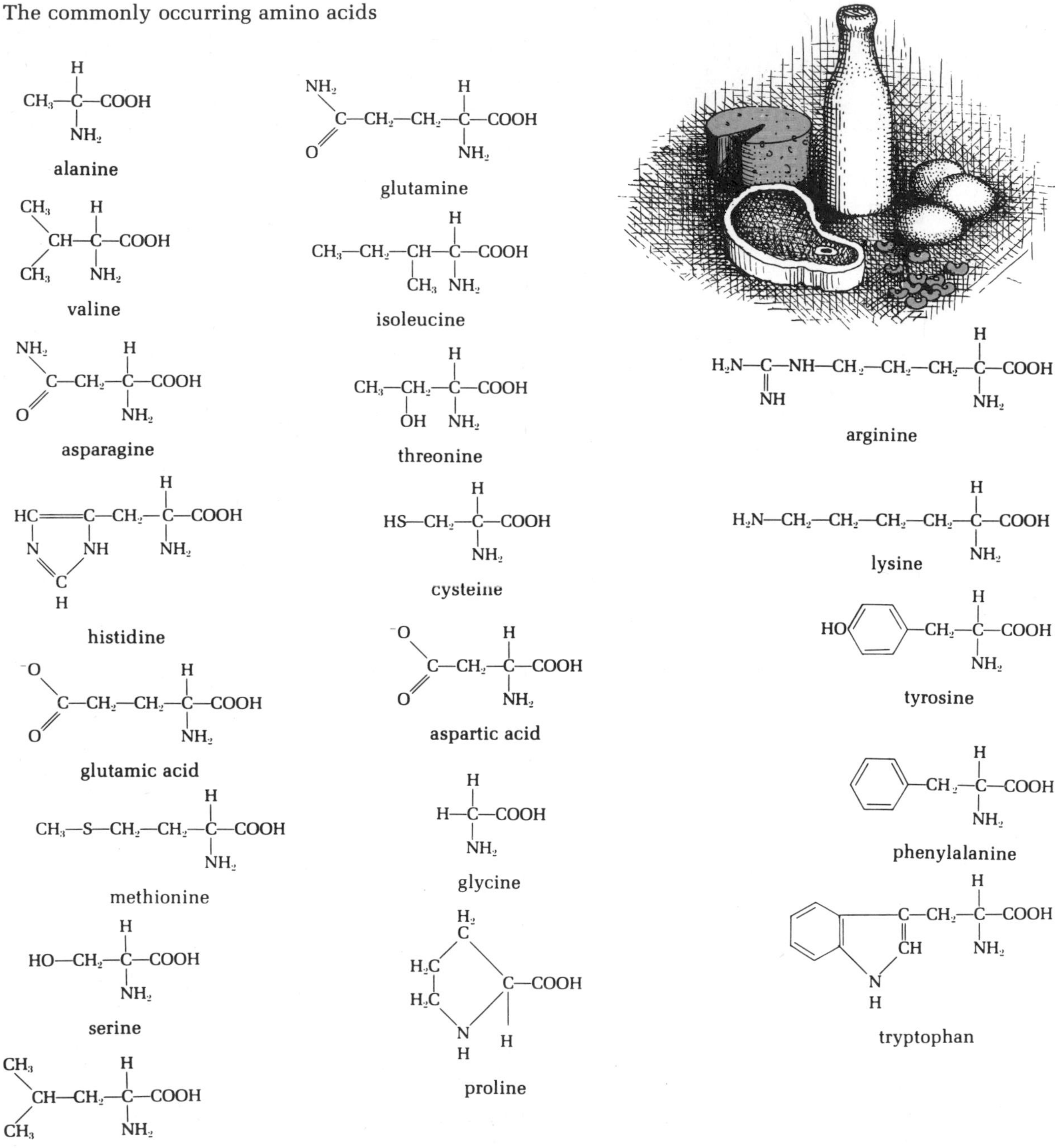

alanine

valine

asparagine

histidine

glutamic acid

methionine

serine

leucine

glutamine

isoleucine

threonine

cysteine

aspartic acid

glycine

proline

arginine

lysine

tyrosine

phenylalanine

tryptophan

Figure 7–1 Proteins. Amino acids are the structural subunits of proteins.

Each species of organism contains certain proteins specific to it and to it alone.

that contain oxygen, hydrogen, carbon, nitrogen, and often sulfur. The proteins in the food we eat are composed of 20 commonly occurring amino acids and 2 relatively rare ones (Figure 7–1). Each protein is made up of a number of amino acids bonded together in a precise sequence. Insulin, a relatively simple protein, contains 51 amino acid units bonded into one long chain. As large as this molecule is, it shrinks beside one of the truly big proteins. Glutamate dehydrogenase, a protein found in cow's liver, has 8,300 amino acid units bonded into 40 linked chains. Since there are 22 amino acids and since any number of them can be linked together in any order, there is no practical limit to the number of possible proteins. Many, many different kinds of proteins are found in each kind of organism. Even a simple bacterium contains some 3,000 different kinds of proteins. An organism as complex as man may possibly contain hundreds of thousands of different varieties of protein, each of which serves a different purpose. Each species of organism contains certain proteins specific to it and to it alone.

The key to the biological usefulness and uniqueness of a protein is its shape. The sequence of amino acids in a protein is known as its *primary structure*. Bonds form between every fourth amino acid in the primary structure, forming the helical, or spiral, shape known as the *secondary structure*. Some proteins with a secondary structure bind together into tough sheets or ropes that give them the name *fibrous proteins*. In other proteins, chemical attractions between parts of the secondary structure folds the helix back on itself, forming the *tertiary structure* of the molecule (Figure 7–2). The resulting twisted compound is known as a *globular protein*. It's important to note that the secondary and tertiary structures are direct consequences of the primary structure. Change one amino acid in the primary structure, and you change the shape of the whole molecule.

Proteins: builders and catalysts

The fibrous proteins are tough and insoluble in water, making them good, stable structural components. The fibrous protein keratin is a prin-

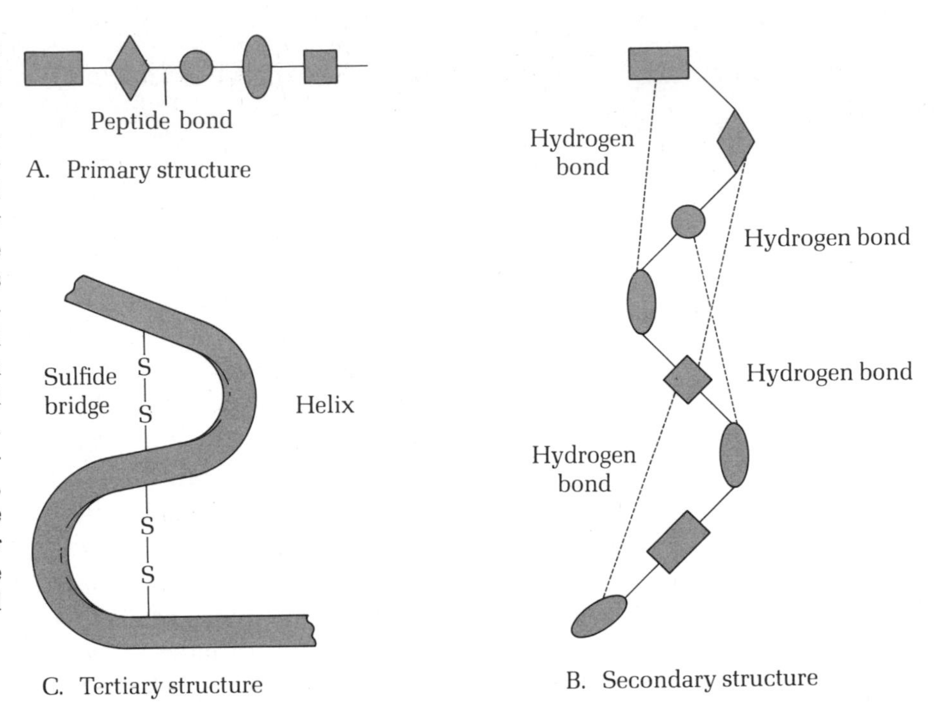

Figure 7–2 The three levels of protein structure. A. The primary structure is determined by the sequence of individual amino acids linked by peptide bonds. B. The secondary structure is determined by the coiling of the basic strand of amino acids into a spiral shape secured by hydrogen bonds. C. The tertiary structure, the most complex, is determined by the position of sulfur bridges that bind the spiral into a fixed coiled shape.

cipal ingredient of hair, nails, and skin. Collagen, the most common structural protein in the human body, makes up a great portion of our skeleton.

Certain globular proteins are the cell's "solution" to a difficult chemical "problem." To maintain itself, a cell must carry out a great many chemical reactions, some of which tend to occur very slowly. In a chemistry laboratory, a reaction can be sped up by adding energy in the form of heat, but the amount of heat needed to promote many necessary biological reactions would destroy the cell. Reactions, though, can also be hurried along by chemicals that act as *catalysts*, accelerating a reaction without actually entering into it. Globular proteins that act as catalysts

In a chemistry laboratory, a reaction can be sped up by adding energy in the form of heat, but the amount of heat needed to promote many necessary biological reactions would destroy the cell.

in living systems are known as *enzymes*. Over 1,500 different enzymes have been identified, each acting as a catalyst for a different reaction.

Enzymes can catalyze reactions because of the shapes produced by their tertiary structures. The reactant molecules fit into a portion of the enzyme molecule in much the same way that a key fits into a lock. The proximity of the reactant molecules promotes the reaction. Once the reaction is completed, the product molecule or molecules leave the enzyme, freeing it to be used again. Enzymes can work so fast that the space of one second is enough time for certain enzymes to catalyze several hundred thousand reactions (Figure 7–3).

Proteins play roles other than structure and catalysis. Globular proteins are soluble in water, and they can serve as transport molecules. Hemoglobin is an example. Some proteins act as stores of amino acids. The proteins in egg white and plant seeds are such storage forms; they supply nutrition to the organism as it develops. Proteins may also be poisons. Snake venoms are proteins, as are the bacterial toxins that cause such serious diseases as diphtheria and botulism.

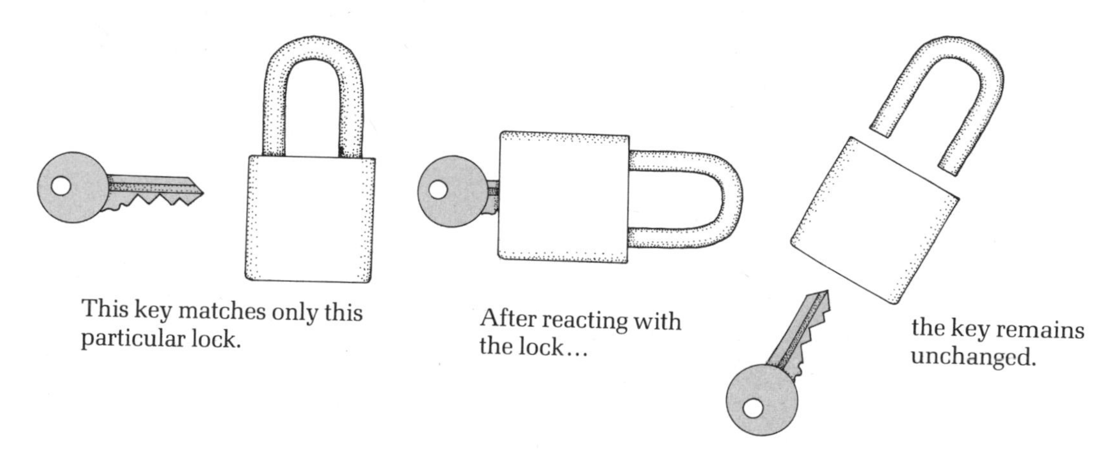

Figure 7–3 The specificity of enzyme activity can be compared to the match between a lock and a key. Each time the enzyme-key causes a reaction in the specific lock for which it has the proper conformation, it emerges unchanged, ready to perform the job again.

Enzymes can work so fast that the space of one second is enough time for certain enzymes to catalyze several hundred thousand reactions.

The supply of proteins

Because many proteins are specific to particular species, all organisms assemble their own, in a process we'll detail in Chapter 8. Obviously, for a cell to synthesize the proteins it needs, it requires a ready supply of the necessary amino acids. Plants can manufacture all the amino acids from carbon, hydrogen, oxygen, and nitrogen in the form of nitrates and nitrites. Animals are able to synthesize only some of their amino acids and have to depend on plants for the rest. Man can synthesize 14 of the amino acids; the remaining 8 must be included in his diet. Without them all the necessary proteins cannot be produced by the cells. The essential amino acids are found in animal protein sources such as meat, milk, and eggs. Most plant foods, except for certain beans and nuts, are missing one or more of the essential amino acids. Generally, an adequate human diet has to include at least some animal protein.

Besides providing amino acids needed for protein synthesis, food proteins can also fuel cellular respiration. The protein is broken into its component amino acids. The amino acids are then converted into acetyl CoA, which enters the energy-releasing reactions of the Krebs cycle and the electron transport chain.

Carbohydrates: chemical energy

We met the carbohydrates earlier, as the class of compounds in which the energy captured by photosynthesis is stored. All carbohydrates comprise carbon, hydrogen, and oxygen, usually in the ratio of 1:2:1. The smallest carbohydrate molecules are the *simple sugars*, or *monosaccharides*. Examples of simple sugars are glucose and fructose. Curiously, both glucose and fructose have the same chemical formula, $C_6H_{12}O_6$ (note the 1:2:1 ratio). The difference between the two compounds is the particular carbons to which the oxygens and hydrogens are attached (Figure 7–4).

Two simple sugar molecules put together produce a *disaccharide*, the form in which many common sugars are found in nature. Sucrose, the common table sugar extracted from sugar cane and sugar beets, is a combination of glucose and fructose. Lactose, a sugar present only in milk, is a combination of glucose and galactose.

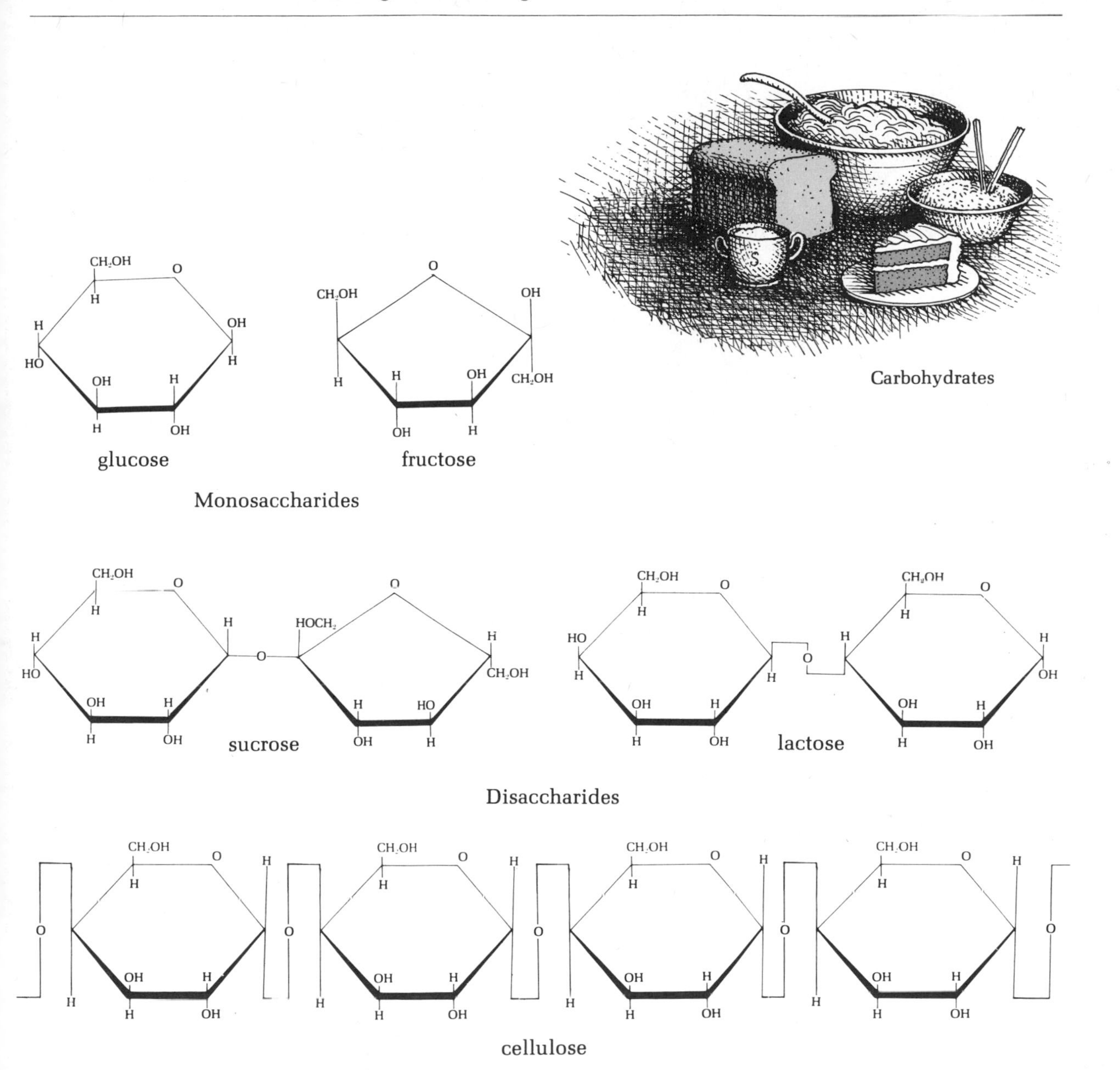

Figure 7–4 Carbohydrates. The sugars and starches are formed of single or multiple monosaccharide units.

A series of simple sugars linked together in a long chain produces a *polysaccharide.* Generally, polysaccharides are the chemical form in which sugar is stored. Starch is the storage polysaccharide for plants, and glycogen the one for animals. Both polysaccharides are made of repeating units of glucose. Certain polysaccharides play a structural role similar to that of proteins. The hard external coating of insects is composed of chitin, a polysaccharide. Another polysaccharide, *cellulose,* is the major structural material of plants and the most abundant organic material in the world. Cellulose is made of units of glucose, as is starch, but the chains of simple sugar are arranged in such a way that the molecule is very rigid. The page these words are printed on is made of fibers of cellulose extracted from tree trunks.

The most common foodstuff

Carbohydrates make up a great proportion of our diet, particularly in the form of starch and sugar. Bread, potatoes, rice, and flour are examples of foods that are mostly carbohydrate. Broken into their monosaccharide units, the carbohydrates are a major source of fuel for cellular respiration. And the carbon, hydrogen, and oxygen atoms they contain are used as building blocks for complex molecules like proteins synthesized in the cell (Figure 7–5).

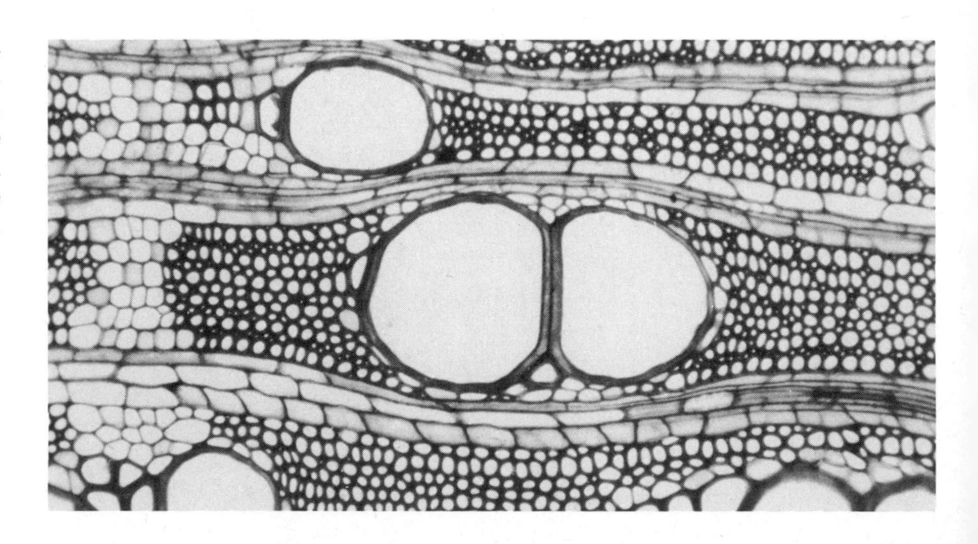

Figure 7–5 Cellulose is the most common carbohydrate, forming much of the structural material in plants, such as the fibers of this plant stem.

Even though cellulose is the most abundant carbohydrate, we humans can't use it as a food source, nor can most animals. Only certain microorganisms, such as fungi, and a few animals, such as silverfish and marine worms, can digest cellulose directly. Grass- or wood-eating animals like deer and termites can eat it only because their stomachs contain cellulose-consuming microorganisms. A product known as "edible" cellulose is produced by the food industry for use in diet foods and as a substitute for milk and ice cream in the milk shakes peddled by cheap drive-ins. "Edible" cellulose is simply cellulose molecules that have been broken mechanically into smaller pieces. It's still cellulose. Despite its name, "edible" cellulose cannot be broken down into its monosaccharide units and used by the body. It provides no energy and no building-block atoms. Nutritionally, eating "edible" cellulose is the same thing as eating paper.

Lipids: energy in reserve

Fatty, oily, and waxy substances fall into the category of *lipids*. Like the carbohydrates, the lipids contain hydrogen, carbon, and oxygen, but the amount of oxygen is typically much less than in the carbohydrates. Lipids may also contain other elements such as nitrogen, phosphorus, and sulfur.

What purpose do lipids serve in living systems? The lipids' roles are built around their two principal characteristics. First, since the lipids are nonpolar, they usually do not dissolve in water. As a result, they are an important component of cellular membranes, which separate the watery media of the different parts of the cell and which divide one cell from another. Second, lipids contain a larger proportion of carbon-hydrogen bonds than other organic molecules. When broken, these bonds release a considerable amount of energy. A given quantity of lipid contains about twice the calories as the same amount of protein or carbohydrate. Thus lipids serve as a concentrated source of reserve energy ready to be converted into acetyl CoA and used in cellular respiration (Figure 7–6).

In our food, most of the lipids are in the form of fats and oils. The *saturated fats* are ones in which every carbon is joined with two hydrogens, completing the bonding possibilities of the carbon. Saturated fats

Figure 7–6 Lipids. An example of the formation of the fat tristearin from the alcohol glycerol and three stearic acid units.

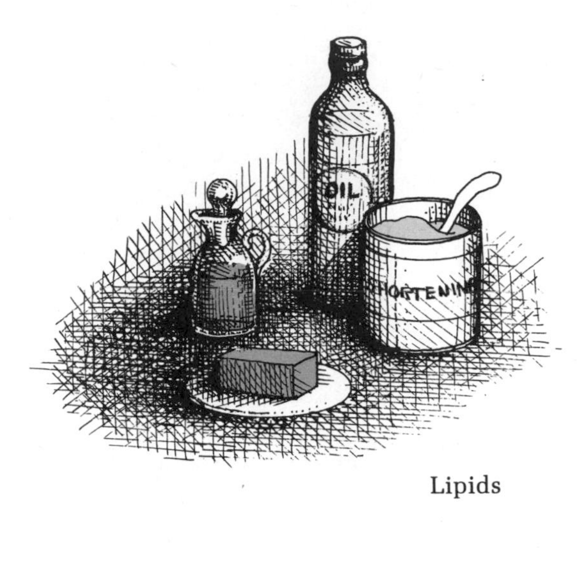

$$
\begin{array}{l}
CH_2\text{—}OH \\
\;| \\
CH\text{—}OH \\
\;| \\
CH_2\text{—}OH
\end{array}
$$

glycerol

$$
\begin{array}{c}
CH_3 \\ | \\ CH_2 \\ | \\ CH_2 \\ | \\ CH_2 \\ | \\ (CH_2)_8 \\ | \\ CH_2 \\ | \\ CH_2 \\ | \\ CH_2 \\ | \\ CH_2 \\ | \\ CH_2 \\ | \\ COOH
\end{array}
$$

stearic acid

$$
\begin{array}{c}
CH_3 \\ | \\ CH_2 \\ | \\ CH_2 \\ | \\ CH_2 \\ | \\ (CH_2)_8 \\ | \\ CH_2 \\ | \\ CH_2 \\ | \\ CH_2 \\ | \\ COOH
\end{array}
$$

palmitic acid

Fatty acids

3 acid units

$$
\begin{array}{l}
CH_3(CH_2)_{16}\text{—}\overset{\overset{\large O}{\|}}{C}\text{—}O\text{—}CH_2 \\
\qquad\qquad\qquad\qquad\quad\;\; | \\
CH_3(CH_2)_{16}\text{—}\overset{\overset{\large O}{\|}}{C}\text{—}O\text{—}CH \\
\qquad\qquad\qquad\qquad\quad\;\; | \\
CH_3(CH_2)_{16}\text{—}\overset{\overset{\large O}{\|}}{C}\text{—}O\text{—}CH_2
\end{array}
$$

tristearin

A fat

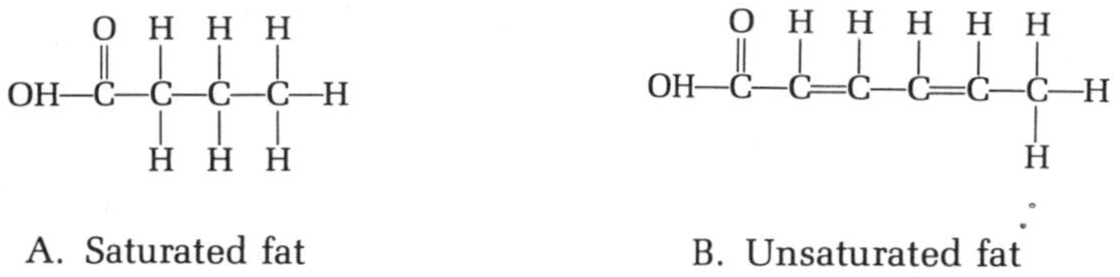

```
   O  H  H  H
   ‖  |  |  |
OH—C—C—C—C—H
      |  |  |
      H  H  H
```

A. Saturated fat

```
   O  H  H  H  H  H
   ‖  |  |  |  |  |
OH—C—C=C—C=C—C—H
                  |
                  H
```

B. Unsaturated fat

Figure 7–7 The difference between a saturated and an unsaturated fat. In a saturated fat the maximum number of carbon–hydrogen bonds has been formed. In an unsaturated fat, however, some of the carbon–carbon links are double bonds; that is, there are fewer carbon electrons available for bonding with hydrogen atoms.

are solid at room temperature. Examples include the animal fats such as lard or beef fat. In the *unsaturated fats,* the carbons can still bond with other atoms. The unsaturated fats are liquid at room temperature and are more common among plants than animals. Soy oil, safflower oil, and corn oil are unsaturated fats (Figure 7–7). The lipids we eat, whether saturated or unsaturated, may serve as building blocks for synthesis of cell membranes, be converted into acetyl CoA for immediate use in cellular respiration, or be stored in special fat cells until they are needed as an energy source at some later date.

Apparently the kinds of fats included in a diet can influence health. The proportion of saturated fats in the diet is one of the factors related to the frequency of heart disease. The Japanese, who eat very small amounts of saturated fat, suffer only about 10 percent the number of heart attacks that befall Americans, whose diet contains a large quantity of saturated fat.

Vitamins and minerals: essentials on a small scale

The proteins, carbohydrates, and lipids make up the greatest proportion of our diet, and we need to consume them in relatively great amounts. Another group of compounds, known as *vitamins,* are needed only in small amounts, but if they are missing from the diet, the results can be drastic.

For many centuries, the disease scurvy was the curse of sailors on long voyages. Cut off from the land and a supply of fresh vegetables and fruits, the sailors lived mainly on salted meat and bread. After a few weeks of such a diet, the often-fatal scurvy would appear in the crew. Physicians of the British Navy found that a daily ration of lime juice eliminated the threat of scurvy. However, it wasn't until 1930 that scurvy was found to result from a lack of vitamin C, or ascorbic acid. Another deficiency dis-

Quantity is not the only factor in the nutrition of a living system. The substances needed in small quantities—such as vitamins and minerals—are as essential to the organism as the smallest bolt in a complex machine.

ease, known as black tongue among dogs and pellagra among humans, is due to a lack of the vitamin nicotinic acid.

Altogether, there are 14 vitamins essential to man's nutrition, divided into two groups: water soluble and fat soluble. These vitamins are listed in Table 7–1. Most of these vitamins are found in the cells of all plants and animals, but not all of them need be present in the diet of these other organisms. For example, most animals can synthesize their own ascorbic acid from simple precursors. Man cannot, and he must take it in ready-made. The other organisms that must have ascorbic acid regularly are the monkeys, the guinea pig, and the Indian fruit bat.

Why is it that compounds needed in such small amounts should be so important? Research has shown that most of the vitamins are important building blocks for essential enzymes or carrier molecules. If the vitamin isn't present, the enzyme or carrier isn't synthesized, and an important reaction cannot occur. For example, nicotinic acid converted to the form of nicotinamide is a principal component of NAD, which acts as a carrier in cellular respiration (see Chapter 5). A deficiency in nicotinic acid means that the respiratory activities of the body's cells are severely hampered.

Table 7–1 Vitamins needed by man

Water soluble		*Fat soluble*
Thiamine: vitamin B_1	Biotin	Vitamin A
Riboflavin: vitamin B_2	Folic acid	Vitamin D
Nicotinic acid	Lipoic acid	Vitamin E
Pantothenic acid	Vitamin B_{12}	Vitamin K
Pyridoxal: vitamin B_6	Ascorbic acid: vitamin C	

The *minerals* are the micronutrients mentioned in Chapter 2, elements that are needed only in small amounts but that are necessary to the body. Iron, for example, provides the core for the four protein chains of the hemoglobin molecule, while sodium in its ionic form plays a role in sending impulses along nerve fibers.

Digestion: simplifying and absorbing

To the series of organs known as the *digestive system* falls the task of first breaking the proteins, carbohydrates, fats, vitamins, and trace minerals of food into a form that the body can use and then absorbing them. Figure 7–8 provides a diagram of the human digestive system,

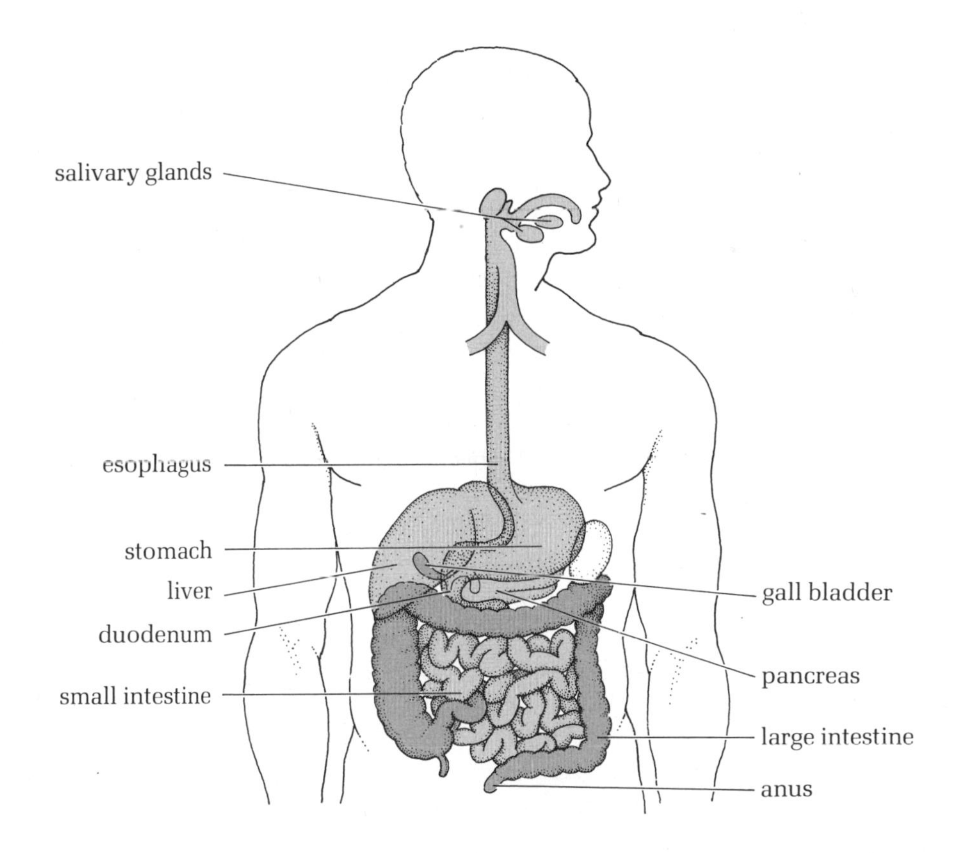

Figure 7–8 The human digestive system. Within this long, twisted tube, food is chemically and physically processed until it can be absorbed by the blood vessels in the intestines.

which is much like that found in all omnivores. In essence, the digestive system is a simple tube that runs from the mouth to the anus. Food moves down the tube by means of rhythmic muscular contractions known as *peristalsis.* The digestive system is equipped with several glands that secrete enzymes and other substances that speed the breakdown of the food molecules. Chemically and physically, the long tube of the digestive tract is "outside" the body. That is, the lining of the system is continuous with the skin, and the products of digestion don't actually enter the body until they are small enough to be absorbed through the lining of the digestive tract.

How does digestion work? How is food broken down and taken in as it passes down the long tube?

Step one: the mouth

We humans process some of our food before we eat by grinding, cutting, soaking, cooking, and so forth. But if we had to, we could digest most foods, even meat, as is. Digestion starts as soon as the food enters the mouth. The teeth tear and grind the food into bits small enough to be easily swallowed, and the tongue keeps the food moving through the mouth.

As the food is being chewed, the *salivary glands* secrete considerable amounts of *saliva.* The saliva coats, moistens, and softens the food, allowing relatively rough or fibrous materials to be swallowed without damaging the walls of the digestive tract. Human saliva also contains *ptyalin,* an enzyme that begins to break down starch into its component simple sugars. This is the reason that well-chewed bread or potato tastes sweet. Although digestive enzymes are found in the saliva of most mammals, they are not present in some species. Dogs are an example; they swallow almost immediately and additional chewing would do nothing to accelerate digestion.

In swallowing, the food is moved from the mouth into the pharynx toward the stomach. Swallowing begins voluntarily, but once the food reaches the pharynx, involuntary peristalsis takes over, pushing the food into the *esophagus,* a tube that connects the pharynx with the stomach. Although the pharynx is part of both the respiratory and the digestive systems, the food is kept from going the wrong way by the *epiglottis,* an elastic flap that covers the trachea during swallowing.

Step two: the stomach

Softened and moistened, with some of its carbohydrates broken into simple sugars, the ball of food enters the *stomach*. The stomach is an elastic bag that can hold as much as 4 quarts of food at a time. The food is held in the stomach by a valve at each end, and the rhythmic churning of the stomach continues the physical breakdown of the food begun by the teeth.

The cells of the stomach wall secrete mucus, *pepsinogen*, and *hydrochloric acid*. The hydrochloric acid, which is highly acidic, kills practically all living matter in the food. It also reacts with the inactive pepsinogen to produce *pepsin*, an enzyme that breaks down proteins. After spending 3 or 4 hours in this acidic, protein-breaking environment, the once solid food has become more or less liquid.

If the stomach is so acidic, why doesn't it digest itself? The mucus produced by the stomach wall acts like a protective barrier against the acid. Apparently, though, prolonged nervous tension inhibits the production of the mucus. If there is too little mucus for too long, the hydrochloric acid may burn the stomach wall, producing an ulcer.

Almost no food is absorbed into the bloodstream through the stomach, with the exception of glucose and alcohol. This is the reason that honey has a reputation as a quick-energy food and that alcohol can exert its effects in so little time.

Step three: the duodenum

From the stomach, the food moves into the *small intestine*. The upper portion of this long and looping organ is called the *duodenum*, and it is here that most of the breakdown of food molecules takes place. Some necessary enzymes are produced by the duodenum itself. Others are produced by the *liver* and the *pancreas*, special digestive organs connected to the duodenum by ducts. These two organs produce enzymes and other digestive secretions. From the pancreas comes a fluid that neutralizes the acidity of the juices pouring from the stomach into the duodenum. The liver produces *bile*, which is stored in the *gall bladder* until needed. Bile breaks down fats into droplets small enough to be carried away by the blood. Human excrement gets its brown to green color from bile. If the duct leading from the gall bladder is blocked by a stone or if the liver is diseased and produces no bile, excrement turns white.

Step four: the lower intestines

After passing through the duodenum and its barrage of enzymes and digestive juices, the food has become a watery broth of nutrients in a form that can be absorbed. That absorption takes place in the lower part of the small intestine. The walls of the intestine are lined with outpocketings called *villi,* which increase the surface area available for absorption, in the same way that the alveoli increase the surface area of the lungs. Food molecules move through the walls of the intestine and into the beds of capillaries within them. As digestion has proceeded, several quarts of water have moved into the stomach and intestines. This water is taken up through the intestinal walls along with the food, for it is too valuable to lose.

From the small intestine, the remaining food mass moves into the *large intestine.* Any leftover water is absorbed here. The large intestine is the dwelling place of a great number of bacteria that live on the food remains. In return they manufacture vitamin K, necessary for the clotting of blood. What the bacteria can't use passes on into the end of the large intestine, where it is stored in the form of *feces,* and excreted to the outside through the *anus.*

Note that the feces are not the waste products of digestion. They are instead those portions of food that cannot be digested, such as the natural cellulose in a serving of green beans or the "edible" cellulose in a low-cost milkshake.

Digestive specialists: eating what others can't

Practically all organic matter can serve as food as long as one can get to it, digest it, and dispose of any poisons it might contain. Much of the marvelous diversity produced by evolution is shown in the various ways animals have of getting food and then digesting it.

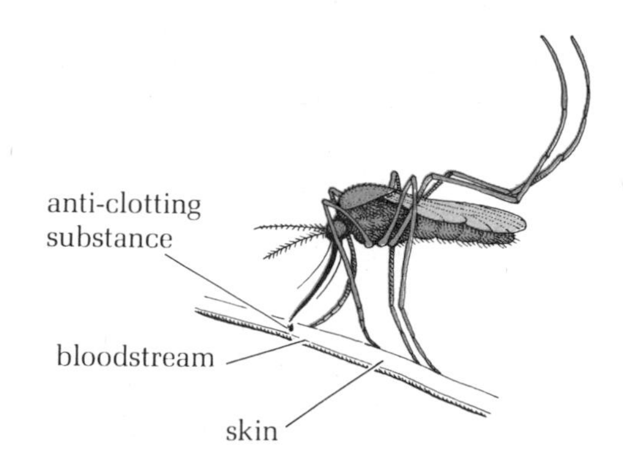

Figure 7–9 The mosquito can pierce the skin of a higher animal and feed on the host's bloodstream.

The carnivores have little problem with their diet, for it contains large amounts of the proteins needed for growth and maintenance and of the fats that are such a good energy source. Among the carnivores, there is specialization not so much in how they digest their food but how they catch it. The superb eyesight and sharp beak of the eagle, the sticky tongue of the frog, and the strength of the lion all serve in catching and killing food animals. Those animals that feed on the fluids of plants and animals —among them the mosquito, vampire bat, aphid, leech, and lamprey eel— also have an easily digested diet. They too have evolved special ways of getting to their food sources. Mosquitoes have a mouth built like a needle for piercing the skin, and they squirt into the wound a substance that keeps the blood from clotting, thus ensuring a steady flow (Figure 7–9).

Cellulose, although the most ubiquitous organic material, is available as food to only a few animals. The class of mammals known as the *ruminants*, which includes cattle, antelope, and camels, are able to utilize cellulose because of a specialized digestive system that is an internal "farm" of microorganisms. All of the ruminants are equipped with four stomachs. A grazing deer, for example, eats a large amount of food very quickly, filling the first stomach, then retreats to a safe place to continue its digestion. There it regurgitates the food in a ball-like mass called the *cud*. The deer chews the cud slowly and thoroughly. The processed cud moves into the stomach called the *rumen*, from which this class of mammals takes its name. The rumen contains a rich growth of microorganisms that break the cellulose into its components and give off carbon dioxide and methane. The deer rids itself of the gases by burping; the other component molecules pass into the bloodstream and are transported through the body to be used as fuel for cellular respiration. The microorganisms

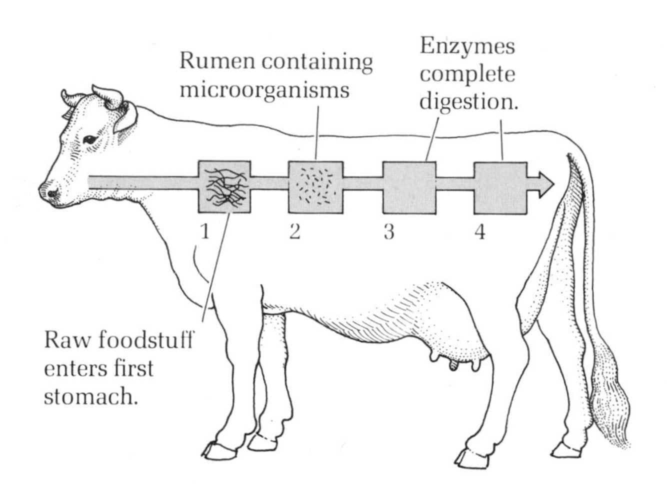

Figure 7–10 The complex stomach system of the ruminant. The microorganisms in the rumen "digest" cellulose for the host animal, breaking the complex molecule into simpler substances.

of the rumen are swept into the two final stomachs, which contain enzymes able to break down the microorganisms' protein. The amino acids are then absorbed through the stomach wall. The protein that is such a large proportion of a beef steak or a venison chop is ultimately produced by microorganisms nourished by grass supplied by the cow or deer (Figure 7–10).

The kidney: filter and regulator

Most of the food consumed by animals is used as fuel in cellular respiration to produce usable energy in the form of ATP. ATP, though, is not the only product of cellular respiration. The breakdown of carbohydrates and fats produces carbon dioxide and water. The carbon dioxide diffuses into the blood to be disposed of through the lungs. Since man lives in a dry, land environment and always faces the problem of retaining enough moisture, the water is a gain. But when proteins are broken down, nitrogen is produced as a by-product, usually in the form of ammonia. Unlike carbon dioxide or water, ammonia is very poisonous, and the body must get rid of it quickly and efficiently. How can this be done?

In birds and insects, the ammonia is converted to the form of *uric acid* (Figure 7–11). Uric acid is the hard, cementlike, wax-eating ingredient in

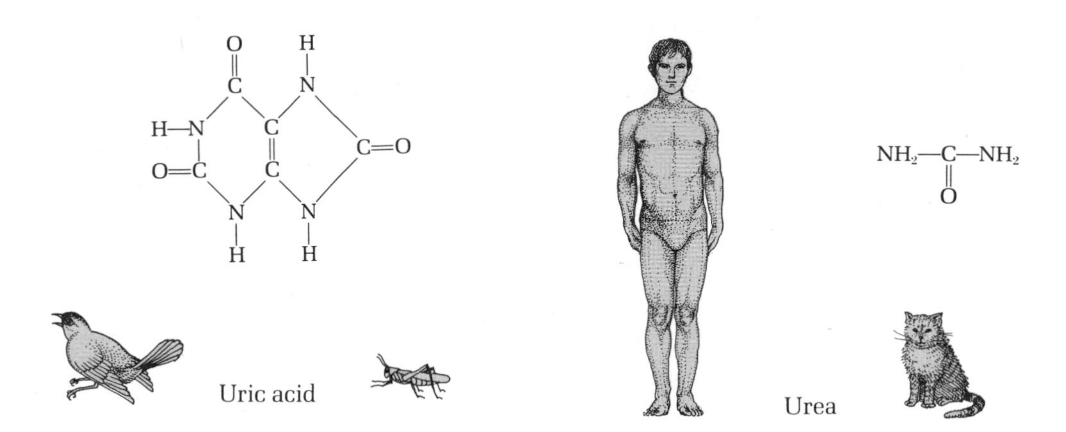

Figure 7–11 Excreting nitrogen. Birds and insects eliminate nitrogen wastes by excreting uric acid in semisolid form. Most mammals excrete a dilute solution of urea. Both methods relate directly to the water retention needs of an organism.

bird excrement that can make washing a car or sitting on a park bench something less than pleasant. In man, however, as in all mammals, the ammonia is converted to the form of *urea* in the liver. Urea must be excreted in solution with water.

Blood carries with it chemicals other than urea. The blood and the cell fluids contain a number of ionic salts taken in with food or produced by the chemical activities of the cells. The hydrogen ion (H^+), the sodium ion (Na^+), the potassium ion (K^+), the magnesium ion (Mg^{2+}), and the chloride ion (Cl^-) are among them. A certain concentration of these salts is needed for vital chemical reactions, but at higher concentrations they too are toxic. The ions can also be excreted in solution with water.

But a land animal like man can't use great amounts of water to carry away wastes like urea and excess ions. He has to retain water to keep from drying out. Thus man faces three excretory problems: to dispose of urea; to maintain the proper concentration of salts, getting rid of the excess but retaining any that are in short supply; and to make sure that the body loses no more water than it can afford. These chemical functions are carried out by the *kidney*.

The workings of the kidney

The kidneys are a pair of fist-sized, bean-shaped organs situated about halfway up the back. The kidneys, along with the bladder and the

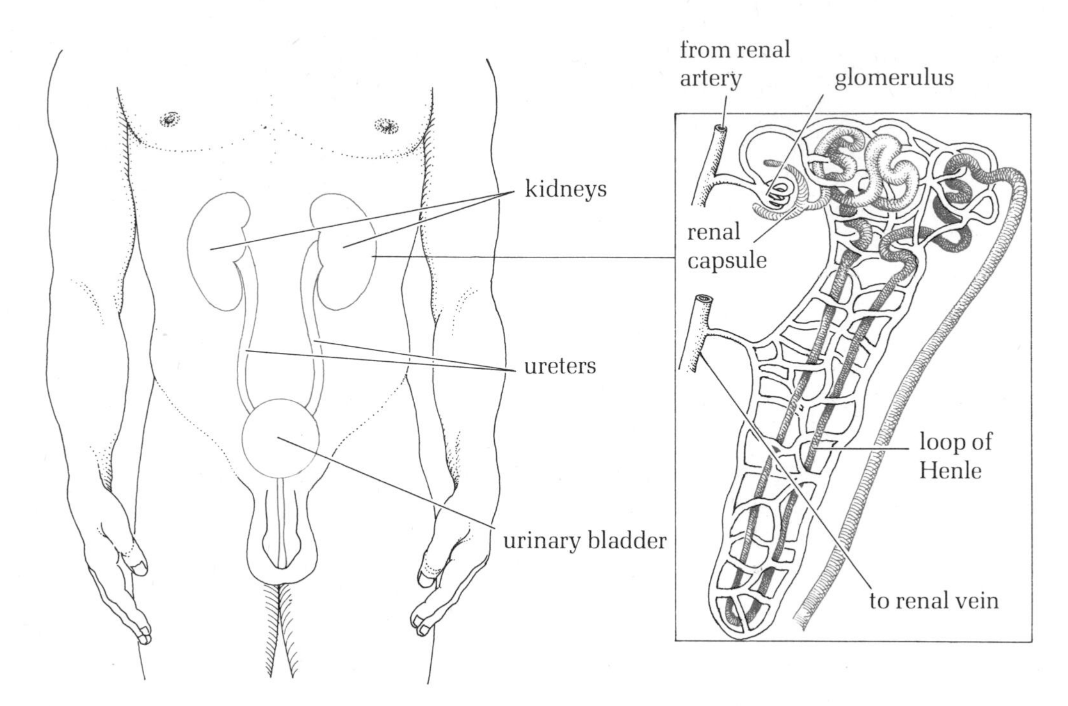

Figure 7–12 The human urinary system. Within the kidneys are over one million nephron (see inset). These minute filtering units selectively process the circulating blood, monitoring the body's chemical balance and removing liquid wastes.

ducts that connect kidney with bladder and bladder with the outside, constitute the *urinary system* (Figure 7–12).

How is it possible for an organ to select and remove harmful materials and to leave beneficial ones in a mixture as complex as the blood? The kidney first filters out all of the small molecules in the blood and then takes back only the valuable ones. Instead of having to identify a multitude of harmful substances, the kidney needs to recognize only a handful of good ones.

How is it possible for an organ to select and remove harmful materials and to leave beneficial ones in a mixture as complex as the blood?

The actual unit of the filtration is the *nephron*, shown in Figure 7–12. Each kidney contains about 1 million nephrons, providing the average adult with some 50 miles of filtration surface! A branch of the aorta, the *renal artery*, which supplies the kidney with blood, branches and divides and finally ends in the tufted bundle of capillaries called the *glomerulus*. Because of the relatively high pressure of the blood, the plasma and the materials pass through the walls of the glomerulus and into the *renal capsule*. The openings in the capillary walls are too small for blood cells and blood proteins to pass through, so these components of the blood remain behind. From the renal capsule, the plasma flows through a long and convoluted tubule surrounded by a bed of small arteries and veins. In this tubule, the contents of the blood are analyzed and sorted. Glucose, amino acids, and vitamins are returned to the blood. Waste materials like urea remain in the tubule. The fate of the ions depends on the chemical status of the body. If there is too much Na^+, for example, it remains; but if it is in short supply, it is returned to the blood. A specialized portion of the tubule known as the *loop of Henle* recovers most of the water in the filtered plasma. The amount of water recovered also depends on the water needs of the body. The kidneys of a man suffering from thirst on a hot hike in the desert will recover all the water except for that amount needed to keep the urea from building up to a toxic level, while the kidneys of someone who has been drinking lemonade for several hours will produce a very watery urine. The kidney tubules connect into larger and larger structures, ending in the *ureters* that connect the kidney to the *urinary bladder*, where the urine collects until it can be excreted.

The kidneys act something like chemical watchdogs for the body, analyzing the blood and checking its contents against the body's needs. The daily flow of blood through the kidneys is about 50 gallons, yet the adult human contains only some 6 quarts of blood. Thus every drop of blood in the human body is checked and filtered in the nephrons of the kidneys some 33 times a day.

Adapting the kidney to the environment

All the various kinds of kidneys, found in the vertebrate organisms ranging from fish to man, work in roughly the same way, as blood analysts and blood filters. But different animals living in different habitats face different excretory problems, and the kidney has adjusted accordingly.

The bodies of freshwater fish contain a far greater concentration of ionic salts than the waters they live in. As a result, water tends to move into their bodies by diffusion. The kidneys of these fish serve to pump out water and keep salt; they release great quantities of very dilute urine. The saltwater fish have just the opposite problem. The relatively primitive hagfish and sharks have body fluids with roughly the same salt content as seawater. The bony fishes, which include most varieties of fish apart from the sharks and rays and which first evolved in fresh water and then moved into the sea, have body fluids less salty than the sea. Thus they lose water through their skins. These fish drink seawater, remove most of the salts and keep the water, and then excrete the salts through special glands in the gills.

Land animals face the problem of losing water steadily (see Figure 3–3). One of the major sources of water loss is the urine. Animals that live in particularly dry climates have developed the ability to secrete a very salt-concentrated urine. The desert-dwelling kangaroo rat has the most concentrated urine found in any animal. It can live its whole life without ever drinking liquid water, subsisting entirely on the water produced as a by-product of the breakdown of fats in the seeds it feeds on. If it had to, the kangaroo rat could maintain the amount of water it needs by drinking seawater. Why can't a man adrift at sea do the same? The kidneys of various species differ in the salt concentrations of the urine they produce. Man, for example, cannot produce urine that is more than about 2 percent salt. Seawater contains about 3.5 percent salt. Thus a man drinking seawater would have to use some of the water in his body fluids just to get rid of the excess salt. As a result, drinking seawater would result in a loss, not a gain, of water.

A word about cycles

In Chapter 3, we saw how the interrelationship between the nonliving and the living worlds could be described in terms of nutrient cycles.

The desert-dwelling kangaroo rat can live its whole life without ever drinking liquid water.

Necessary materials like hydrogen and carbon are picked up in the physical environment by living organisms, passed on to other life forms, and finally returned to the surroundings to be cycled again. Assembled into compounds on their way through the trophic levels of a given ecosystem, they serve two purposes: as a source of energy that can be converted to the usable form of ATP, and as a supply of the molecules needed to provide for maintenance and growth. That was the big picture.

In looking at the digestive and excretory systems of animals, we've looked at the little picture. The digestive system provides the animal organism with a way of taking in complex organic molecules and breaking them down into a form that can be used for energy or for maintenance and growth. And through the excretory system, the waste and excess products of these processes are returned to the environment. The same cyclic movement that described the great mass of organisms also holds true for the individual human being.

Summary

The molecules of food can be classified into a few groups. Proteins are large, complex molecules made of chains of connected amino acids, whose sequence is the primary structure of the protein. Bonding within the primary structure produces the helix of the secondary structure, a characteristic of fibrous proteins. Foldings of the helix back on itself give rise to the twistings of the tertiary structure of the globular proteins. Fibrous proteins are important components of such structures as skin and bone, while globular proteins serve primarily as enzymes. Although man synthesizes his own proteins, certain component amino acids must be supplied by his diet. Proteins are also used as a fuel for cellular respiration.

Carbohydrates are compounds of carbon, hydrogen, and oxygen. The simplest carbohydrates are the monosaccharides. Two linked monosaccharides produce a disaccharide. A chain of linked monosaccharides results in a polysaccharide like starch or cellulose. Carbohydrates are important as an energy source. Cellulose is the major structural molecule in plants.

Lipids are waxy, fatty, and oily substances. They are not soluble in water, and play a structural role in cell membranes. They also provide great amounts of energy when broken down. Unsaturated fats are generally plant oils, while solid animal fats are saturated.

The vitamins are needed only in small amounts, but they are crucial components of important enzymes and carriers. The trace minerals are equally essential materials also required in small amounts.

The function of the digestive system is to break down food and to absorb the useful molecules in it. In man, digestion begins in the mouth, where the food is broken mechanically into smaller pieces and carbohydrates are treated with a salivary enzyme. In the stomach, pepsin and hydrochloric acid further reduce the food and prepare it for the main stage of digestion in the duodenum. Secretions from the liver and pancreas play an important role at this point. As the food moves into the lower reaches of the small intestine, its useful molecules and much of the water used in digestion is absorbed. The indigestible remains of the food are stored in the large intestine until they are excreted through the anus.

Evolution has produced a range of specializations for catching food and digesting it. The carnivores and fluid feeders are specialized in how they get to their easily digested diets. The ruminant herbivores are able to digest cellulose with the aid of microorganisms living in their intestines. The microorganisms then act as a source of amino acids for the ruminant.

The kidneys dispose of urea and excess salt and restrict water loss. The unit of filtration is the nephron. Blood plasma flows through the walls of the glomerulus and into the renal capsule. As it passes down the tubules of the nephron, needed items are returned to the blood, while wastes are retained. Most of the water is reclaimed in the loop of Henle before the urine passes into the ureters on its way to the urinary bladder.

Different kidneys fit different habits. The kidneys of freshwater fish serve to pump out water and keep salt. In most marine species, little water but much salt is excreted. Desert dwellers are able to secrete highly concentrated urine that minimizes water loss.

Questions to consider

1. a. What are the two primary roles played by proteins in cells?
   b. What is it about the structures of proteins that fits them for such roles?
2. a. How do monosaccharides, disaccharides, and polysaccharides differ from each other?

b. What principal roles do polysaccharides play in cells?

3. a. The lipids have two significant chemical characteristics. What are they?
   b. How are these characteristics evidenced in the functions lipids serve in cells?

4. How can it be that the vitamins and trace minerals, needed in such small amounts, should be so crucial to animal life?

5. What are the major occurrences in digestion from the time the food enters the mouth until its remains pass out through the anus?

6. The cells of the stomach wall produce and secrete hydrochloric acid and pepsinogen. Inside the stomach the pepsinogen and hydrochloric acid react to produce pepsin, which breaks down proteins. Isn't this a rather roundabout procedure? Why don't the cells of the stomach wall simply manufacture pepsin in the first place?

7. a. What are the three principal tasks performed by the kidney?
   b. An engineer designing a filter, say for crankcase oil in a car, would very likely try to find some way to hold impurities back and let the clean oil pass through. Blood is more complex than oil, and it contains a far greater number of potential impurities. How does the body solve this complex filtration problem?
   c. What are the components of the nephron?
   d. What role does each one play?

8. On the basis of how the kidneys of the bony fishes work, most biologists agree that these species first evolved in fresh water and then moved into the sea. Why does this seem a reasonable assumption?

9. The breakdown of 1 ounce of fat yields 1.1 ounces of water and no ammonia. The breakdown of 1 ounce of protein yields about 0.3 ounce of water and some ammonia. A kangaroo rat fed a diet of fatty seeds needs no extra water. But if it is given only seeds high in protein, such as soybeans, the rat will die of thirst. Part of the reason is, of course, the different amount of water produced by each diet. But that is only part of the answer. What is the rest of it?

Glossary

amino acid one of twenty-two simple nitrogen-containing compounds that are the chemical subunits of proteins

anus the opening at the end of the digestive tract through which solid food wastes are excreted

bile a secretion of the liver which breaks down fats during digestion

catalyst any substance which speeds the rate of a chemical reaction without being altered by the reaction

cellulose a complex polysaccharide that is the major structural substance in plants

cud a food mass regurgitated by a ruminant for further chewing

digestive system the organ system which chemically and physically processes food for absorption into the bloodstream

disaccharide a compound of two simple sugars

duodenum the upper portion of the small intestine

enzyme a globular protein that acts as a biological catalyst

epiglottis the elastic flap that folds over the trachea during swallowing

esophagus the long muscular tube that connects the pharynx and the stomach

feces solid animal waste products of digestion

fibrous protein a type of protein characterized by a ropelike molecular structure; important for structural tissues in living systems

gall bladder the sac-like structure of the digestive system that regulates the secretion of bile

globular protein a type of protein characterized by a compact and coiled structure; important for the chemical regulation of living systems

glomerulus the tufted collection of capillaries in the nephron of the kidney

hydrochloric acid an acid secretion in the stomach which partially sterilizes incoming foods and reacts with pepsinogen to produce the enzyme pepsin *kidney* one of the two fist-sized organs which are the major portion of the urinary system

large intestine the thick, coiled tubular system which connects the small intestine and the anus

lipid the class of nutrient compounds that includes oils and fats; important for structural functions in living things

liver the large organ which secretes numerous enzymes and bile into the digestive tract

loop of Henle the portion of the nephron tubule that is the site of the reabsorption of water

minerals the inorganic substances needed by living systems in very small amounts

monosaccharide a simple sugar unit, the subunit of complex carbohydrate molecules

nephron the filtration unit of the kidney

pancreas the digestive organ which secretes digestive enzymes and regulates the acidity of the stomach

pepsin a stomach digestive enzyme which breaks down protein molecules

pepsinogen an inactive stomach enzyme which is activated by hydrochloric acid to release pepsin

peristalsis the rhythmic muscular movement of the digestive system which controls the flow of food materials

polysaccharide a large carbohydrate molecule composed of repeating monosaccharide units

primary structure the sequence of amino acids in a protein

protein the class of nutrient compounds composed of large complex molecules of linked amino acid units

ptyalin a salivary enzyme that begins the digestion of sugars

renal artery the artery supplying the kidneys

renal capsule the portion of the nephron that collects the plasma escaping the glomerulus and channels it into the tubule

rumen one of the four stomachs found in the ruminants

ruminant a mammal characterized by four stomachs and the ability to digest cellulose with the aid of microorganisms

saliva the enzymatic secretion of the mouth

salivary glands the digestive glands that secrete saliva

saturated fats a group of fats whose carbon atoms have formed the maximum number of possible bonds

secondary structure the twisted form of a protein molecule formed by linkages between the amino acids of the primary structure

simple sugar a monosaccharide

small intestine the long narrow tube system which connects the duodenum and the large intestine

stomach the sac-like muscular digestive organ that processes food before releasing it to the duodenum

tertiary structure the folding of the secondary structure of a protein to form a coiled and complex molecule

unsaturated fats a group of fats whose carbon atoms have not all formed the maximum possible number of bonds

urea the nitrogen-containing end product of protein digestion which some animals excrete in a dilute solution

ureter one of the two tubes leading from the kidney to the bladder

uric acid the nitrogen-containing end product of protein digestion which can be excreted without loss of water by the organism

urinary bladder the sac-like organ in which urine is stored until it is excreted

urinary system the organ system which filters the blood and removes liquid wastes

vitamins the relatively simple organic compounds required by organisms in small amounts

8 Inheriting

If a backyard gardener bought a package of radish seeds and planted them, and then later found zucchini growing in place of the radishes, how would he explain what had happened? Very likely, he would assume that the seed company had made an error and mistakenly put zucchini seeds into a package intended for radishes. Never would he think that what he'd planted were radish seeds in the first place. The very suggestion that zucchini could grow from radish seeds is mind-boggling. We all know from common sense and common experience that each species produces its own kind and only its own kind. Radishes seed radishes, and cats have kittens; never do the two become confused. This point about organisms seems so self-evident that we take it for granted.

Yet it really is quite a lot to take for granted. All living organisms are able to copy themselves, to pass on to another generation precise instructions about how to develop properly, to ensure that radishes do indeed become radishes and kittens cats. No human engineer has been able to design a machine able to pass on instructions for self-assembly to another generation of machines. Such things are the stuff of science fiction. But this self-copying ability, so far beyond the creative capabilities of the best human minds, is possessed by even the simplest life forms.

But another curious thing about life's ability to copy itself is that the copies aren't perfect. The children produced by one couple may strongly

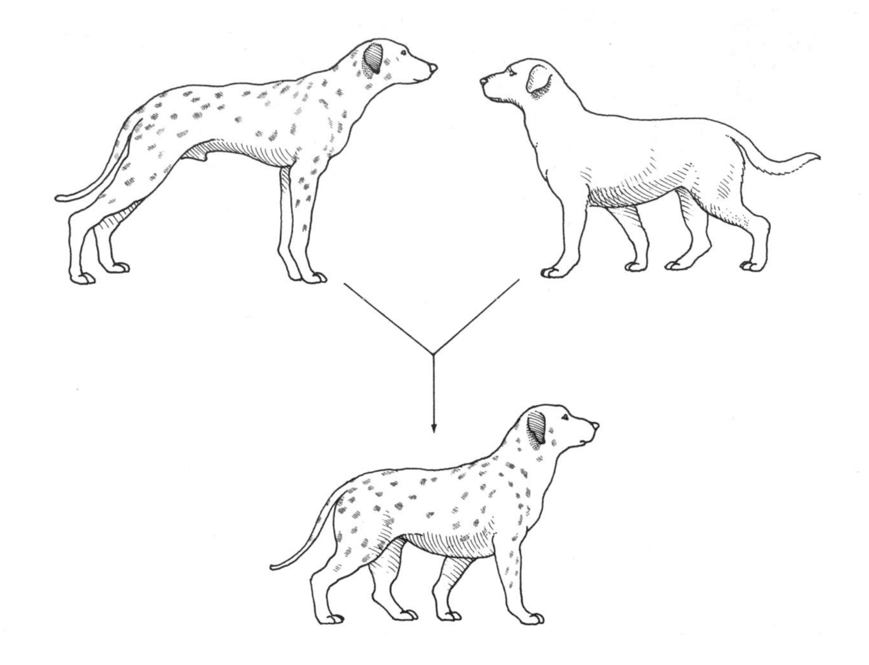

Figure 8–1 Heredity provides a predictable link between generations. Living things produce offspring of the same species, with many of the traits of the parent generation.

resemble each other, but they also represent considerable variety. Except in the case of identical twins, none of them look exactly like each other or like either of the parents. Thus, while inheritance from one generation of humans to another ensures that each generation is indeed human, totally new individuals are produced (Figure 8–1).

Genetics is the field of biology concerned with the hereditary instructions passed from one generation to the next. In this chapter we'll be looking at some of the chief questions of genetics. For example, how is the inherited information passed and received? What form does this message take? What is the genetic basis for the many different forms of individuals within a species? And does genetics contribute anything to our understanding of evolution?

The cell's blueprints

Since all organisms are composed of one or more cells, it's not surprising to find that the carriers of hereditary information are found within

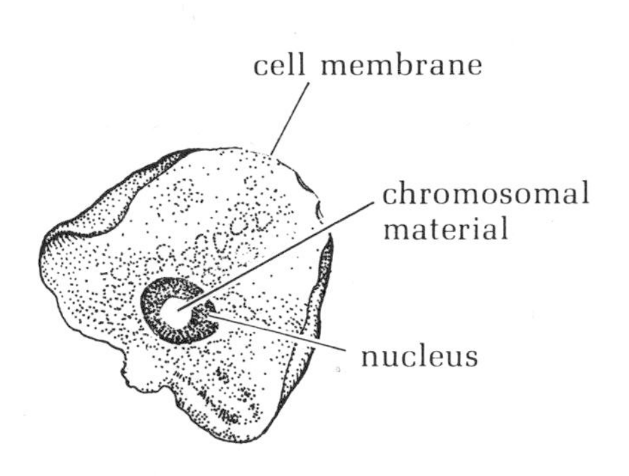

Figure 8–2 A. A cell from the lining of a human cheek. Note the dark material coiled in the nucleus; these are the chromosomes of the cell. B. A "map" for the photograph, identifying the major parts of the cell.

the cell. In fact, these carriers are present within each and every cell of our bodies. Figure 8–2a shows a typical human cell stained with dye to reveal its various components. Compare it with the schematic drawing of Figure 8–2b, which acts as a sort of map for the photograph. You'll notice that within the nucleus of the cell are a number of short, stubby, and dark bodies. Because these bodies readily absorb the dyes used to stain cells for examination under the microscope, they are called *chromosomes,* a combination of Greek words that means "colored bodies." This photograph shows 46 chromosomes arranged into 23 pairs, the number characteristic for humans. The exact number of chromosomes varies from species to species, from 2 (one pair) to 400 (200 pairs); for example, dogs have 78, frogs 26, and cabbages 18. Chromosome number is not necessarily related to complexity. For example, the dog and the horse, which are lower on the evolutionary scale than man, have more chromosomes than we do.

Cells, whether single cells or components of complex organisms, reproduce by dividing; that is, one cell divides into two, the two into four, the four into eight, and so forth. This division follows one of two definite and predictable patterns. The major difference between these two patterns is what happens to the chromosomes.

The dog and the horse, which are lower on the evolutionary scale than man, have more chromosomes than we do.

Mitosis: duplicating information

The first pattern of cell division is known as *mitosis*. In mitosis, the nonnuclear contents of the parent cell are divided more or less equally between two daughter cells. But, even more importantly, the chromosomes are first duplicated and then divided precisely, a complete and full set going to each of the daughter cells. Figure 8–3 diagrams the several stages of the process.

Before mitosis and before the chromosomes can be seen within the nucleus, the chromosomal material is duplicated. As mitosis starts, the chromosomal material condenses into thicker and thicker strands and finally becomes visible as individual chromosomes. If you look carefully, you can see that each chromosome is made up of two parts, called *chromatids*, joined at their centers. Each chromatid of the chromosome is identical to the other. The point of attachment between the chromatids is called the *centromere*.

While the chromosomes are condensing and thickening, the nuclear envelope disappears, and a fan-shaped bundle of fibers, the *spindle*, arises. Some of these fibers attach themselves to the centromeres of the chromosomes. The chromosomes arrange themselves along the midline of the cell, and then the centromeres replicate, separating the chromatids of each pair. Pulled by the spindle fibers, the separated chromatids travel

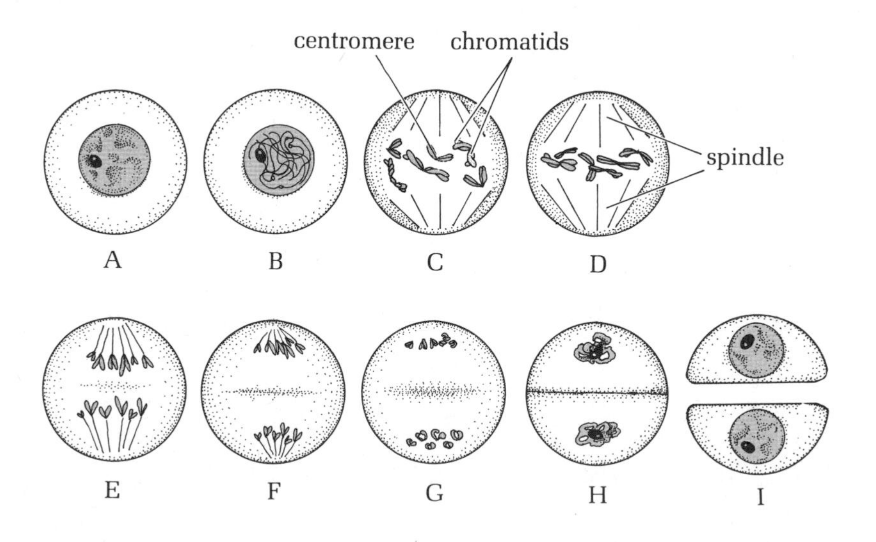

Figure 8–3 The process of mitosis. Through this simple doubling and dividing process, a cell is divided into two smaller—but genetically identical—cells, each with the diploid number of chromosomes. From the "resting cell" stage (A), the chromosomes in the nucleus thicken (B). Chromosome pairs become visible (C). The spindle fibers attach to the centromeres and separate the chromatids (D through F). Nuclear membranes form around the separated portions of nuclear material (H), and the process is complete with the formation of the new cell membranes (I).

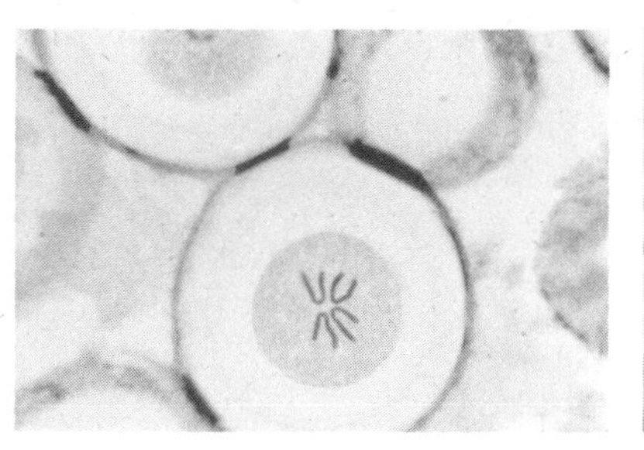
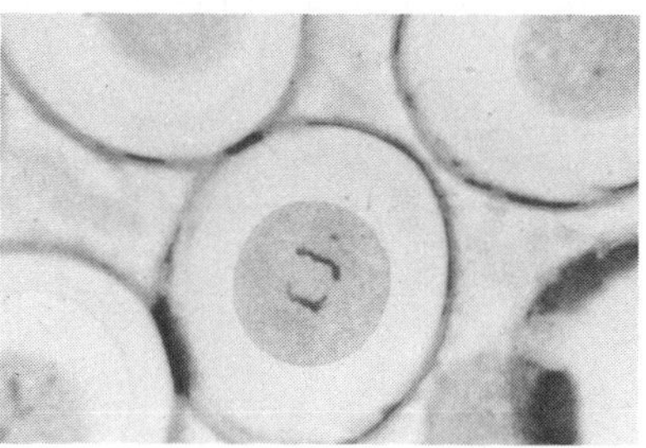
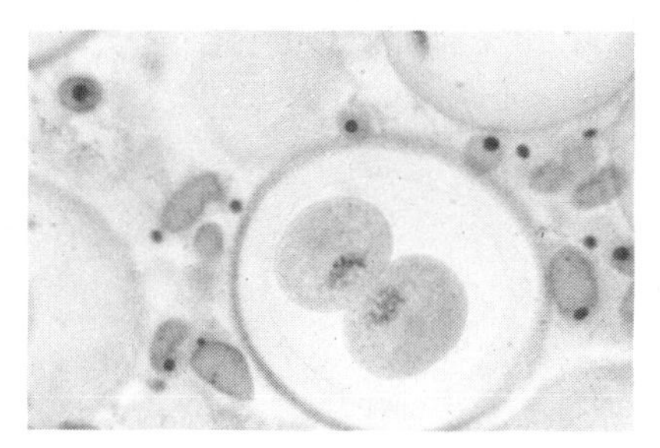

Three of the phases of mitosis in the roundworm, *Ascaris*. In metaphase (A) the duplicated chromosomes line up in pairs along the center axis of the cell; in anaphase (B) the spindle fibers separate the chromatids; in telophase (C) the development of two separate, genetically identical cells is evident.

to opposite ends of the cell, so that each end has an identical set of chromosomes. Nuclear envelopes develop about the chromosomes. Cell division is complete when the cytoplasm is apportioned between the two new cells. In animal cells, the nuclear membrane pinches inward, almost as if the original cell were being squeezed in half. In plants, a plate appears in the middle of the cell and grows until it touches the cell membrane and cuts the original cell in two.

The result of mitosis is that one parent cell has given rise to two daughter cells, and each of these daughter cells has exactly the same genetic information as the other. Also, each of these cells contains the full number of chromosomes characteristic of the species. A cell with the complete set of chromosomes is called *diploid*. The diploid number in man, for example, is 46. By means of the events of mitosis, a diploid parent cell has given rise to two diploid daughter cells.

Meiosis: dividing information

The second procedure of cell division produces a result quite different from that of mitosis. Instead of two diploid daughter cells, the offspring are four half-diploid, or *haploid*, cells. Each contains exactly one-half the number of chromosomes typical of the species. Figure 8–4 depicts the series of occurrences that produces this result.

The first stages of meiosis look much like those of mitosis; the chromosomal material duplicates, condenses, and thickens as the nuclear envelope disappears. As we noted, the chromosomes come in pairs; the two chromosomes of a pair are called *homologues*. As the chromosomes

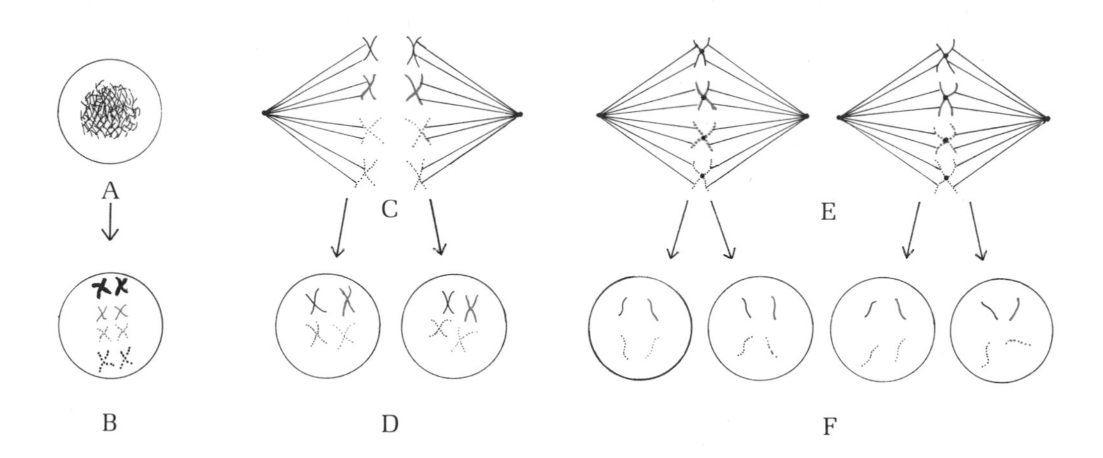

Figure 8–4 The process of meiosis. The chromosomal material in the diploid cell condenses and thickens (A). Chromosome pairs become evident (B), and spindle fibers are attached (C). The homologues are separated and move to opposite ends of the cell (D). Two diploid cells result (E). In the next division, spindle fibers are again attached to the chromosomes, but separation occurs without replication of the genetic material. Thus, four haploid cells are produced.

appear, the homologues, each consisting of two chromatids connected by a centromere, move very close together. Now occurs an event characteristic of meiosis and meiosis alone. The chromatids of the two homologues intertwine with each other and exchange bits of chromosomal material. This event is called *crossing-over*, and in many species it occurs at almost every meiotic division.

The homologous, intertwined pairs arrange themselves along the midline of the cell. Spindle fibers attach themselves to each of the homologues. Unlike the case in mitosis, the centromeres do not divide, and the chromatids of each chromosome remain together. The spindle fibers pull the homologues apart, sending one of the pair to one end of the cell and the other to the opposite end. There the chromosomal material breaks down and again becomes diffuse, completing the first stage of meiosis.

The second stage of meiosis begins with the condensation and thickening of the chromosomal material into chromosomes. This material has not been duplicated, as happened at the beginning of the first stage. Spindle fibers appear and attach themselves to the centromeres of the chromosomes. The centromeres replicate, separating the chromatids from each other, and the spindle fibers pull the separated chromatids, or single chromosomes, toward opposite sides of the cell. Nuclear envelopes and cell membranes appear, dividing the one cell with four clusters of single chromosomes into four separate cells.

The result of meiosis is four cells with a haploid number of chromosomes; in man, the number is 23. And, because of crossing-over, these chromosomes are not necessarily identical to each other.

The purpose of the difference

Why should there be two modes of cell division—particularly when one of them is as complex as meiosis seems to be? The answer is that each process serves a different purpose.

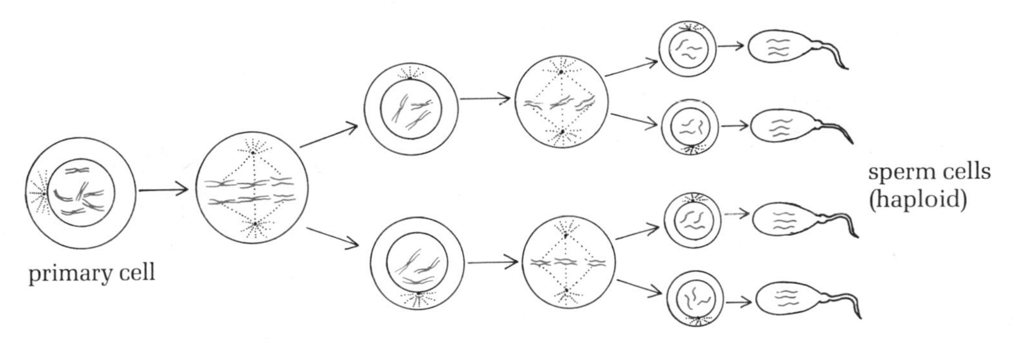

Spermatogenesis—production of sperm cells

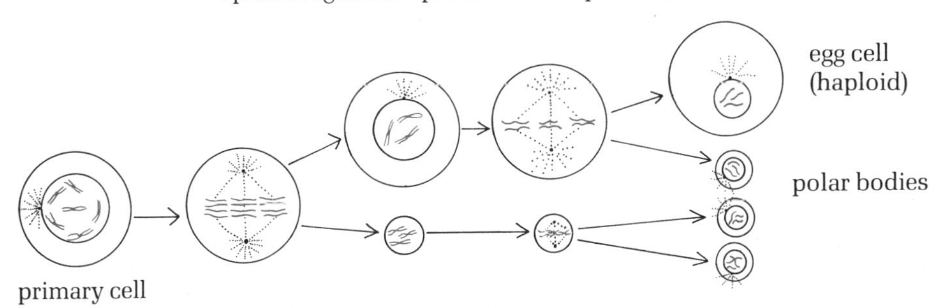

Oogenesis—production of egg cells

Figure 8–5 Gamete production. The processes of sperm production and egg production involve meiotic divisions that reduce diploid primary cells to four haploid gametes.

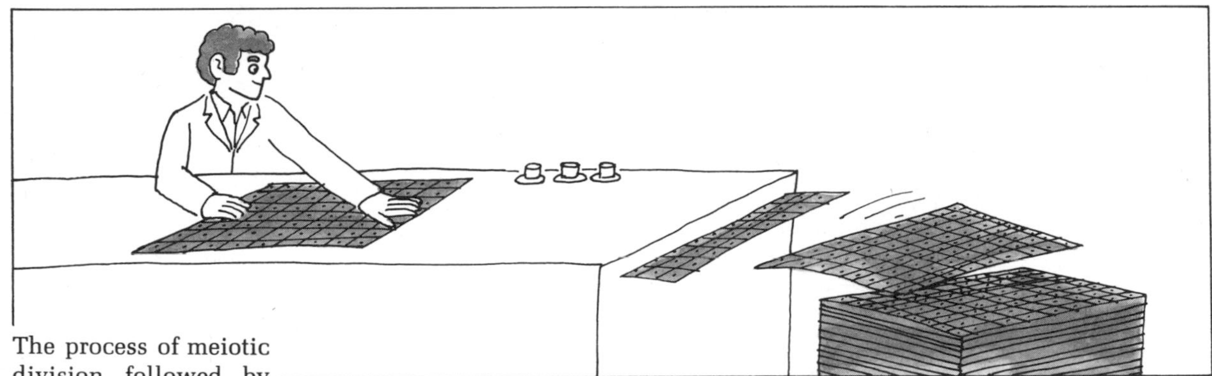

The process of meiotic division followed by the fusion of gametes can be compared to the cutting and splicing of two different sheets of paper. The sum total of material has not increased, but the distribution of the pattern has been altered, so the final set has pieces of both original patterns. Mitosis, on the other hand, provides a series of identical copies of the same material—analogous to running the same sheet through a copying machine.

Mitosis is the way that many one-celled microorganisms reproduce. It is also found among the body cells of plants and animals. For example, flowering plants grow throughout their life cycles, and it is mitosis, mostly in the stems and roots, that adds to the number of cells in the plant and therefore to its size. Human skin cells go through a similar, constant process of division. Mitosis ensures that every one of the cells in the body of the organism contains precisely the same genetic information as all the other cells.

Meiosis produces the specialized reproductive cells called *gametes*, the eggs of the female and the sperm of the male. Gametes are haploid. By fusing, a male gamete and a female gamete contribute their haploid sets of chromosomes to form one diploid cell, called a *zygote*, which will develop into an offspring. A male gamete contains one homologue of a chromosome pair, and the female cell carries the corresponding homologue. When the two gametes fuse into the zygote, homologues are reformed into a pair. Thus the chromosomal material in the zygote is derived equally from each parent. Without meiosis, two parent, or sexual, reproduction would be impossible (Figure 8–5).

Meiosis also makes possible very considerable differences in gametes. Half the chromosomes of any organism come from each parent. Meiosis does not necessarily divide the paternal and the maternal chromosomes equally among all the gametes. Instead, they are apportioned independently. The number of possible mixes of maternal and paternal chromosomes is determined by the number of pairs in the diploid cell. Mathematically, the total number of possible combinations is 2^n, where n is the number of chromosome pairs in the organism. In an organism with four chromosomes, or two pairs, $2^n = 2^2$, or 4. Thus such an organism can produce four kinds of gametes. With 8 chromosomes, the number is 2^4, or 16; with 10, it's 2^5, or 32. In the case of man, with 46 chromosomes, the number is 2^{23}, or 8,388,608! This means that any one human being can produce three times more combinations of his chromosomes than there are people in Los Angeles. Multiply the 8,388,608 by the human world population of almost 4,000,000,000, and you can get some sense of the astronomical number of genetic variations made possible by meiosis.

Meiosis ensures endless alternative combinations of the same genetic material in gametes, while mitosis guarantees that once two gametes come together to form a cell, the chromosomes contained in this new diploid cell will be replicated faithfully as the organism develops.

The expression of inheritance

After two human gametes have come together, the cell that will develop into a child contains a half-complement of chromosomes from

Any one human being can produce three times more combinations of his chromosomes than there are people in Los Angeles.

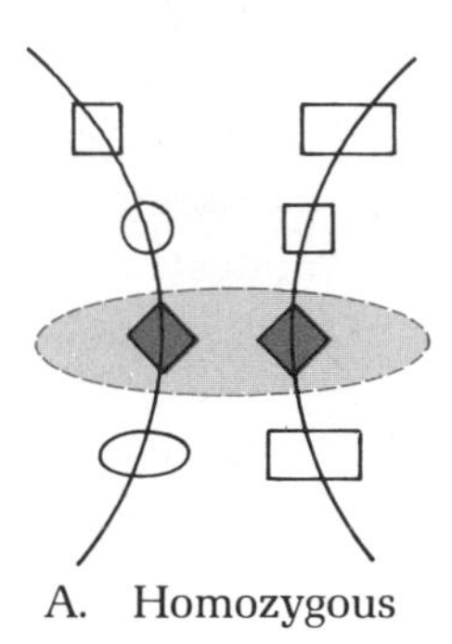

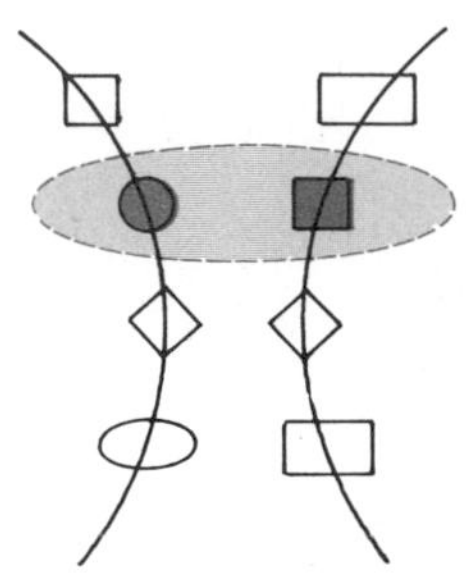

Figure 8–6 The difference between homozygous and heterozygous genes. In each of these two figures, there are eight alleles for four genes, each gene determining a single trait. At the shaded portion in A, the alleles are the same; thus the chromosome pair is homozygous for that trait. In B, at a different site on the chromosome pair, the alleles in the shaded portion are different, making the pair heterozygous for that trait.

the mother and a half-complement from the father. How will the information on these chromosomes be expressed in the individual? In other words, what are the rules to determine whether a child looks more like mom than dad or vice versa?

Information in the chromosome

The portion of a chromosome that influences or determines a particular trait is called a *gene*. Chromosomes are like a series of genes laid end to end and bound together. In the diploid cell, there are two genes for any one trait, one located on each homologue of a chromosome pair. These two genes can be the same, or they may be different. Different forms of the same gene are called *alleles*. A cell containing the same allele of a certain gene is said to be *homozygous* for that particular trait. If the alleles are different, then the cell is *heterozygous* (Figure 8–6).

In a cell homozygous for a certain trait, obviously only the one form of the trait can be expressed. But what happens in the case of the heterozygote, where there are two alternatives for the same trait? Generally, only one allele is expressed. The other form is masked and unexpressed, even though it remains part of the genetic makeup of the organism. The expressed gene is said to be *dominant*, and the masked one *recessive*.

Dominant and recessive: an example

To show how this interaction works, let's take as an example the flower color of the common garden pea. This gene has two alleles, white

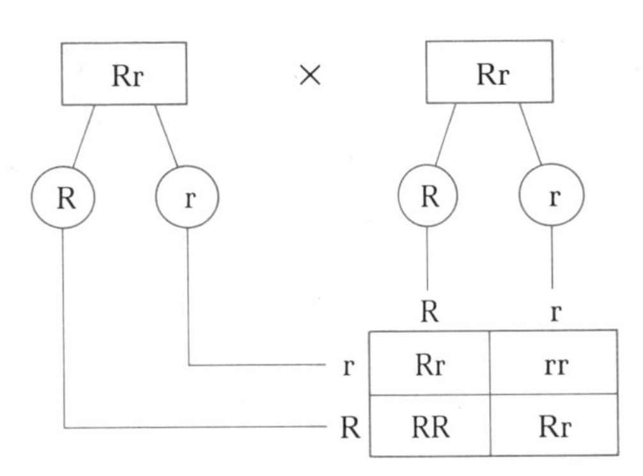

Figure 8–7 This gridwork is a form of Punnett square, named after a British biologist. A Punnett square is a handy way of indicating the expected genotypes of the offspring of a given cross. The ratio shown here also helps determine phenotype—there will be three red-flowered plants and one white-flowered plant.

and red, and red is dominant. By the conventions of the geneticist's shorthand, traits are assigned a single-letter abbreviation, with the capital letter going to the dominant allele. Thus, in this case, r indicates white, and R red. The genetic makeup, or *genotype,* of a homozygous red-flowered pea is RR. The genotype of a white-flowered plant is rr. Each homozygous plant can produce only one kind of gamete so far as flower color is concerned—R or r. If RR is crossed with rr, then all the offspring must be Rr, the heterozygous form. Since R is dominant, the outward expression, or *phenotype,* of Rr is red. However, the Rr plant can produce two kinds of gametes, R and r. If the Rr plants are crossed with each other, some offspring will be white, and some red, as shown in Figure 8–7.

For a recessive to appear in the phenotype, the offspring must inherit that particular gene from both parents. In this example, the only crosses **that can produce white flowers are Rr × Rr, Rr × rr, and rr × rr. No white plants can come from a RR × rr cross.**

Also, note that one phenotype may result for more than one genotype, or, conversely, two genotypes can produce the same phenotype. A red-flowered pea can be either RR or Rr; the phenotype is precisely the same in both cases even though the genotypes differ.

Many genes and many traits

The fact that two genotypes produce the same phenotype is but the barest indication of just how complex the relationship between genetic

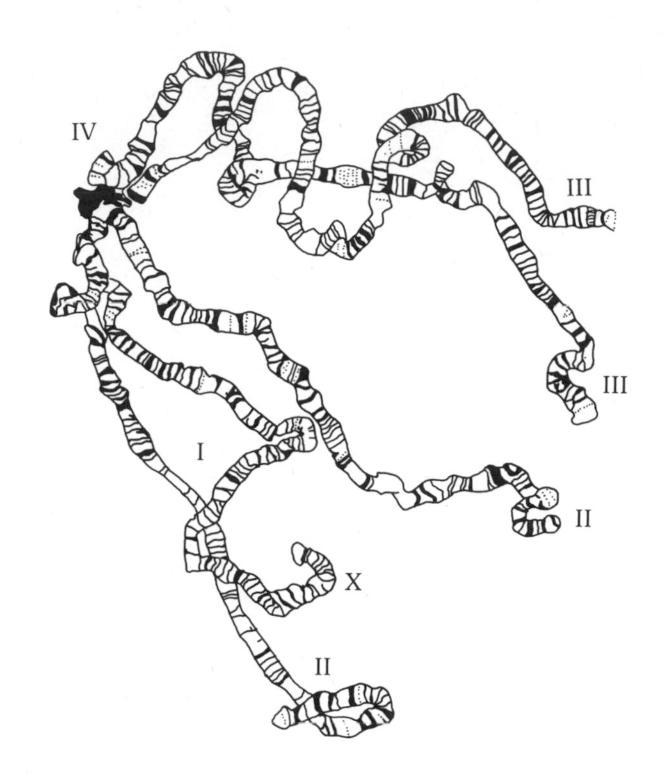

Figure 8–8 An early drawing of the chromosomes of the salivary gland cells of the fruit fly. The common fruit fly has been an important research tool of modern genetics throughout this century. As far back as 1915 the prominent geneticist Thomas Hunt Morgan had identified 85 genes, apparently organized into four distinct groups within the nucleus of the fruit fly's cells. Morgan correctly determined that the fruit fly has four pairs of chromosomes.

information and outward appearance can be. To begin with, the chromosomes carry a great number of genes. The chromosomes of the fruit fly have been the subject of extensive study and mapping. It is thought that the 8 chromosomes of this small insect contain some 10,000 genes! By no means have all of them been identified (Figure 8–8).

The relationship between two alleles in a heterozygote is not always a cut-and-dried situation of dominant and recessive. In some cases, neither gene dominates and both are partially expressed. This phenomenon occurs, for example, in certain flowers like snapdragons and carnations, where crossing a red-flowered individual with a white-flowered one results in pink offspring. This is called *incomplete dominance* (Figure 8–9).

In some cases, too, what we may think of as a single trait is in fact influenced by a number of genes. In such a situation, the trait is the cumulative product of a number of different genes. In man, for example, there is no single gene for height. What we consider one trait—a certain num-

Chromosome I

Name	*Region Affected*
Abnormal	Abdomen
Bar	Eye
Bifid	Venation
Bow	Wing
Cherry	Eye color
Chrome	Body color
Cleft	Venation
Club	Wing
Depressed	Wing
Dotted	Thorax
Eosin	Eye color
Facet	Ommatidia
Forked	Spines
Furrowed	Eye
Fused	Venation
Green	Body color
Jaunty	Wing
Lemon	Body color
Lethals, 13	Die
Miniature	Wing
Notch	Venation
Reduplicated	Eye color
Ruby	Legs
Rudimentary	Wings
Sable	Body color
Shifted	Venation
Short	Wing
Skee	Wing
Spoon	Wing
Spot	Body color
Tan	Antenna
Truncate	Wing
Vermilion	Eye color
White	Eye color
Yellow	Body color

Chromosome II

Name	*Region Affected*
Antlered	Wing
Apterous	Wing
Arc	Wing
Balloon	Venation
Black	Body color
Blistered	Wing
Comma	Thorax mark
Confluent	Venation
Cream II	Eye color
Curved	Wing
Dachs	Legs
Extra vein	Venation
Fringed	Wing
Jaunty	Wing
Limited	Abdominal band
Little crossover	II chromosome
Morula	Ommatidia
Olive	Body color
Plexus	Venation
Purple	Eye color
Speck	Thorax mark
Strap	Wing
Streak	Pattern
Trefoil	Pattern
Truncate	Wing
Vestigial	Wing

Chromosome III

Name	*Region Affected*
Band	Pattern
Beaded	Wing
Cream III	Eye color
Deformed	Eye
Dwarf	Size of body
Ebony	Body color
Giant	Size of body
Kidney	Eye
Low crossing over	III chromosome
Maroon	Eye color
Peach	Eye color
Pink	Eye color
Rough	Eye
Safranin	Eye color
Sepia	Eye color
Sooty	Body color
Spineless	Spines
Spread	Wing
Trident	Pattern
Truncate intensf.	Wing
Whitehead	Pattern
White ocelli	Simple eye

Chromosome IV

Name	*Region Affected*
Bent	Wing
Eyeless	Eye

ber of feet and inches—is actually the result of a great number of separate traits having to do primarily with the building of the skeleton, and each of these traits is influenced by a different gene. Another example of a human trait affected by several genes is skin color.

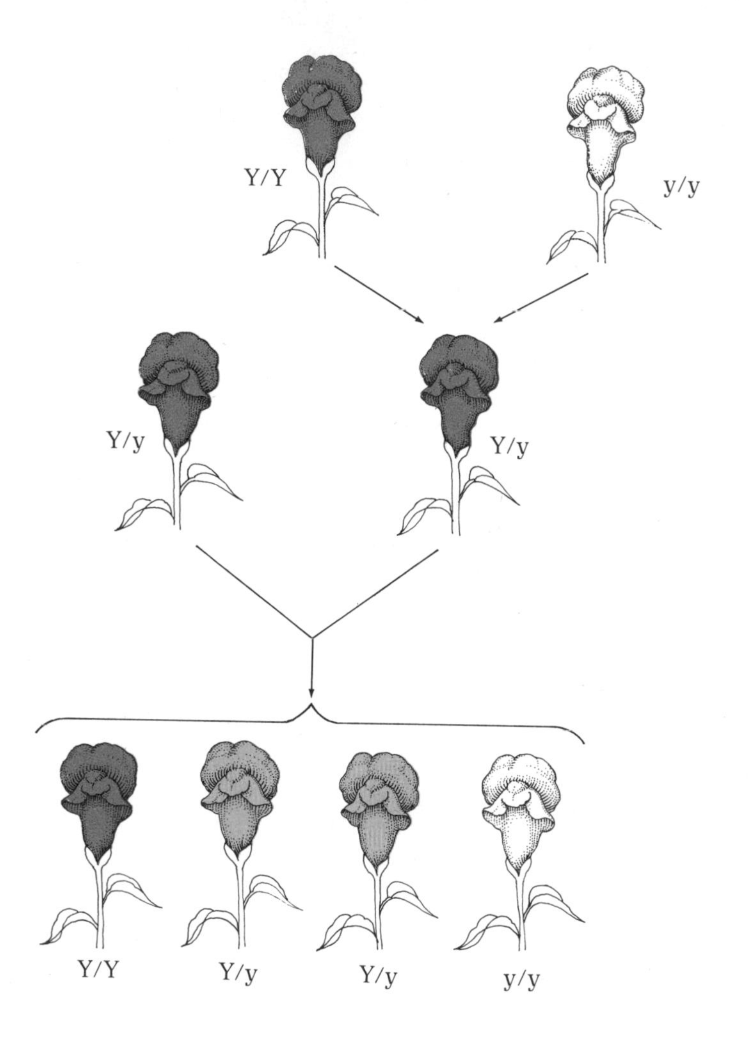

Figure 8–9 An example of incomplete dominance in snapdragons. A cross between a homozygous dominant and a homozygous recessive will produce a blended color effect in the second generation.

In much the same way that many genes may influence one trait, one gene may affect several traits. The many parts of an organism are so interrelated that a seemingly insignificant trait can have very pronounced effects throughout the organism. In rats, for example, a single gene affect-

The many parts of an organism are so interrelated that a seemingly insignificant trait can have very pronounced effects.

ing the makeup of cartilage, an important component in the skeleton, produces offspring with thick ribs, narrow tracheas, blunt snouts, and thickened heart walls. In chickens, a single gene produces frizzle, unusually structured feathers that give the chicken the appearance of having been the unfortunate victim of a mad beautician. Because of the misshapen feathers, a chicken with frizzle loses body heat more rapidly than a normal chicken and must burn food more quickly to keep its body temperature up. Because of the larger volume of food, the kidneys have to work harder at their filtering tasks. The overall rise in the body's chemical activity caused by the need to maintain body heat forces the heart to pump faster and work harder. The heart grows in size and changes in shape. All these changes can be traced to the altered feathers produced by a single gene!

Determining sex

As complex as the relationship between some traits and some genes is, the one that determines sex stands out as remarkably simple. Figure 8–10 shows an enlarged photograph of the human chromosomes arranged in homologous pairs. As you can see, the homologues of each of the first 22 pairs are pretty much the same size and thickness. This is not necessarily the case with the twenty-third pair. This pair determines sex. If it consists of an X chromosome and a Y chromosome, then the person is male. If there are two X's, then the person is female.

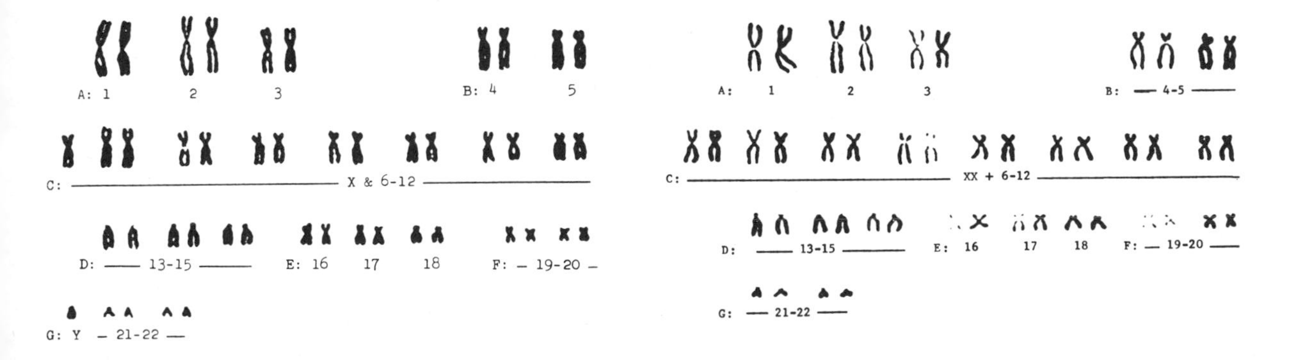

Figure 8–10 A. The chromosomes of a normal human male, showing XY chromosomes. B. The chromosomes of a normal human female. Note the XX chromosomes.

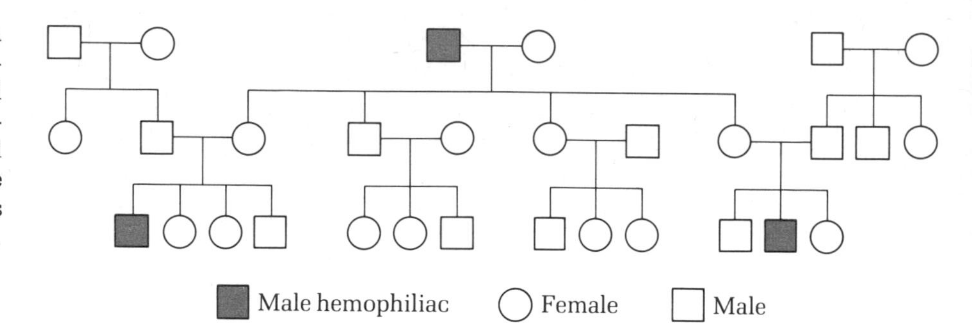

Figure 8–11 A pedigree of hemophilia, a sex-linked trait. The two hemophiliacs in the second generation are both the offspring of daughters of a male hemophiliac.

At meiosis, when the homologues of each pair are sent their separate ways, the sex chromosomes are divided up just like the rest. Since a woman carries only X chromosomes, all her eggs have an X. Since the male has both X and Y, half of his sperm cells have an X chromosome, and the other half a Y. The fusion of the egg with an X sperm results in an XX nucleus, which will develop into a female. The union of the egg with a Y sperm results in an XY combination, which will become a male. Thus the male's sperm determines the sex of the offspring, and the dividing up of chromosomes at meiosis keeps the number of males and females in the population approximately equal.

The Y chromosome carries almost no genetic information other than sex. The X chromosome, on the other hand, does have additional, *sex-linked* genes. In the XY chromosome of the male, the Y chromosome provides no corresponding alleles for the X chromosome. Thus, all of the alleles, recessive as well as dominant, of the X chromosome are expressed. In the female, the recessive trait can appear in the phenotype only if the genotype is homozygous for that trait. However, if the woman is heterozygous for the trait, it will appear in half of her gametes. Sex-linkage is the reason that men may inherit certain genetic abnormalities that are rarely found in women but that can be passed on by them (Figure 8–11).

An example of such a sex-linked recessive trait is the disease hemophilia. A hemophiliac's blood lacks the factors needed to promote normal

The dividing up of chromosomes at meiosis keeps the number of males and females in the population approximately equal.

clotting. As a result, even the slightest bump or bruise can result in severe, possibly fatal bleeding. Hemophilia results from a recessive allele on the X chromosome. A woman carrying the gene produces two kinds of gametes in equal number: $X_{hemophilia}$, or X_h, and X_{normal}, or X_n. Assuming that her husband is not a hemophiliac and carries the normal gene on his X chromosome, the combinations with his sperm result in the following genotypes and phenotypes:

X_nY = normal son
X_hY = hemophiliac son
X_hX_n = normal daughter, but a carrier
X_nX_n = normal daughter, not a carrier

It is possible for a woman to suffer from hemophilia, but only if she inherits the gene from both parents. To meet this requirement, the woman's mother would have to be a carrier and her father a hemophiliac. The odds against such a pairing are very high to begin with, and, to boot, hemophiliacs usually don't live long enough to reproduce.

Another example of sex-linkage, fortunately a more benign one, is color blindness, one type of which is the inability to distinguish red from green. This condition appears most commonly in men and rarely in women. However, daughters of color-blind men will carry the gene, and half of their sons will probably be color-blind.

An excess of information

The purpose of meiosis is to divide homologous pairs and apportion them equally, so that at the union of two gametes the right diploid number of chromosomes results. What would happen if a cell received too few or too many chromosomes?

The results are disastrous. Occasionally meiosis goes awry and produces gametes with the wrong haploid number. Fused gametes with too few chromosomes do not develop, but those with too many sometimes can. Too many chromosomes is one cause of the form of mental retardation known popularly as mongolism and medically as Down's syndrome. People suffering from Down's syndrome often have three twenty-first chromosomes instead of the usual two. Apparently, meiosis of

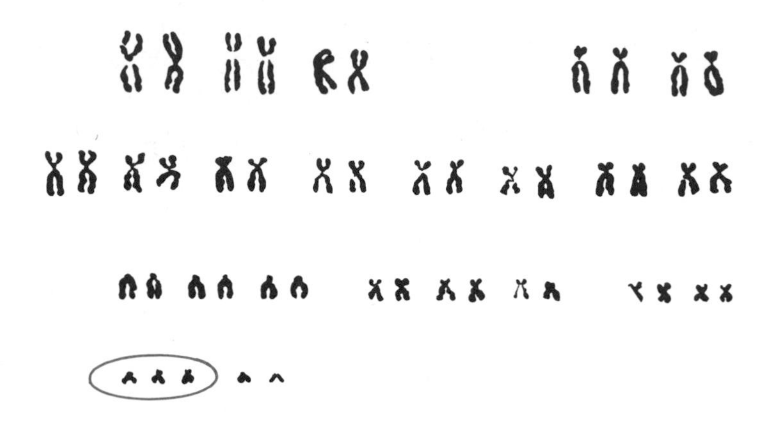

Figure 8–12 The chromosomes of a female victim of Down's syndrome. Note the extra chromosomes aligned with the twenty-first pair.

the gametes of one or the other parent failed to divide the twenty-first pair, resulting in an extra chromosome in the gamete and then in the zygote. It is interesting, but as yet unexplained, that this phenomenon is more common in offspring of parents over 35 years of age.

Down's syndrome can also result when part of an extra twenty-first chromosome is attached to a larger chromosome, often number 15. It often turns out that one of the parents, usually the mother, of such people has one too few separate chromosomes, because one of the homologues of the twenty-first pair is linked to one of the fifteenth. Although such a person has a normal phenotype, meiosis cannot proceed properly, and most gametes are unusual. In fact, of any six children produced by the pairing of such a parent with a normal mate, three will die before birth, one will suffer from Down's syndrome, one will be an outwardly normal carrier of the defect, and only one will be both genotypically and phenotypically normal (Figure 8–12).

The chemistry of the gene

Given the fact that seemingly minor counting errors in the chromosomes can have such marked effects on the final form of the organism, it is obvious that we are dealing with a powerful cellular structure. But it is one thing to note the power of inheritance, and quite another to ask why and how. What is the form of the genetic message? How is this message expressed?

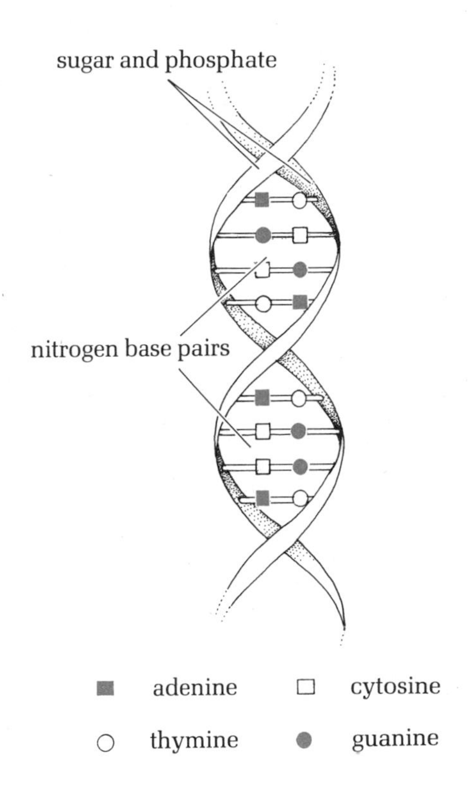

Figure 8–13 The DNA molecule. Within this relatively simple structure is the genetic code in chemical form. The spiral-staircase shape is made of outer bands of phosphate and sugar molecules with crossbars composed of linked nitrogen-base pairs. Notice that adenine and thymine are always linked with each other, and that cytosine and guanine also form repeating pairs.

The double helix

Chromosomes comprise two main components, protein and DNA. DNA is the abbreviation for *deoxyribonucleic acid*. DNA is a very large molecule, which assumes the shape of a ladder twisted into a spiral, or a double helix, as shown in Figure 8–13. Despite its size though, DNA is a relatively simple compound, made up of repeating units of a phosphate, a sugar, and one of the four bases adenine (A), guanine (G), thymine (T), and cytosine (C). The bases on one strand of the double helix bond with those on the other. However, because of the shapes of the molecules, A can bond only with T, and G with C. Therefore an A on one strand must bond with a T on the other, and a G on the one with C on the other.

Early researchers interested in the chemical nature of the genetic message were convinced that the protein had to be the bearer of the code of inheritance. All organisms are extremely complex, and it seemed logi-

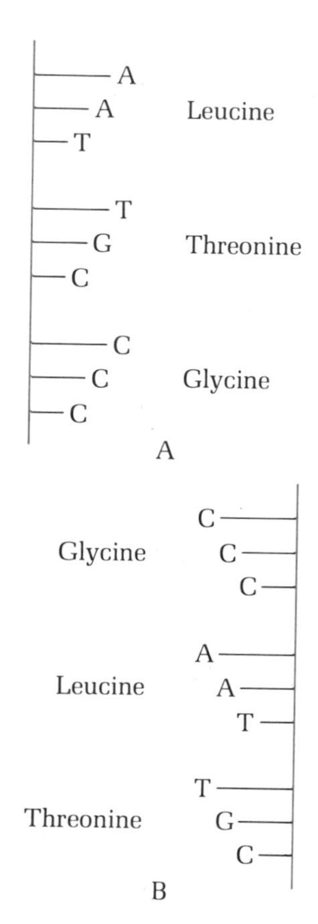

Figure 8-14 An example of triplet codes. The specific nitrogen bases and the sequence in which they are arranged—in groups of three—within the DNA helix encode specific amino acids.

cal to expect that the code for this complexity was equally sophisticated. The proteins offered that complexity, with their many amino acids capable of being arranged in an infinite variety of combinations. Since the only variable part of the DNA molecule was its four bases, DNA appeared far too simple to bear such complex information.

As is so often the case, the logical deduction turned out to be wrong. Subsequent research—and a goodly dose of creative insight on the part of two young scientists, James Watson and Francis Crick—showed that DNA was indeed the bearer of genetic information. But how can such a simple four-base molecule as DNA possibly transmit complex information? The answer lies not in the bases themselves, but in the sequence of bases arranged in threes, or triplets. The genetic code is composed not of A, G, T, C, G, T, but of the triplets AGT or CGT or any other possible combination of the four bases. Still, it seems a long way from a sequence of bases to the outward expression of gene. How is the base code expressed in the cell?

By proteins. You will recall that the proteins play two primary roles in cells, as structural components and as enzymes. Enzymes are particularly important to the function of a cell because the presence or absence of a specific enzyme determines whether the chemical reaction it catalyzes can occur in the cell. The protein's shape, which determines its chemical role, is the result of the sequence of amino acids in its primary structure. Each triplet of bases on a DNA molecule codes for a specific amino acid, and the sequence of triplets codes for the order of amino acids. For example, the amino acid leucine is represented by AAT, while TGC and CCC stand for threonine and glycine. The sequence AAT/TGC/CCC would mean leucine/threonine/glycine, while CCC/AAT/TGC would result in glycine/leucine/threonine (Figure 8-14).

The DNA code is a rather remarkable example of the economy and simplicity of nature. The essence of the code is simplicity, a matter of four possible bases arranged in triplets. Yet this simple thing can serve as a way of transmitting great complexity. The bases can follow any sequence, and there is no end to the number of triplets that can be tied to-

The DNA code is a rather remarkable example of the economy and simplicity of nature.

gether. The chromosomes of man contain in the neighborhood of 10 billion bases, and the possible combinations are practically limitless. The other remarkable thing about the genetic code is its universality. The same triplet of bases codes for the same amino acid in all organisms. In a bacterium, a cornstalk, a snake, and a man, AAT means leucine. At this most basic chemical level, all life is the same.

Making a protein

Although the chromosomal DNA is located in the nucleus, protein is usually synthesized outside the nucleus, in the cytoplasm of the cell, and the genetic information has to be passed from one spot to the other. First, the two strands of the DNA molecule separate, almost as if they had been unzipped. The DNA bases face outward and are available to make new bonds. These bonds are taken up by molecules of a compound known as *messenger RNA*, or *mRNA*. RNA (ribonucleic acid) is a single-strand molecule that, like DNA, is made of phosphates, sugar, and bases. RNA's bases are A, C, G, and uracil (U), which behaves chemically much like T. The RNA bases bond with the bases of the separated DNA strand. A bonds with U, C with G, and so on. This bonding forms a linked molecule of hundreds or thousands of base units—mRNA—that is a chemical mirror-image, so to speak, of the DNA.

Then the newly formed mRNA separates from the DNA. The nuclear envelope contains pores large enough for the mRNA to pass through into the cytoplasm. Once into the cytoplasm, the mRNA becomes attached to one of the small cellular organelles known as *ribosomes*, which act as the site of the remaining steps of protein synthesis.

Now that the information coded onto the DNA has been carried into the cytoplasm, the next task is translating that code into the proper sequence of amino acids. A second ribonucleic acid, known as *transfer RNA*, or *tRNA*, enters the picture. In man, there are 22 kinds of tRNA, each one specific for a particular amino acid. One end of the tRNA molecule bonds with the amino acid, while the other end of the molecule contains a triplet of bases. This triplet chemically complements mRNA. Figure 8–15 depicts the assembly of the protein. The mRNA moves lengthwise across the face of the ribosome. As it does, each successive base triplet bonds with the corresponding bases on a tRNA molecule carrying its

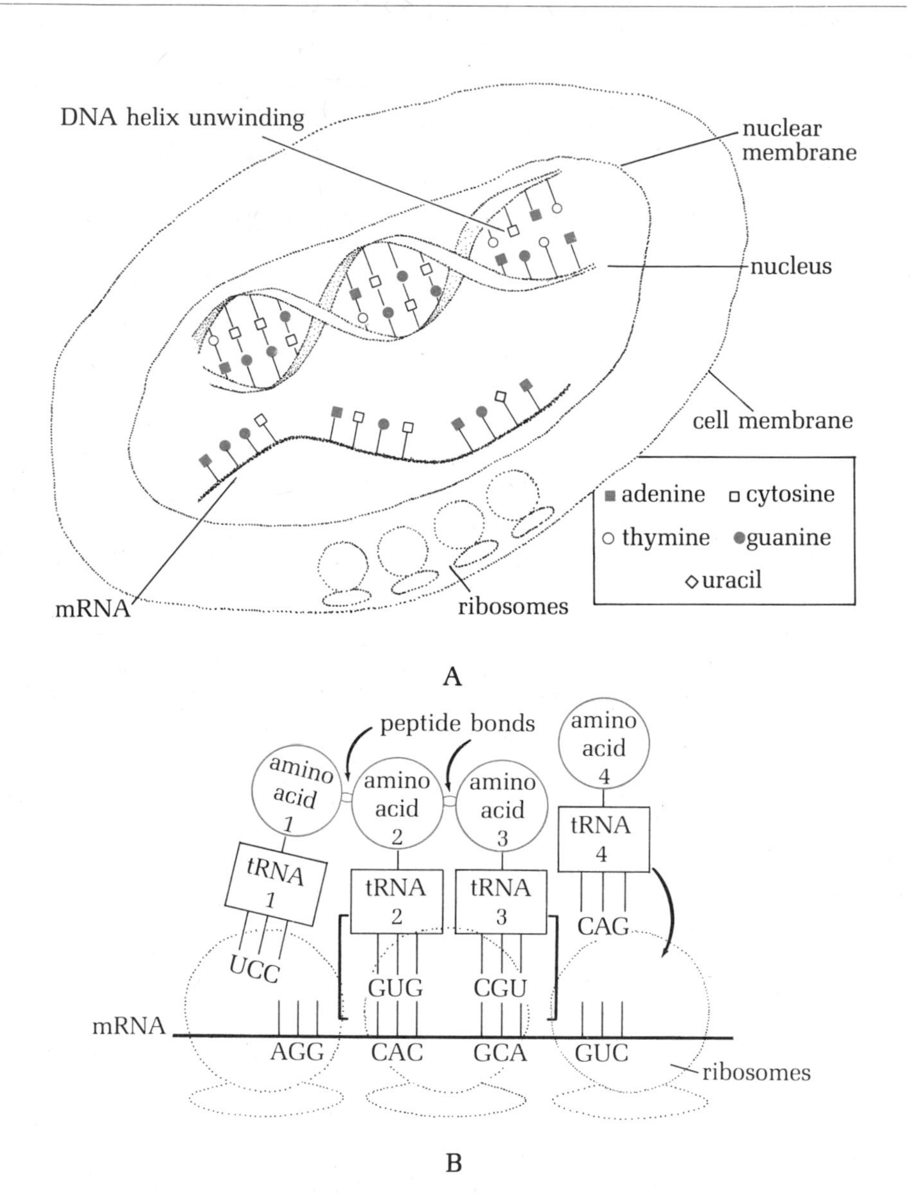

Figure 8–15 A. The first phase of protein synthesis is the transfer of the genetic information within the nucleus to the ribosomes in the cytoplasm, through the action of messenger RNA (mRNA), it copies the triplet codes from the unwinding DNA helix and moves across the nuclear membrane into the cytoplasm. In the cytoplasm it forms an mRNA–ribosome complex, which serves as the template for protein synthesis.

B. In the second phase, transfer RNA units, each specific for a single amino acid, carry amino acids in the cytoplasm to the appropriate place (as indicated by the triplet code) on the ribosomal surface. Peptide bonds are formed between the amino acids, thus forming a polypeptide, or protein. As the polypeptide chain is formed, the tRNA units are released, recirculating into the cytoplasm.

specific amino acid. The amino acids bond together, and the chain of the protein comes off the ribosome piecemeal. When the whole sequence has been completed, the protein separates from the ribosome and becomes an independent molecule.

The essence of the gene

Although the specifics of the translation of a gene into a protein are involved, the outline and implication of the process are clear. The gene is a certain sequence of base triplets on the DNA molecule. Through the mechanisms of mRNA and tRNA, the sequence of triplets is translated into the primary structure of a particular protein. Simple and efficient as the process is, it is also capable of infinite variation.

Genes and natural selection

Darwin postulated that the traits culled by natural selection were the raw material of evolution. But Darwin had to make this assumption simply on faith, because at the time of his work there was no science of genetics. He was unable to show how traits are inherited or how new ones could appear. But now that we do know a good deal about inheritance, we can profitably ask how genetics adds to our understanding of evolution.

Gene pools and hidden traits

If you take all the alleles of all the genes found in any given population of organisms and lump them together, the result is the *gene pool* of that population. Evolution affects the frequency with which particular alleles appear in the gene pool. Those that contribute to phenotypes favored by natural selection become more common, while the less-favored phenotypes and their genes become increasingly infrequent (Figure 8–16).

An important point to remember is that the amount of information contained in any gene pool far exceeds that expressed in the existing phenotypes. The reason is that most genes have at least two alleles and that each individual contains a relatively high proportion of heterozygous combinations. Among humans, a minimum of 16 percent heterozygous gene pairs in each individual is the general rule. Thus, not only does a gene pool contain a large amount of genetic information simply because

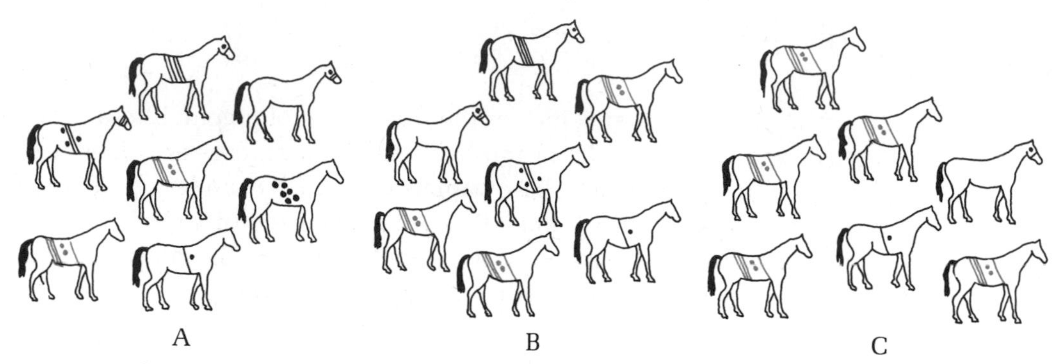

Figure 8–16 The concept of gene pool. The gene pool of a population is the sum of the genotypes of all the individuals within that population. Thus the percentage of occurrence of those traits will increase within the gene pool, even if the total number of organisms in the population does not change.

so many individuals are represented, but also each individual's genotype contains more information than appears in his phenotype.

But how is this genetic material made available to natural selection?

Trading parts of the chromosome

Although the number of genes in a human cell is great, all of them are assembled on the 23 pairs of chromosomes. It would appear that all the alleles of the genes arranged on a particular homologue would have to travel together and that there would be a limit to the number of alternative combinations of these alleles. We have seen how such a *linkage* of genes works in the case of the X chromosome.

However, the sequence of alleles does change, a phenomenon called *recombination*. The mechanism underlying recombination is the crossing-over of meiosis, when the chromatids exchange bits of themselves with each other. After crossing-over, the chromatids are different from what they were before, carrying substitute pieces from their homologue. Crossing-over seems to occur at practically every meiosis.

The effect of recombination is to present a variety of alternative arrangements of the same set of alleles. We noted that meiosis can produce more than 8 million different assortments of maternal and paternal chromosomes in the gametes of a single individual. Because of additional variation through crossing-over, the maternal and paternal traits can be sorted out in an endless variety of possible combinations.

Changing the gene

Recombination allows for many alternative arrangements of the same set of genes, but it doesn't account for new genes. Yet any genetic explanation of evolution must account for new genes. For there to be change, there must be novelty.

So far we've been talking about genes as if they were immutable, like hard, glassy beads strung on the binding twine of a chromosome. For the most part, this is an accurate picture. But sometimes genes do change, in what is known as a *mutation*. Chemically, a mutation is simply a change in the base triplets of the DNA of the gene. Most mutations are of no value. As precise as is the chemical balance of the cell, it's not surprising that random changes in the proteins are generally lethal or meaningless. But some beneficial mutations do occur, and they are the raw material of evolution.

How often do mutations appear? Not often, but the rate varies markedly from one gene to another. For example, in field corn, the gene for seed color mutates at a rate 400 times greater than the gene for shrunken seed.

Genetics and adaptation: the case of sickle cell anemia

Genetically, evolution is a culling and sorting of the gene pool of a given species. Recombination offers new assortments of these genes in each generation, and mutation provides a small but steady source of new alleles. Each generation, then, is very nearly like a series of new models, recombined and perhaps slightly mutated from the ones that went before. From this number the environment makes its selection. The fit individuals live into adulthood and pass their genes—again recombined and possibly

Genetically, evolution is a culling and sorting of the gene pool of a given species.

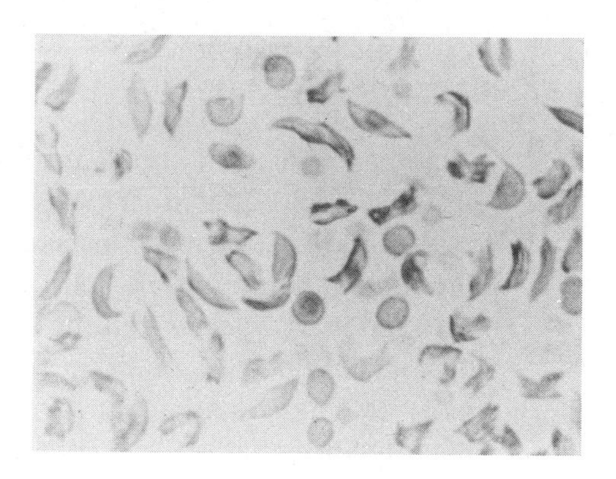

Figure 8–17 Sickled red blood cells. The debilitating effects of this disease are coupled with an important adaptive advantage, immunity from malaria.

mutated—to the next generation from which the environment again picks and chooses.

Research into a serious human disease, sickle cell anemia, has provided us with a rather intriguing and dramatic case of an evolutionary process affecting the human gene pool. Sickle cell anemia is a serious, inherited condition affecting hemoglobin. The hemoglobin molecule, a protein, consists of four chains, two identical pairs of two types called the alpha and the beta. Each chain numbers about 150 amino acids. In normal hemoglobin, the sixth amino acid of the beta chains is glutamic acid. In sickle cell anemia, the glutamic acid is replaced by valine. The change of 2 amino acids in 600 completely alters the behavior of the hemoglobin. When oxygen is lacking, the abnormal hemoglobins tend to form long, rigid rods that twist the red blood cells from their usual concave shape to the crescent shape of a sickle (Figure 8–17). If present in sufficient number, these abnormal cells clog small vessels and cut off blood to internal organs.

A person heterozygous for the sickle cell trait produces both normal and abnormal hemoglobin. Generally, he suffers from the effects of sickling only when he is hard-pressed to maintain enough oxygen, say by being at a high altitude or by engaging in long-distance running. Otherwise the heterozygote can lead a normal life. But a person homozygous for the sickle trait is severely debilitated because all his hemoglobin is abnormal. With careful medical attention, a sickle cell anemic may live to adulthood. Without it, he will almost surely die as a child.

In this country, sickle cell anemia is restricted almost entirely to black people. It is also found in Africa, where many of the ancestors of

America's blacks were captured and sold into slavery. In Africa, the sickle cell gene is very common in certain tribes. As much as 40 percent of the population has the gene, mostly in the heterozygote. Among American black people, about 9 percent are heterozygous, and 0.25 percent homozygous. Why should such a dangerous gene be so common in Africa? And why should black Americans, less than 200 years distant from their African origins, be so much less subject to the disease than their ancestors?

The answer lies in the fact that the sickle cell gene, both in the heterozygous and in the homozygous state, makes a person almost completely immune to malaria. In Africa, malaria is a constant and serious threat to life. Most likely the sickle cell gene arose as a mutation. Anyone who carried the gene was immune to malaria and produced more offspring, also carriers, than did noncarriers. Although those people unfortunate enough to inherit the trait from both parents died at an early age, malaria took a heavy toll among those who didn't carry the trait at all. Over time, the gene became common in the population. But in America, where malaria hasn't been a serious threat to health since the end of the last century, the sickle cell trait conferred no particular advantage to those who carried it, and they did not reproduce more successfully than those missing the gene. Thus the gene became less and less frequent over time.

Man and mutagens

The original cell, from which a complex, multicelled organism develops, inherits its chromosomes directly from its parents, half from each. In a chemical sense, what one generation contributes to another is DNA. If you look back into the history of life on earth, what you see is essentially a long, unbroken strand of DNA tying one generation to another, from the present to some unknown point close to the debut of life.

If you look back along the history of life on earth, what you see is essentially a long, unbroken strand of DNA tying one generation to another.

We also carry in our bodies the genetic instructions for the human generations that will follow us.

Mutations may occur spontaneously, but genes can also be altered by a number of physical and chemical agents called *mutagens*. Unfortunately, man's technology has made some of these mutagens comparatively common. X-rays in sufficient dosage, for example, can alter genes and produce mutations. The radiation given off by the atomic bombs detonated over Hiroshima and Nagasaki, Japan, caused mutations, many of them grotesque and crippling, in some of the offspring of survivors of the blasts. Methyl mercury, an industrial waste, is also a mutagen. This chemical is commonly dumped into bodies of water where it enters the food chain and accumulates in fish often eaten by man. Fruit flies fed a diet containing a very small amount of methyl mercury have an extra chromosome. Recently, several aerosol adhesives were taken off the market after they were shown to cause chromosome breaks responsible for several abnormal infants.

This is an area where man has to proceed with extreme caution, because tampering with chromosomes is tampering with the very basis of human life. The DNA connection of one generation to the next is the weakest link in the chain of life, for there we are subject to the greatest destruction of the very stuff we are made of. In keeping the environment as free as possible of mutagens, we are protecting not only our own welfare but that of all the human generations to follow.

Summary

The carriers of genetic information are the chromosomes, passed from one cell to another during cell division. Cell division follows two pathways, each distinguished by how the chromosomes of the parent cell are apportioned between the daughter cells. In mitosis, the diploid number of chromosomes is first duplicated, and then a full diploid set is given to each of the two daughter cells. In meiosis, the diploid number is duplicated, but then divided twice, with the result that one parent diploid gives rise to four daughter haploids. Meiosis is also distinguished by crossing-over between the chromatids of homologues. Mitosis is the mode of cell division in many one-celled organisms and in the body cells of higher plants and animals. It results in precise genetic copies. Meiosis produces gametes needed for sexual reproduction and guarantees great genetic variation.

The actual units of inheritance on the chromosomes are genes. The two genes for the same trait in the diploid number of chromosomes may be either homozygous or heterozygous. The expressed allele of the heterozygote is the dominant, and the recessive is the unexpressed one. A heterozygous genotype usually produces the same phenotype as the homozygous dominant. The recessive phenotype can appear only when the genotype is homozygous recessive. Some traits are the result of many genes, while in other cases one gene may be responsible for a great many related traits. Examples of the first are height and skin color in humans; of the second, frizzle in chickens. In humans, sex is determined by the twenty-third chromosome pair. An XX is female, and an XY is male. The sex of a zygote is determined by the male gamete. The X chromosome carries sex-linked genes that generally appear in the phenotypes of men only, but are transmitted through the female's gametes; an example is hemophilia. In some cases, zygotes with too many chromosomes result from faulty meiosis of one gamete. One possible result is Down's syndrome, which can also be inherited from a parent with abnormally connected chromosomes from two different pairs.

The chemical form of the expression of a gene is the proteins of the cell. The code for these proteins is carried in the bases of the DNA molecules of the chromosomes. A triplet of DNA bases codes for each amino acid, and the sequence of triplets determines the primary structure of the protein. At the beginning of the making of a protein, the DNA double helix uncoils so that the bases face outward. Molecules of mRNA form on the DNA strand, then break away and travel to a ribosome in the cytoplasm. Molecules of tRNA transport amino acids to the ribosome, where they are linked together according to the sequence coded onto the mRNA. When the primary sequence is complete, the protein leaves the ribosome.

In genetic terms, evolution is the selection and culling of the gene pool of a population. Because of heterozygous individuals in a population, a gene pool contains considerable hidden variability. Recombination resulting from crossing-over produces new combinations of genes. Mutations provide new genes. Sickle cell anemia is a good case of how the pressures of natural selection have affected the frequencies of the same gene in two different gene pools.

Mutations appear both at random and as the result of mutagens such as X-rays, methyl mercury, and atomic radiation. These mutagens pose a definite threat to the hereditary material of the human race.

Questions to consider

1. What are the principal events of mitosis?
2. Mitotic division in the skin cell of a dog, which has 78 chromosomes, produces two daughter cells.
   a. How many chromosomes does each daughter cell contain?
   b. Are the equivalent chromosomes in each daughter cell, say the seventeenth pair, identical or different?
3. What are the principal events of meiosis?
4. Meiotic division in the testes of a dog produces four sperm cells from one body cell. The diploid number in the dog is 78.
   a. How many chromosomes does each sperm contain?
   b. How many combinations of the dog's chromosomes can theoretically appear in its sperm? (Simply state this as an exponent.)
5. A gardener has two sweet pea plants, one white flowered and one red flowered. He crosses the two and gets four offspring, all of them red. In the subsequent breeding season, he crosses the all-red generation and gets eight offspring. Six are red and two white. What are the genotypes of each member of each generation of this line of sweet pea?
6. A blue-eyed man and a brown-eyed woman produce one daughter, who is also brown eyed. She subsequently marries a brown-eyed man and has two children, one with brown eyes and one with blue.
   a. Which allele of the eye-color gene is dominant, brown or blue?
   b. If the daughter has two more children, what color eyes are they likely to have?
7. a. How is the sex of a human offspring determined genetically?
   b. In humans, which parent determines sex?
8. Direct male-to-male inheritance of hemophilia is impossible. Why?
9. a. What is the basic chemical structure of the DNA molecule?
   b. What chemical form does the DNA code take?
   c. How is the code expressed in the cell?
10. How is a protein assembled in a cell?
11. If you describe all the known phenotypes of the entire population

of gypsy moths, you could not include all the information carried in the gypsy moth gene pool. Why is this so?

12. a. What is the evolutionary significance of recombination?
    b. What is the source of new genes?

13. a. What are some common mutagens?
    b. Is there any reason to be particularly concerned about mutagens present in the environment, or should they be treated just like any other group of pollutants?

Glossary

allele one of the alternative forms of a gene for a specific trait

centromere the point of attachment for the two chromatids of a chromosome

chromatid one of the two strands of a single chromosome held together by a centromere

chromosome a threadlike structure in the nucleus of cells which carries genetic information

crossing-over the exchange of genetic material between the chromatids of homologous chromosomes during meiosis

diploid a full complement of chromosomes, composed of both homologues of each pair of chromosomes

DNA deoxyribonucleic acid; a complex molecule, usually found in the nucleus of cells, whose nitrogen base structure carries the code of genetic information

dominant the allele for a trait which always is expressed in the phenotype even when the alternative form of the gene is also present

gamete a haploid reproductive cell (designated either male or female)

gene the unit of heredity; a portion of a chromosome which contains the specifications for a single trait

gene pool all of the alleles of all of the genes within a population

genetics the branch of biology which specializes in the study of heredity

genotype the complete genetic information of an organism including both expressed and unexpressed traits *haploid* a half complement of chromosomes composed of one of each homologous pair of chromosomes

heterozygous having different alleles of a specific gene

homologue one of a pair of chromosomes

homozygous having the same alleles of a specific gene

incomplete dominance the expression of both alleles for a specific gene with neither being dominant

linkage the transmission of two or more genes on the same chromosome

meiosis a process of cell division in which one diploid cell produces four haploid cells

messenger RNA (mRNA) a form of ribonucleic acid which is produced in the nucleus of the cell and which carries the information of the triplet codes from the nucleus to the ribosome and directs the synthesis of proteins

mitosis a process of cell division in which one diploid cell produces two diploid cells that are genetically identical

mutagen a chemical or physical factor that causes a permanent change in genes

mutation a change in a gene

phenotype the expressed genetic traits of an organism

recessive the allele for a trait that is masked by the dominant and is therefore unexpressed

recombination the alteration of the sequence and combinations of the alleles of a chromatid resulting from crossing-over

ribosome the cytoplasmic organelle in which amino acids are assembled into proteins

sex-linkage the transmission of certain genes on the sex chromosomes

spindle the bundle of fibers that separates chromosomes during meiosis and mitosis

transfer RNA (tRNA) a form of ribonucleic acid that carries amino acids to the ribosome and aids in the assembly of proteins

zygote the diploid cell produced by the fusion of two gametes

9 Reproducing

If the various activities of any organism were ranked in terms of importance to its species, reproduction would have to be given first position. The reason is simple enough: Death comes to us all, whether amoeba or man. A species can survive for longer than a single lifetime only if it is able to reproduce—to bring forth living offspring themselves capable of reproducing. Reproduction is the ultimate biological necessity.

In genetic terms, reproduction is the passing of the hereditary message coded in the DNA of the chromosomes from one generation to the next. But, given the many forms of life on earth, it's hardly surprising to find that there are a great many different ways of passing that information on. Were we to study reproduction solely in man, we would quickly be confused by the cultural and social aspects of our sexuality. To help put ourselves in perspective, we will first sample the reproductive diversity of the world. Then we will take a closer look at the sexual apparatus and behavior of the human species and examine the special problem of birth control.

A species can survive for longer than one lifetime only if it is able to reproduce.

Figure 9–1 One of the simplest forms of reproduction is the asexual division of a single *Amoeba* into two separate individuals.

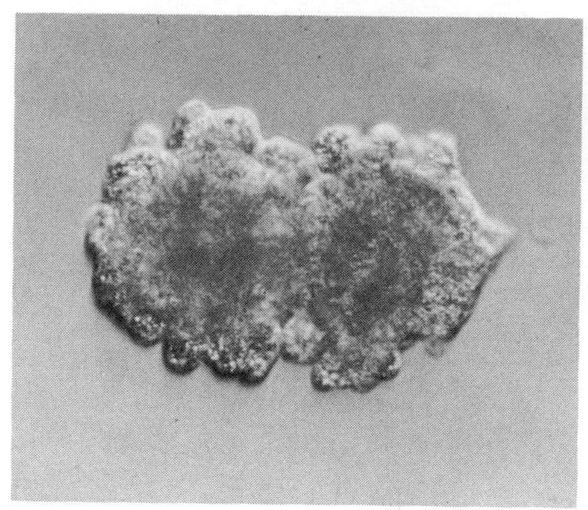

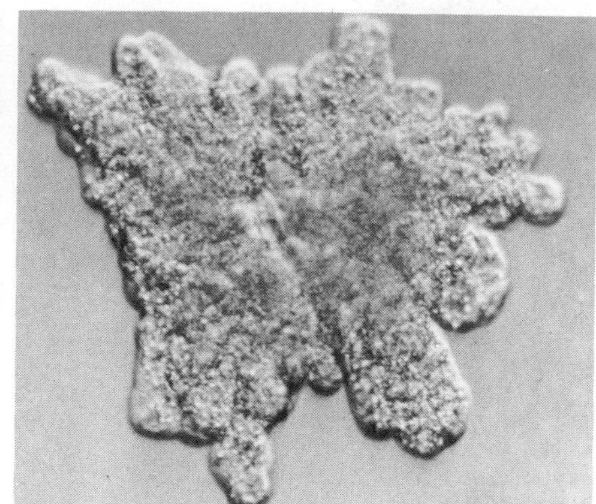

Courtesy Carolina Biological Supply Company

Reproduction without sex

Because we tend to think in terms of ourselves alone, we may assume that offspring always result from two parents. But this is not the case at all. In some species, offspring are produced at least part of the time by only one parent. Such reproduction is called *asexual*; that is, it involves genetic material from only one parent organism.

The simplest form of asexual reproduction is found among such primitive, one-celled organisms as the amoeba, shown in Figure 9–1. To reproduce, the amoeba simply divides in two. In more complicated life forms, offspring are produced from a portion of the parent's body rather than from a division into two. An example is a hydra, a small, freshwater relative of the jellyfish. Buds appear near the base of the hydra's body. Each develops into a fully formed animal before dropping off to begin life on its own (Figure 9–2).

In some cases, though, the offspring remain attached to the parent. This kind of reproduction is suited to organisms that stay in one place throughout their life cycles. The coral is a good example of an animal that reproduces in this fashion. Corals form colonies of many, many thousands of connected individuals—parents and offspring and offspring of offspring. When the corals die, the soft parts of their bodies decompose, exposing the hard community skeleton. A coral reef of the sort shown in Figure 9–3 is a community skeleton built up by the combined forces of reproduction and death over the course of many centuries.

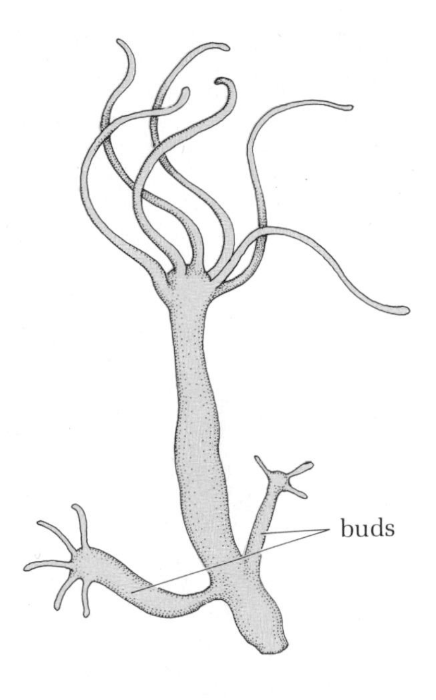

Figure 9–2 A hydra reproducing through the simple procedure of budding.

Figure 9–3 A portion of the Great Australian Barrier reef seen from the air. This massive reef is actually composed of the remains of millions of corals.

Figure 9–4 The common plague of the gardener, crabgrass, is actually a colony of linked, individual plants.

As any weekend gardener can attest, asexual reproduction by organisms that live in colonies can provide a very effective form of life insurance. Crabgrass is a coarse, unsightly grass that tends to invade a lawn and displace the grasses already growing there. Crabgrass reproduces by sending runners out under the surface of the soil; offspring bud from these runners (Figure 9–4). A clump of crabgrass is not just a bunch of separate plants but a colony of interconnected individuals. And this is the reason that crabgrass is so hard to stop once it gets started. If even one plant survives an onslaught of weeding and herbicide, that one plant can give rise to a whole new colony of crabgrass.

The organisms we have discussed so far produce their offspring from portions of the parent. In some organisms, asexual reproduction involves single, specialized cells called *spores*. Spores are so small and light that the slightest air currents can carry them long distances. Thus spores provide immobile plants and fungi with a way of traveling to new, favorable habitats by hitchhiking on the wind. The spores that happen to fall on suitable soil—usually a very small percentage of the total number of spores—grow and develop into adult fungi or plants. The mushroom, for

Asexual reproduction by organisms that live in colonies can provide a very effective form of life insurance.

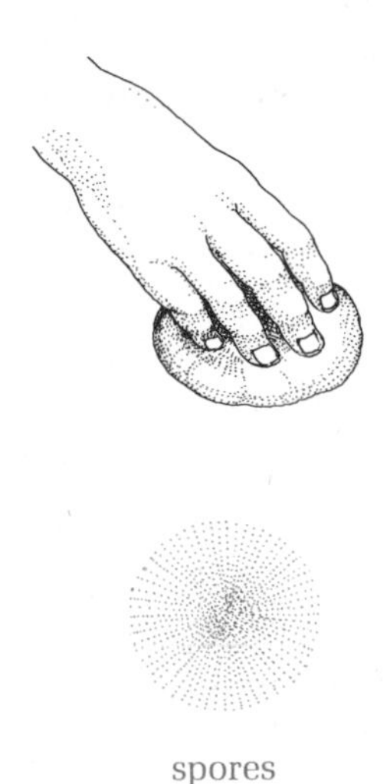

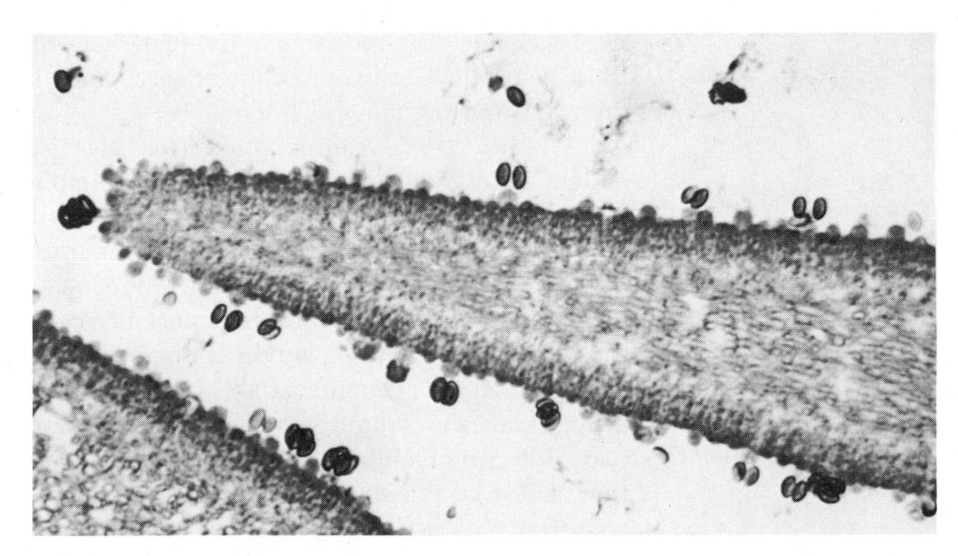

Figure 9–5 A. The spores within a mushroom cap are organized in a distinctive pattern. B. A common example of a spore-producing organism.

example, is the spore-forming body of a particular class of fungi. If you took the cap from a fresh mushroom and placed it on a sheet of paper, the spores would settle on the paper in the characteristic pattern shown in Figure 9–5. Every one of these spores, which are too small to be seen individually without the aid of a high-power magnifying glass, can grow and give rise to a new colony of mushrooms. Although the actual number of spores that do mature and form new colonies is very small, a spore-producing plant releases so many spores that the continued existence of the species is practically guaranteed. This efficient mode of reproduction can sometimes cause us humans inconvenience, as happens with the black mold that grows on stored bread.

Reproduction with sex

Like spore formation, sexual reproduction has to do with the production of specialized cells, the gametes. But, unlike spore formation, sexual reproduction involves the union of the nuclei of two separate gametes—an action called *fertilization*—to produce the zygote from which the offspring will develop. Although some species produce both kinds of gametes in the same individual, many other species have specialists who produce only one or the other. One of the specialists is known as the male, and the other as the female.

Why did sexual reproduction ever arise? Asexual reproduction has certain definite advantages. It's relatively simple, and it allows a species to reproduce even if only one individual is around. It's also very fast; given enough food, most bacteria can bring forth a new generation every 20 minutes! However, any offspring produced asexually is a carbon copy of its parent. The only source of change is genetic mutation. Since sexual reproduction involves the union of gametes produced by meiosis, it makes for a great store of genetic variability. Sexual reproduction, though, exacts its price. It is a complex process, involving a great number of "moving parts." It is more subject to disruption and error than asexual reproduction, and it takes more time for each generation to appear. Because of the complexity of sexual reproduction, a sexual organism has to expend a great amount of energy in reproducing. Thus there is a trade-off: simplicity and speed versus variability. Among the less complex forms, the simplicity of asexuality has proven more adaptive. But the variability of sexual reproduction has added to the fitness of the more complex plants and animals.

A simple variation

Even though sexual reproduction is characteristic of higher organisms, a sort of precursor of sex can be seen among some of the simplest organisms. For example, bacteria, which are one celled and obviously cannot produce gametes, generally reproduce asexually by a simple division process called *fission*. However, genetic variability is ensured by a form of reproduction called *conjugation*, which occurs periodically. A long, hairlike tube called the *sex pilus* is formed between the cells. Through this the bacteria exchange genetic information in the form of DNA. After conjugation, the bacterium has a different genetic makeup than it did before, and cell division now produces new, variant offspring.

Sexual reproduction is more subject to disruption and error than asexual reproduction, and it takes more time for each generation to appear.

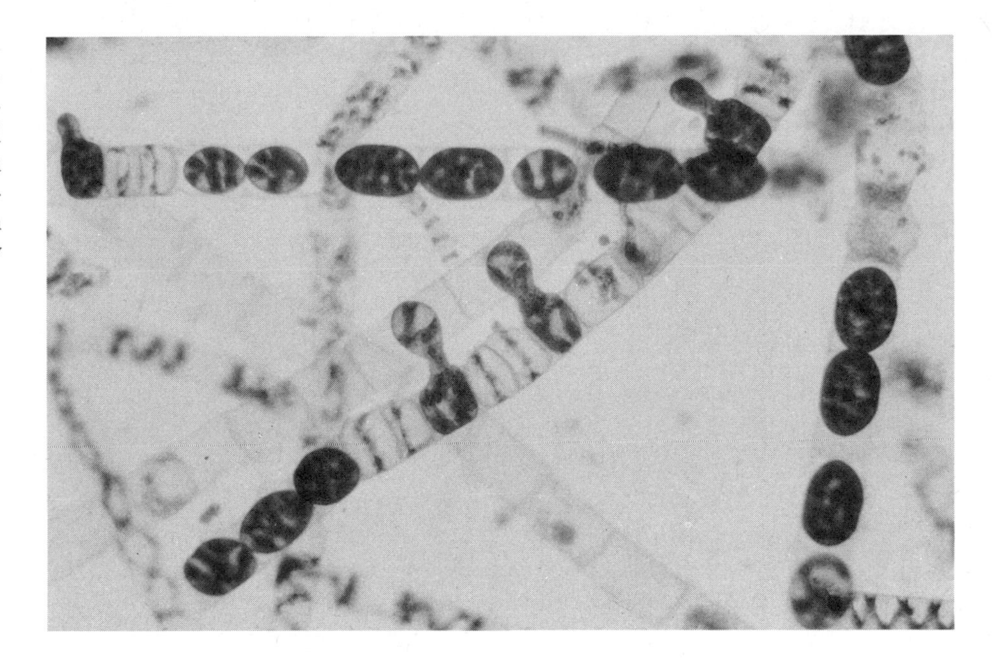

Figure 9–6 Conjugation in the alga *Spirogyra*. Genetic information is transferred between cells through the tubelike connections. The cells then divide but with a new genetic makeup.

In the absence of male and female bacteria, conjugation accomplishes roughly the same genetic purpose as sex. In *Spirogyra*, a simple green alga, a more complex form of conjugation is found (Figure 9–6).

The sex life of plants

The reproductive cycle of the plants is characterized by a phenomenon called the *alternation of generations*. That is, a diploid generation is followed by a haploid, which is followed by a diploid, which is followed by a haploid, and so on. In some species, the alternation is very pronounced, while in others it is subtle, apparent only to someone who knows what he is looking for.

The sea lettuce

The sea lettuce is a marine alga that earned its name from its striking resemblance to the common salad vegetable. The remarkable thing about the sea lettuce is its pronounced pattern of alternate generations.

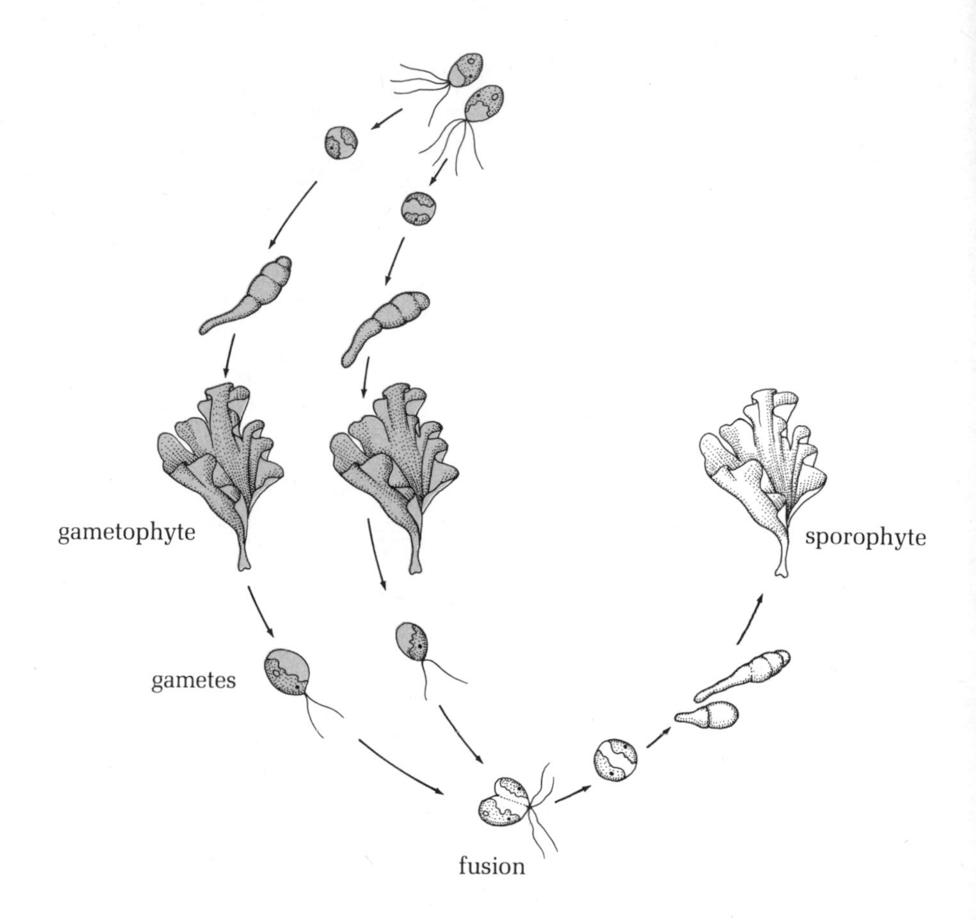

Figure 9–7 The life cycle of the sea lettuce. The sexual phase of the cycle is shown in color.

Sea lettuce comes in both a haploid and a diploid form, which can be distinguished from each other only by counting the number of chromosomes in the cells. The haploid form produces gametes mitotically, which are, of course, haploid, and it is therefore called the *gametophyte*. Although all the gametes look alike, certain ones will unite only with certain other ones, so they can be divided into males and females. After a male gamete and a female gamete meet and fuse, the resulting diploid zygote develops into a diploid sea lettuce. The diploid sea lettuce produces haploid spores through meiotic division, and it is called a *sporophyte*. These spores develop into haploid gametophytes. And so the process goes, from gametophyte to sporophyte and back again (Figure 9–7).

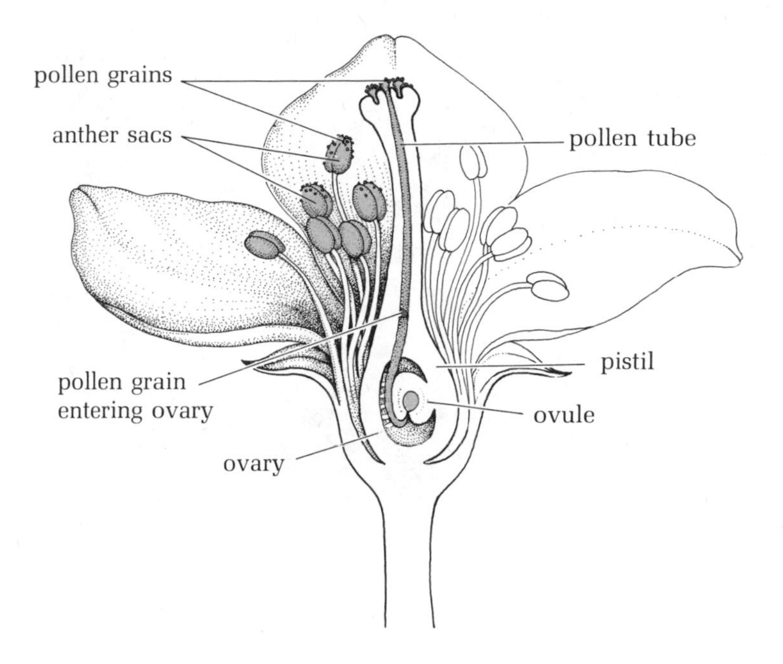

Figure 9–8 The reproductive organs of a flowering plant.

The flowering plants

The flowering plants, which make up the great majority of land-dwelling plant species, also alternate generations, but one of them is relatively inconspicuous. The plant itself is diploid, and is therefore a sporophyte. The gametophytes are tiny structures nourished by the sporophyte and dependent on it.

The key to the alternation of generations in the flowering plants is the flower itself. Figure 9–8 depicts a typical flower and points out its principal structures. The *anther sacs* near the top of the flower are the site where the male gametophytes are produced. These gametophytes are called *pollen grains*. Each mature pollen grain contains three haploid nuclei. The female gametophytes form within the chamberlike *ovules* clustered at the bottom of the *pistil*. Meiosis of a diploid cell in the ovule results in four haploid nuclei within a single cell membrane. Three of these nuclei are reabsorbed into the cell and disappear. The remaining nucleus divides three times mitotically, resulting in eight identical haploid nuclei, all of them still enclosed within the same cell membrane. This cell is the female gametophyte.

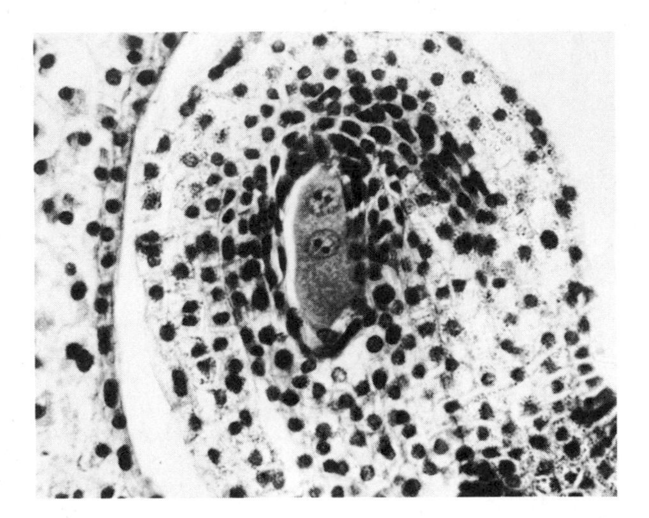

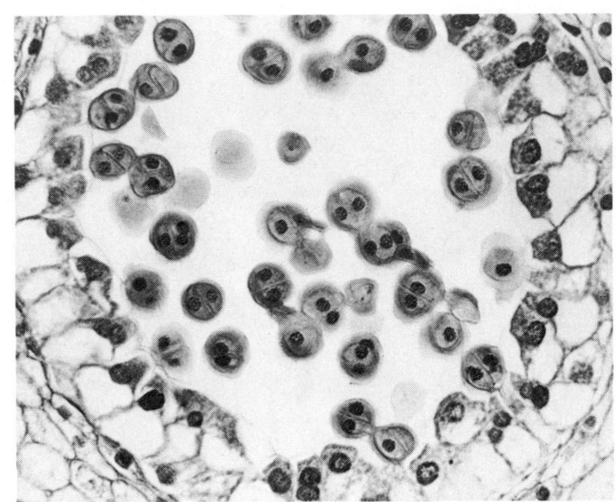

Figure 9–9 The gametophytes of the lily. A. A cross section of the ovule of the lily, showing the developing female gametophyte. B. Pollen grains of the lily, showing the developing male gametophyte.

At maturity, the anther sacs split open and release the pollen grains. The top of the pistil is covered with a wet, sticky substance that traps pollen much the same way that flypaper catches flies. Figure 9–8 shows what happens next. A pollen grain stuck to this surface breaks open, and a long tube that contains one of the pollen's nuclei grows down into the pistil. Guided by a navigational system that is still not understood, the pollen tube eventually reaches an ovule with a female gametophyte and releases its two remaining nuclei, the sperm nuclei. One of the sperm nuclei unites with an egg nucleus in the gametophyte, resulting in the zygote, the first cell of the sporophyte generation. The other sperm nucleus fuses with two of the gametophyte nuclei to produce a single cell that contains three times the haploid, or a *triploid*, number of chromosomes. This triploid cell develops into the layer of the seed known as the *endosperm*, which provides nutrition for the sporophyte zygote as it develops.

Two features of reproduction in flowering plants are found among the flowering plants alone. The first is the double fertilization of sperm cells and egg cells, one to become the sporophyte zygote and the other to become the endosperm. The second is that the endosperm is the only known triploid tissue.

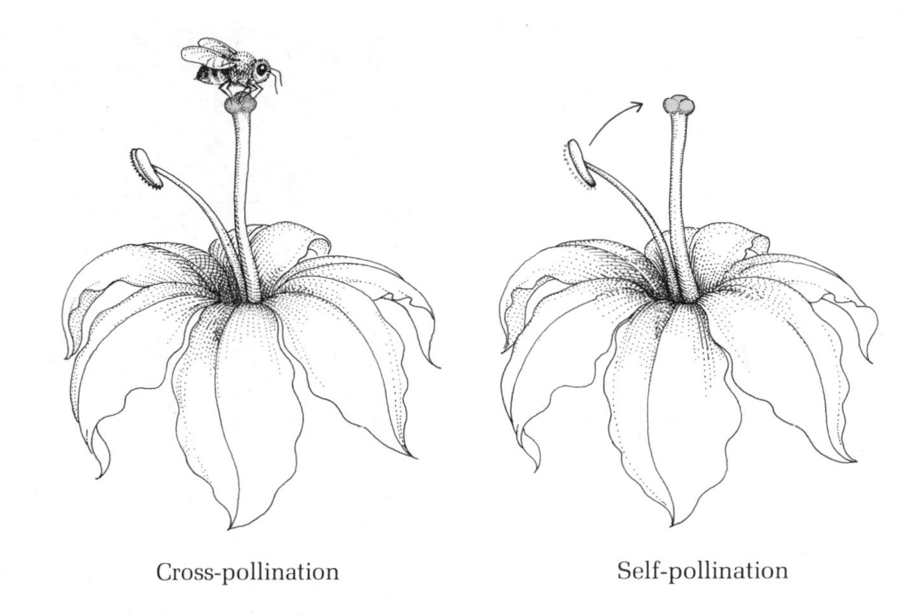

Figure 9–10 Two different methods of pollination. Cross-pollination permits an exchange of genetic material between flowering plants, but it requires some outside agent—an insect, wind, a large animal—to ensure success. Self-pollination has the advantage of simplicity but does not provide new genetic material.

Patterns of pollination

The journey of pollen grains from anther sac to pistil is the process known as *pollination*. Although the distance from anther sac to pistil within a single flower is not great, most plants are not compatible with themselves—that is, a pollen grain from one flower will not be able to fertilize an egg cell in the same flower. In fact, it is also generally true that the pollen from one plant cannot fertilize egg cells from the same plant. Some plants, such as the common garden bean, do fertilize themselves, though, and they are known as *self-pollinators*. But in most species of flowering plants, egg cells in one plant must be fertilized by pollen grains from another plant altogether. This process is given the name *cross-pollination*. Figure 9–10 compares cross-pollination with self-pollination. Since plants are immobile, cross-pollinating plants must find some way of spreading pollen grains from one plant to another. Two methods of cross-pollination are found among the flowering plants. Plants with small and inconspicuous flowers, such as grasses, oak trees, and herbs like ragweed, use wind pollination. The flowers of these plants produce great amounts of pollen and distribute it freely to the air, relying on chance to carry the pollen to any ready and waiting pistils. The

A bee pollinating a thistle flower, an example of cross-pollination.

more pollen grains a plant produces, the greater the chance that at least some of them will find a flower to pollinate.

Other plants, particularly those with large, colorful, and good-smelling flowers, let animals do their pollinating for them. These species excrete a sugary fluid known as nectar from the base of the pistil. The sole purpose of the nectar is to attract animals interested in feeding on it. Some animals eat the pollen as well. Additional pollen grains may adhere to the body of any animal feeding on the flower, and, when the animal moves on, these hitchhiking pollen grains may then fertilize the next plant the animal visits. Numbered among the animals involved in cross-pollination are beetles, flies, bats, hummingbirds, and the well-known honeybee. In one species of orchid, a curious instance of mimicry ensures cross-pollination (Figure 9–11). The flower looks so much like the female of a particular kind of butterfly that the males regularly fall for the imitation and try to mate with the flower. Undaunted by its first failure, the butterfly tries orchid after orchid, cross-pollinating each one in turn.

Figure 9–11 Mimicry, an extraordinary pattern of adaptation in the plant world. The butterfly orchid's resemblance to a female butterfly helps ensure successful pollination.

The reproductive variety of the animal world

As one moves up the evolutionary tree from the lower forms of animal life to the higher, organisms become increasingly complex. That complexity is mirrored in the reproductive methods and systems of animals.

The two-sexed sea hare

Hermaphrodite is the name given to an animal that produces both male and female gametes in the same individual. The common garden snail, for example, is a hermaphrodite, an organism with the reproductive organs of both sexes. A marine relative of the snail known as the sea hare

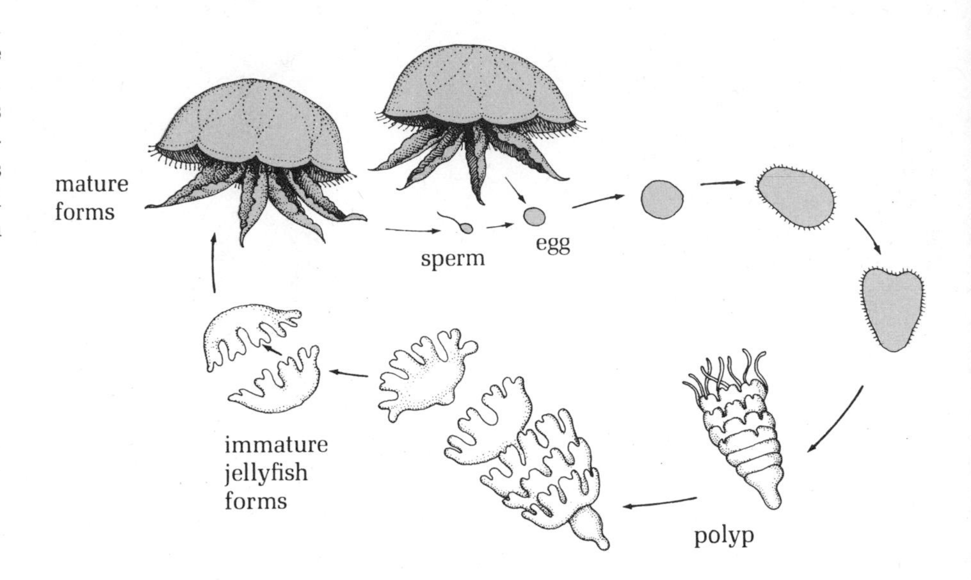

Figure 9–12 The life cycle of the jellyfish. The jellyfish combines sexual and asexual reproduction within its life cycle. The sexual phase is shown in color.

carries hermaphroditism to an extreme. As is nearly always the case with hermaphrodites, one organism cannot fertilize itself, and two individuals are needed to begin reproduction. When two sea hares meet, one approaches the other from behind and mates with it. The sea hare in front is using only "her" female organs to produce eggs, while the sea hare in the rear is using only "his" male organs to produce sperm. To a third sea hare, the rear animal is an available female, and "he" can mate with "her." Of course, the third sea hare's female organs are available to a fourth sea hare "male." As many as six or seven sea hares may get together as simultaneous males and females.

The best of both worlds: the jellyfish

The jellyfish is not a hermaphrodite like the sea hare, but it does employ a unique duality. At alternate stages in its life cycle, the jellyfish reproduces both sexually and asexually. Figure 9–12 diagrams the succession of stages. The familiar sac-shaped jellyfish is only one part of the life cycle of the organism. Each jellyfish is either male or female, producing one or the other kind of gamete. The jellyfish release their gametes into the water, where they fuse to begin the development of a new individual. But this new organism grows not into a jellyfish, but into a bulbous animal with radiating tentacles that lives a sedentary life at-

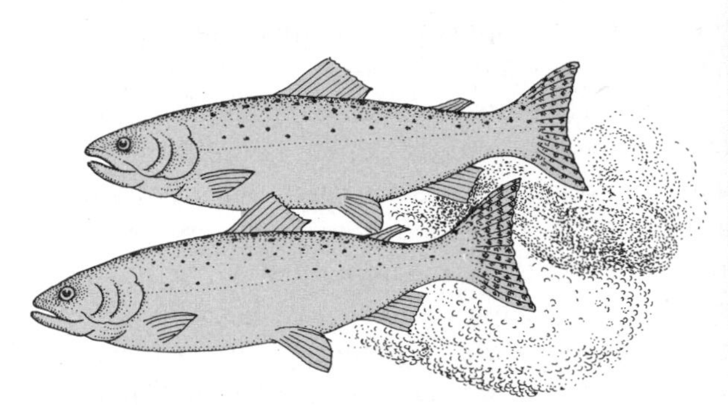

Figure 9–13 A. Mating salmon. The salmon seek a shallow, calm area of the spawning stream. After the female has released her egg cells, the male covers them with a layer of sperm. B. Mating frogs. In a manner similar to the reproduction of salmon, fertilization in frogs takes place outside the parent bodies. The male frog climbs onto the female and releases his gametes as she releases the egg cells.

tached to some fixed object like a coral reef or the pilings of a wharf. This animal, called a *polyp,* reproduces asexually by producing branches which grow into other polyps. In a short time, the original polyp has produced a whole colony. During the autumn and the winter, the polyp's body constricts into what looks like a stack of saucers. The "saucers" break apart from each other, turn over, and change into immature, free-floating jellyfish. With additional development, the adult jellyfish are ready to produce gametes, thus bringing the cycle full circle.

Complex sex in complex animals

Hermaphroditism and combinations of sexual and asexual reproduction are found only in relatively simple organisms. In all vertebrates, reproduction is entirely sexual.

In the great majority of aquatic vertebrates, fertilization takes place outside the body of the parent organisms. These animals do not simply release their male gametes like so many pollen grains and count on the water currents to carry them to free-floating female gametes. Instead, males and females find each other and release their gametes together. Figure 9–13a shows two salmon mating in the shallow waters of a spawning stream. As the female releases her eggs, the male covers them with a cloud of sperm. This basic method of fertilization, common to most fishes, is also found among amphibians. Figure 9–13b shows two mating frogs. The male holds on to the female and sprays sperm over the eggs she releases as a gelatinous mass into the water.

One of the most important factors in successful reproduction is getting a mature male together with a mature female at the proper time. The frogfish, which lives in the relatively unpopulated depths of the sea, has a unique way of making sure that the male is always ready when the female is. When an immature male frogfish finds a female, he bites into her body and buries his head in the female's flesh. His fins and gills disappear, and the male frogfish becomes totally dependent on the female for oxygen and nutrients. In fact, the female even controls the production of sperm. When she matures and lays her eggs, chemical changes in her body trigger the production of his sperm.

The amphibians, which represent a sort of evolutionary halfway house between land and water, generally mate in the water. But those vertebrate animals totally adapted to the land and independent of the water—the reptiles, the birds, and the mammals—don't use water as a medium for bringing eggs and sperm together. Instead, the male introduces the sperm directly into the body of the female. Birds and reptiles accomplish this task in a similar fashion. Both kinds of animals have only one opening to the outside, which is used for both reproduction and excretion. For example, the rooster mounts the hen from the back and bends the rear end of his body down under her tail so that the openings of the two birds join. The rooster's reproductive organ, located just inside the opening, releases sperm into the hen. Inside her body, the sperm can move about in a watery medium.

In some species, reproduction is made easier by a specialized male organ that can be inserted directly into the female's body. This organ is the *penis*. The penis is found among some reptiles, flightless birds, and ducks. It also evolved independently in the insects. But of all the classes of vertebrates, only among the mammals is the penis to be found in every species.

The crux of the matter: fertilization

The whole point of sexual reproduction is to get a male gamete together with a female gamete so that they may fuse to form the first cell of

Figure 9–14 The human gametes.

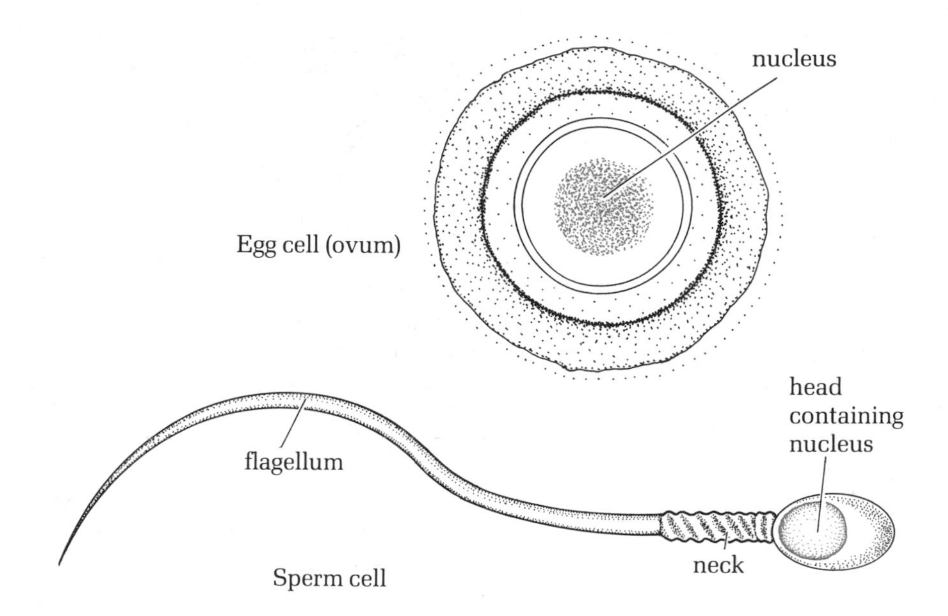

a new individual. All the elaborate anatomy of reproduction and the special behaviors that lead up to it serve that one purpose—to get those two gametes together. What happens then?

As Figure 9–14 shows, male and female gametes are quite different in appearance. That difference serves as a clue to their roles. As cells go, the egg cells of the female are quite large, so large that they can often be seen with the naked eye. The yolk of a chicken egg is actually only one cell. The egg cell of the human female is roughly the size of a printed period on this page. The size of the egg is due to its large cargo of nutrients intended to nourish the zygote as it develops.

By comparison, sperm cells are small. But they make up for their lack of size in speed. Sperm cells are the hot rods of the cellular world, consisting of little more than a cap containing the nucleus and its chromo-

Sperm cells are the hot rods of the cellular world, consisting of little more than a cap containing the nucleus and its chromosomes and a tail that provides locomotion.

Figure 9–15 Adult sea urchins, a species with no apparent physical distinction between male and female.

somes and a tail that provides locomotion. Egg cells cannot move on their own, and the sperm cells must find them, guided by chemicals given off by the eggs.

When a sperm finds an egg, the head of the cell penetrates the egg, leaving the tail behind. What is known as the wave of change passes through the mucuslike substance coating the egg, usually with the effect of preventing any additional sperm from entering. The nucleus of the sperm fuses with the nucleus of the egg, making a diploid cell from the two haploids. In the next chapter, we'll trace the way that the zygote develops into a new individual.

The differences between the sexes

The relative strengths and weaknesses of the two sexes has been a matter of concern and a source of jokes and insults for many thousands of years. In the case of the human species, it is very difficult to determine whether certain emotions or abilities are exclusively male or exclusively female simply because we are dealing not only with the physical facts of sexuality but also with a variety of cultural, social, and psychological influences. But we can learn something of our own differences, or lack of them, by looking at the sexual differences characterizing other organisms.

Among the hermaphrodites, both plant and animal, there are no sex differences. A sea hare is simply a sea hare, neither a "she" nor a "he." Among simple animals that have only one set of sexual organs, it is often

Figure 9-16 The distinctive plumage of the male crown crane. Lavish plumage in birds is common in the male of the species.

impossible to distinguish the male from the female on the basis of outward appearances. Figure 9–15 shows a number of sea urchins in mixed company. Even an expert can't tell a female from a male without opening the sea urchin to see whether it has sperm-producing organs—and thus is male—or egg-producing organs—and thus is female. Such differences in the gamete-producing organs, or *gonads*, are known as *primary sexual characteristics*.

Usually when we think of sexual differences, though, we think of something other than the gender of gametes. All differences between the sexes other than those of the organs of reproduction are known as *secondary sexual characteristics*. Secondary characteristics may be found in both physical appearance and behavior. Oftentimes, the sexes are of different sizes. Among some marine fishes the male is nothing more than a sack of gonads, while among mammals males tend to be larger than females. Males and females of a single species of bird often have very different plumage; Figure 9–16 shows an example. In some species, secondary sexual characteristics are temporary, lasting only as long as the mating season and disappearing afterward. The antlers of male deer fall into this class. Male and female salmon in the ocean are virtually indistinguishable. But before the salmon school returns to fresh water to spawn, the male develops the prominent beak shown in Figure 9–17. At breeding time, the male salmon uses this beak as a weapon to chase scavengers like steelhead trout away from the female while she is laying her eggs.

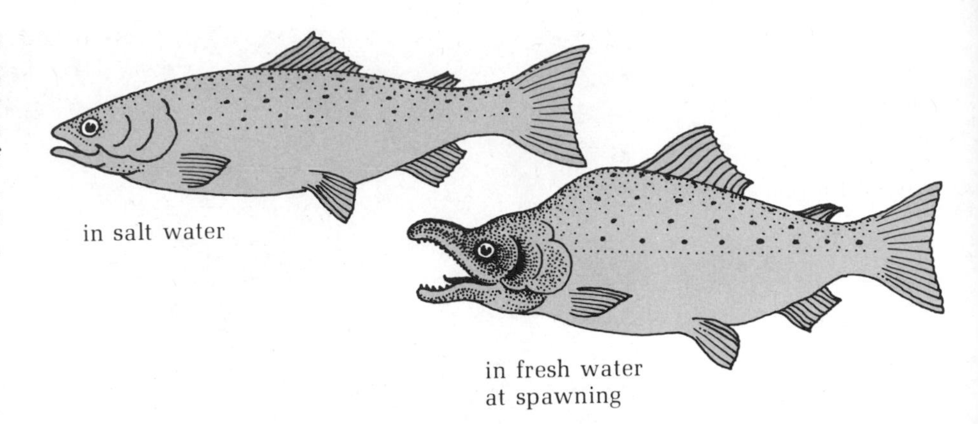

Figure 9–17 The beak of the spawning male salmon. Unlike that of the female, the physical appearance of the male salmon changes drastically as the fish goes upstream to spawn.

Like the salmon's fighting beak, many secondary sexual characteristics serve a definite reproductive purpose. In some species, differences in color and form in one sex draw prospective mates to an animal ready and willing to mate. The male peacock spreads his fan of iridescent feathers to announce his readiness to mate. Oftentimes a behavior is added to the physical display to make the male's intentions obvious and to induce the female to accept him. The elaborate and beautiful courtship dance of the sage grouse is an example.

Sometimes these behaviors involve an element of competition among the members of the same sex. The large and majestic antelope species known as the Uganda kob provides a particularly interesting example. Each herd of kob has a special breeding area some 200 yards across. The breeding ground contains 15 to 18 small, circular areas that look very much like putting greens. The males of the herd fight for ownership of one of these circles, and only those males that get and hold onto a circle breed. In fact, there is even a rank ordering among the males who do win a circle. The females seem to prefer those males closest to the center, and such a king of the mountain may mate three or four times as frequently as a male on the perimeter.

♂—the sexual apparatus of man

As with other species, human sexuality is a blend of anatomy and behavior. We will look first at the primary and secondary sexual charac-

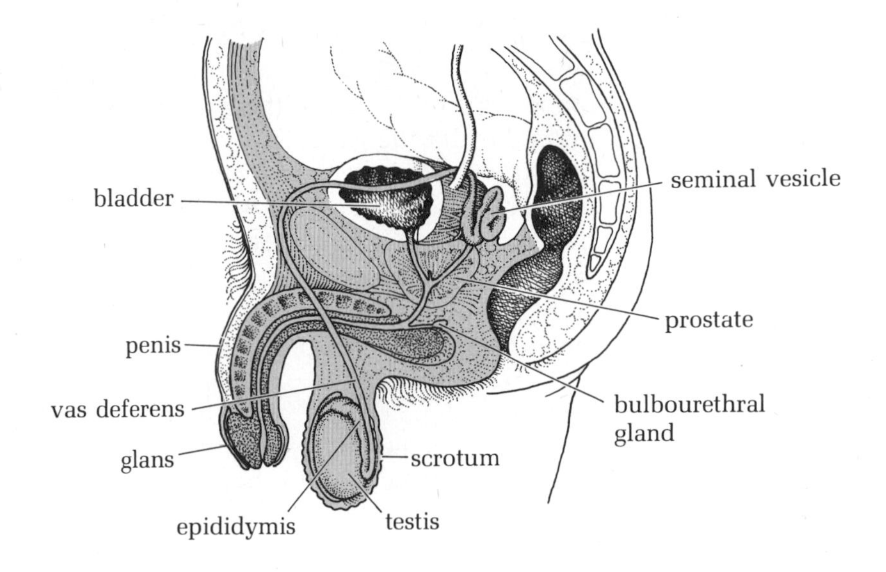

Figure 9–18 The human male reproductive system.

teristics of men and women and then see what happens physically during intercourse.

The male organs

Male gonads are the *testes* (singular; *testis*) located in the saclike *scrotum* hanging from the base of the trunk of the body. The testes serve two functions. One is the production of the male hormones. A *hormone* is a chemical substance produced in one part of the body and transported, usually through the bloodstream, to another part of the body where it exerts its effect. Typically, hormones work in minute amounts. The male hormone *testosterone*, produced by special cells in the testes, is responsible for such secondary sexual characteristics as height, body shape, beard, and deep voice. The second function of the testes is the production of the sperm (Figure 9–18).

After production in the testes, the sperm are stored in the thin, tightly coiled tube called the *epididymis*. The sperm get to the outside by traveling along another thin tube, the *vas deferens*, which is connected to the urethra. Along the way, the sperm receive secretions of nutrients and lubricants from three glands: the *seminal vesicles, the bulbourethral gland*, and the *prostate*. The secretions from these glands plus the sperm cells make up the *semen*, the white fluid expelled, or ejaculated, from the penis at orgasm. The semen moves down the urethra to its opening in the bulbous tip of the penis, the *glans*.

Normally the glans is covered by a flap of tissue, the *foreskin*. The foreskin is often removed surgically from males before they become sexually mature, an operation called *circumcision*. In Judaism and Islam, circumcision is a religious rite akin to Christian baptism. However, religion is not the only reason for circumcision. In this country, male infants are circumcised routinely shortly after birth. In uncircumcised males, dirt and secretions of the skin may build up under the foreskin and cause infection of the glans. Removing the foreskin removes the problem.

The most obvious characteristic of the male penis is its ability to grow large and hard. Within the penis are three masses of spongy tissue capable of retaining blood under pressure. Two of these masses are located above the urethra. The third surrounds the urethra and, at the end of the penis, swells to form the glans. When a man is sexually excited, a small muscle constricts the veins that lead away from the penis. Simultaneously, the arteries leading into it widen, letting more blood in. A second muscle shuts down the flow from the bladder, keeping the urine out of the urethra. The spongy tissue fills with blood and becomes distended, enlarging and hardening the penis. As the penis erects, it becomes increasingly sensitive to touch, particularly on the bottom side of the glans.

The curious place of the testes

The testes develop within the abdominal cavity of the unborn male, and do not actually descend into the scrotum until shortly before the time of birth. After they drop into the scrotum, the holes in the abdominal wall are plugged with tissue, but these spots remain a weak point in the body. Man's upright posture throws much of the weight of the intestines on the wall separating the scrotum from the abdomen. Occasionally the wall splits, and a loop of intestine may push into the scrotum, an injury known as a hernia.

Why are the testes located outside the body where they are so much more vulnerable to injury? Human sperm cells, and the sperm cells of most mammals for that matter, are very sensitive to heat. Human male gametes need a temperature a couple of degrees cooler than the 98.6°F of the body to develop properly. Some men have undescended testes, ones that have remained in the body cavity. These men are completely and normally male, but their semen contains no living sperm. The condition can be corrected surgically. The temperature of normally descended testes is kept constant by a set of muscles in the scrotum. When the air is cold, these muscles contract, drawing the testes up to the warmth of the body. When the air is hot or when body temperature is elevated because of a fever, these muscles relax, dropping the testes away from the body.

Impotence and sterility

Impotence and *sterility* are often used as if they were synonymous, but such use is mistaken. A sterile male is one who does not produce a sufficient number of living sperm to be likely to impregnate a woman. The most extreme case is a complete absence of living sperm. However, a sterile male can have normal intercourse, and he does produce semen. The only difference between him and a normal, sperm-producing male is that the sterile man's semen contains an inadequate amount of living sperm. He cannot father offspring. The impotent male does produce living sperm, but he cannot achieve or maintain an erection. His problem rests not in the product of his testes, but in the function of his penis.

Sterility results from some physical cause, such as undescended testicles or scar tissue sealing the vas deferens. Impotence may also result from physical causes, such as a side effect of surgery on the prostate gland, a common site of cancer in men past middle age. Most of the time, however, impotence has its origin in psychological troubles.

Castration

How is the male affected if the testicles are removed completely, that is, if the man is *castrated?* The man is sterile, but perhaps more importantly he is deprived of a steady supply of testosterone. The exact effect of this loss depends on the age of the man. If he happens to be an immature boy of 12 or 13, he will never develop the deep voice, the body hair, or the mature sex organs of a normal man. Should castration befall a man in his 20s, the effects are not quite so pronounced. His voice will

Figure 9-19 The gelding racehorse Kelso. These creatures are known for their docile temperaments and heavy muscular bodies.

most likely remain the same, but his beard will become softer, and he will gain weight, particularly in his chest and buttocks, so that his figure will become more like a woman's. In a man over 30, castration has fewer physical effects; the man's body will remain pretty much the same. In many animals, castration produces impotence, but this is not always true of a man who has already reached sexual maturity. Castration before sexual maturity, though, does result in impotence.

It has long been common practice for stock farmers to castrate their animals. Castrated animals tend to put on fat, and they are much more docile than normal males. Cattle and pigs raised for meat are almost entirely unbred females and castrated males. A castrated horse, called a gelding, makes a less temperamental mount than a stallion (Figure 9-19). At times, humans have castrated their own kind. Middle Eastern harems were usually guarded by castrated slaves called eunuchs. Since the eunuchs had no sexual interest in the women they guarded, they could be trusted to do their job without availing themselves of the women's charms. This custom was not only confined to the exotic East. For some centuries following the Renaissance, the Church held that it was sinful for women to act on the stage or to sing in public. The soprano voices of Italian choirs were supplied by castrates, or "white voices" as they were called in polite society. Fortunately, this custom is no longer practiced.

♀ – the sexual apparatus of woman

The male's sexual organs serve to produce sperm and to insert that sperm into the body of the female. The female's role is to produce egg

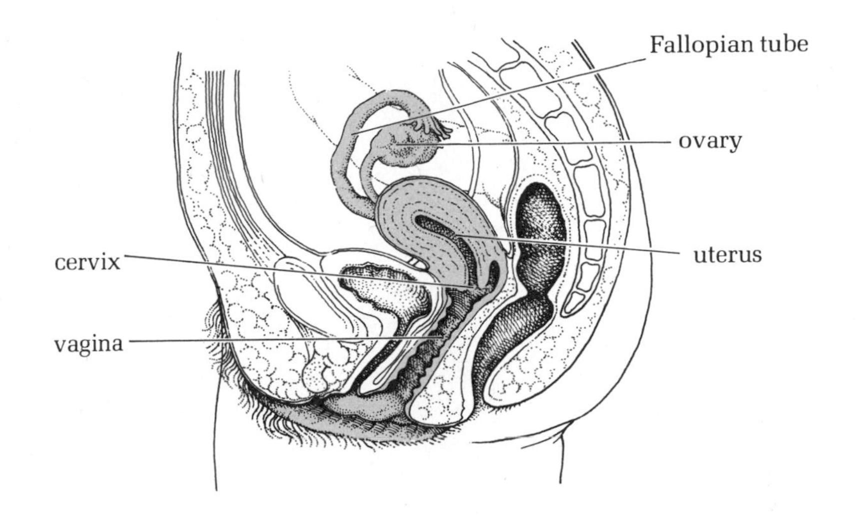

Figure 9–20 The human female reproductive system.

cells, to receive the male gametes, and to provide the site for fertilization and the development of the zygote. Figure 9–20 shows a sketch of the female reproductive system.

The primary sex organs in the woman are the *ovaries*, suspended on the thick cords called ligaments in the lower abdominal cavity. As in the male, these primary organs produce both gametes and hormones responsible for such female characteristics as breasts. A matched pair of tubes, known as the *Fallopian tubes*, or *oviducts*, connect the ovaries to the *uterus*.

The uterus is a chamber with thick, muscular, folded walls that can expand greatly. The normal uterus is only about the size of a pear, but it can distend itself to accommodate a baby weighing 10 pounds or more. Like the ovaries, the uterus is held in place by heavy ligaments. The bottom of the uterus is a tight ring of muscle called the *cervix*. The cervix contains a small opening into the vagina. The *vagina*'s walls are similar to those of the uterus, in that they can expand greatly and contract rhythmically.

The vagina's opening to the outside of the body is partially covered by a thin membrane, whose label, the *hymen*, is derived from the name of the Greek god of marriage. The hymen generally covers most of the vaginal opening and is often broken by first intercourse, sometimes with slight bleeding and pain. An intact hymen has long been considered a sign of virginity, and indeed it can be, but a virgin may have a broken hymen. Some women's hymens are so thin and delicate that they are broken well

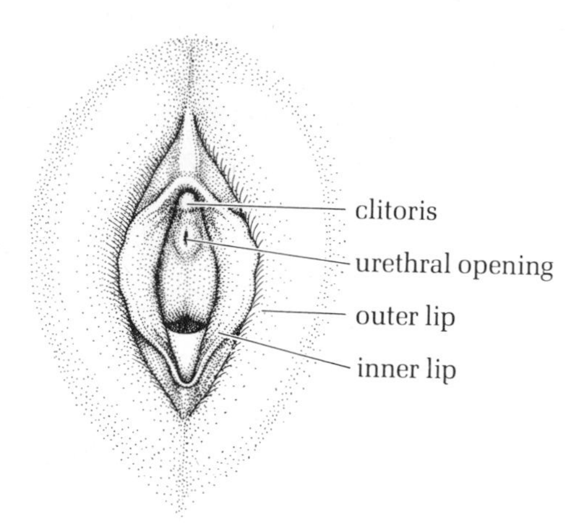

Figure 9–21 The external female sex organs.

before maturity by vigorous exercise like running or jumping rope. Other women are less fortunate, plagued with a hymen so thick that it must be opened surgically before they can enjoy intercourse.

The exterior female sex organs are given the collective name *vulva*. The *outer lips* cover the *inner lips*, visible only when the outer pair is parted as in Figure 9–21. The inner lips cover the entrance to the vagina and what there is of the hymen. Just above the vaginal opening is a small orifice that is the end of the urethra. Although the urethra has a reproductive role in the man's sexual anatomy, in the female it serves only as a pipeline for urine leaving the bladder. Above the urinary opening is the *clitoris*. The clitoris is the primary sensual organ in the female sexual system. It is richly supplied with nerves, and direct or indirect stimulation of the tip of the clitoris can greatly excite a woman sexually. The clitoris contains erectile tissue, and it is the anatomical equivalent of the penis.

The timing of ovulation

A sexually mature man produces sperm continuously, and he is capable of making those cells available for reproduction any time. The woman, though, produces gametes—a process called *ovulation*—only once in a cycle that usually lasts 28 days from beginning to end.

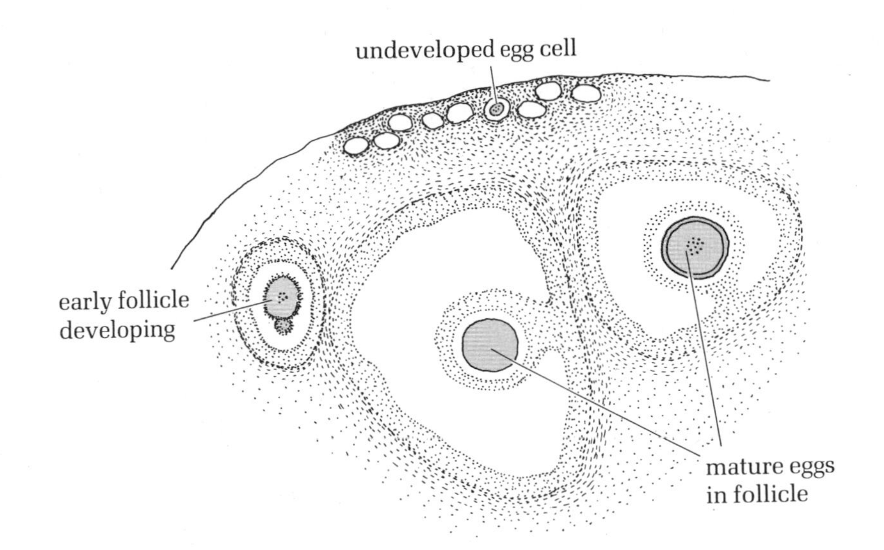

Figure 9–22 A cross-sectional view of a portion of a human ovary. After the egg cell has developed fully within a follicle, it can be released in the process known as ovulation. This entire sequence is under the control of the female hormones.

Each egg in a female ovary develops in a *follicle*, a small saclike structure. Figure 9–22 shows the follicles of an ovary and their eggs in various stages of development. When the egg is mature, the follicle ruptures and sends the egg into the open end of the Fallopian tube. After shedding the egg, the follicle changes into a structure known as the *corpus luteum*.

While the egg has been developing, the uterus has been preparing to receive it. The lining of the uterus thickens, and its blood supply greatly increases. If the woman has had intercourse recently and if she is not practicing birth control, sperm deposited in her upper vagina may have traveled through the cervix and uterus and up the Fallopian tubes. If a sperm meets the egg, fertilization takes place in the upper third of the oviduct, and the egg cell begins to divide and develop. When the fertilized egg cell reaches the uterus, it implants itself in the uterine wall and continues developing. An unfertilized egg does not implant in the uterine wall, and the specially prepared lining breaks down in order to be sloughed off in the bloody flow known as *menstruation*.

The menstrual cycle is usually dated from the first day of discharge. When the menstrual flow stops, a new egg matures, and the lining of the uterus rebuilds. Ovulation usually occurs on the fourteenth day, and, if the egg is not fertilized, the menstrual discharge will begin again on day

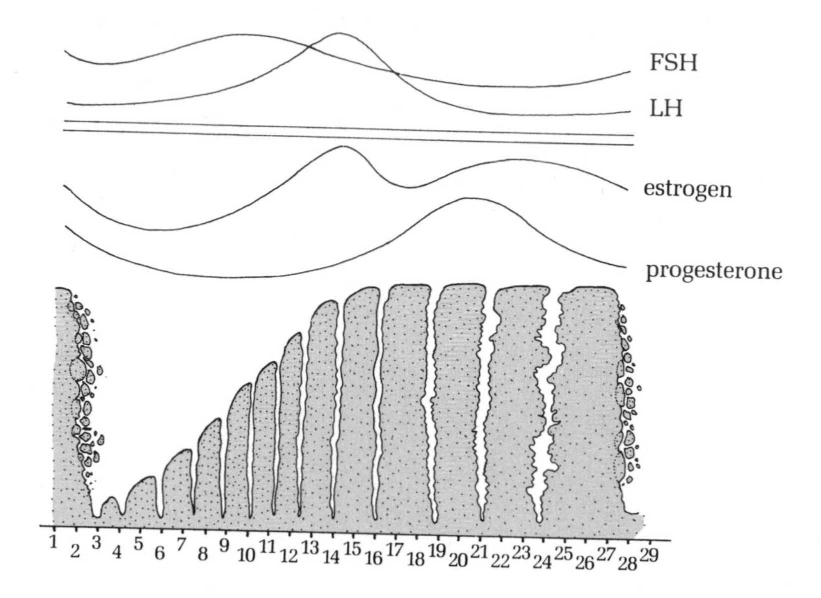

Figure 9–23 Hormonal regulation of the menstrual cycle. Each month the lining of the uterus prepares to receive a fertilized egg cell. A complex combination of female hormones regulates this event over the course of approximately 28 days.

28. These figures are only averages, though. Some women have highly irregular periods spanning anywhere from 18 to 60 days. Other women have regular periods that run on some number other than 28, say 21 or 35. But even the most regular period can be thrown off by sickness, emotional upset, or a change in altitude or climate.

How are the complex events of the menstrual cycle coordinated?

The hormones of the menstrual cycle

While the male produces testosterone at a steady level, the female produces a number of different hormones in varying amounts during the course of the menstrual cycle. The hormones direct the events of the cycle. Figure 9–23 charts the rise and fall of the hormones through a sample menstrual cycle.

The *pituitary gland*, a tiny organ located near the base of the brain, produces a great number of hormones, several of which affect the ovaries. One of these is the *follicle-stimulating hormone*, or FSH. A rising level of FSH in the bloodstream stimulates development of an egg cell in the ovary. The developing follicle produces its own hormone, *estrogen*. As the estrogen level rises, the pituitary produces less and less FSH and begins to release *luteinizing hormone,* or LH. The rising level of LH stimulates rupture of the follicle and ovulation. After ovulation, the level of LH continues to rise, and it promotes the change of the ruptured follicle into the corpus luteum. The amount of estrogen produced by the corpus luteum falls off, and *progesterone* is produced instead. Progesterone causes the walls of the uterus to thicken and to hold water, sometimes adding a few pounds to the woman's weight.

If the egg becomes fertilized, it lodges in the wall of the uterus and releases a hormone that stimulates continued release of progesterone by the corpus luteum. But if fertilization does not take place, the progesterone level falls off, and the thickened wall it helped to build begins to break up. By the time menstruation actually starts, both estrogen and progesterone are at the very lowest levels of the cycle, and the pituitary responds with increased amounts of FSH, thus beginning the cycle anew. The low level of estrogen and progesterone during menstruation is responsible for the depression and irritation some women feel when they have their periods.

Timing and sexuality

The menstrual cycle is truly uncommon in the animal world, being found only in man and the great apes. In most mammals, the sex hormones peak only at certain times, known as *estrus,* or heat. The female ovulates during heat, and only at this time will she mate. The frequency of estrus depends on the species. Horses, cattle, deer, and elk usually mate once a year. Dogs go into heat about every 6 months, while rats become available every few days. In some species, such as rabbits and cats, mating actually stimulates ovulation.

In all species except man, intercourse and reproduction go hand in hand. Even though the human female is likely to ovulate only within a 2- or 3-day span out of every 28 days, she can mate at any time. The reason for this may have something to do with the course of human evolution outlined in Chapter 4. Some anthropologists hypothesize that the female's

The menstrual cycle is truly uncommon in the animal world.

sexual willingness was favored by natural selection. Because of the sexual rewards of staying by his mate, the male was prompted to protect her and her offspring carefully. This gave rise to the strong male-female pair that is the basic unit of human society. In other words, the sexual nature of the human female may be an important key to the structure of civilization.

The biology of intercourse

Although our knowledge of sexual anatomy has been relatively complete for some time, little was known about sexual response until some 10 years ago. The reason for this lack of information was not a paucity of scientific techniques, but a general social taboo on research into the physical aspects of sex. Actually watching a couple engage in intercourse and recording the physical events smacked somewhat of depravity. But Dr. William Masters and Mrs. Virginia Johnson braved criticism to study sexual function with the scientific care it deserved. Their work has given us the first clear picture of how the sexual organs work.

Masters and Johnson divide intercourse into four stages (Figure 9–24). The first, the *excitement phase*, can be triggered by any one of those many stimuli—sound, touch, thought, or smell—that affect people sexually. In the man, an erect penis signals excitement. In the woman, the walls of the vagina secrete a fluid that lubricates the passageway and eases the penetration of the penis.

In the *plateau phase*, the sexual tensions begun in the first stages of excitement mount. The nipples of the woman, and often of the man as well, become erect. Both partners may show a sex flush, a mild rash on the chest, breasts, hips, or other parts of the body. The penis extends to its maximum length, between 6 and 7.5 inches. The vagina widens and elongates, the vulva becomes so engorged with blood that it looks almost purple, and the uterus may shift to a more elevated position.

Finally the sexual tension reaches a peak, catapulting the person into the *climax phase*, or *orgasm*. The man often feels the orgasm approaching as his urethra fills with semen. Violent contractions of the urethra and pelvic muscles expel the semen from the penis against the back wall of the vagina. The actual orgasm lasts only a few seconds. The

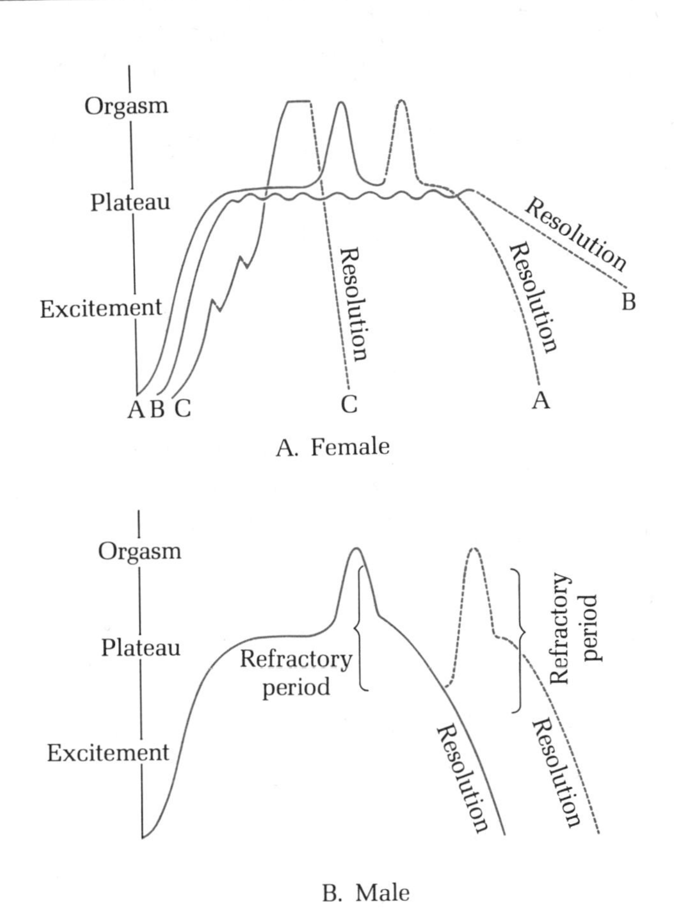

Figure 9–24 The male and female patterns of sexual response, as described by Masters and Johnson. A. Three distinctive female patterns are shown. B. The refractory period in the male patterns refers to the period between the plateau and the orgasmic phases.

woman's orgasm is likewise short-lived but equally intense. She experiences a series of violent contractions in the uterus, often accompanied by a spreading of the sex flush, further erection of the nipples, a suffusion of warmth throughout the pelvis, and a great increase in the sensitivity of the clitoris.

The intensity of orgasm eliminates sexual tension, and both partners relax, entering the *resolution phase*. During resolution, the sex organs return to their normal size and position. The two sexes differ in the speed with which they can return to the orgasmic phase. Most men have to wait 15 to 30 minutes before they can again become erect and have an orgasm. Some women, though, are capable of going from resolution to orgasm in a matter of 30 seconds to a minute time after time, experiencing several orgasms during a single act of intercourse.

How long does the complete cycle of sexual response take? There is no one answer. It can vary from less than a minute to an hour or more. A person who has been looking forward to intercourse with relish all day long may reach orgasm in a very short time. But someone who is physically or emotionally fatigued may require considerable stimulation just to become excited, and more yet to climax.

The myth and mystery of human sexuality

Sexuality is an area of such supercharged emotions that it has long been the subject of a great variety of untruths, distortions, and superstitions. Before the tools of science were turned to a better understanding of sex, there was simply no way to dispel these wrongheaded and often painful notions. The work of Masters and Johnson as well as a great number of psychological studies have given us a somewhat better grasp of the meaning of human sexuality.

The status and nature of the female orgasm has long been a subject of controversy. In much of Victorian Europe, general opinion had it that all well-bred women looked on sex as a joyless obligation to their husbands and that only women of ill repute could take pleasure in intercourse. This attitude prevailed in Vienna when Sigmund Freud pioneered the study of the working of the human mind. Freud saw that the stricture against sex caused many women great psychological distress. But Freud believed that there were two distinct types of female orgasm, clitoral and vaginal. He felt that clitoral orgasm evidenced psychological immaturity. Masters and Johnson showed that not only was Freud wrong about what was mature or immature but that he was also mistaken about the various kinds of orgasms. Their work demonstrated that all orgasms are physically the same, no matter where the stimulation is centered.

On the male side of the fence, there was the prevalent myth that the longer a man's penis, the greater his ability to satisfy a woman. Masters and Johnson found this to be untrue. So far as a woman's excitement is concerned, it's the width, not the length, of the penis that counts, and the woman has a way of making sure that the penis is thick enough. The vagina of a sexually excited woman actually closes down on the penis, thus making up for any variation in size.

A final and particularly ludicrous tale had it that *masturbation,* self-stimulation of the genitals to produce orgasm, was a heinous act resulting in any number of physical afflictions ranging from acne to insanity. In point of fact, the only physical result of masturbation is mild fatigue.

Once one goes beyond the strictly physical aspects of sexual anatomy and function and begins to deal with the "rights" and "wrongs" of sexual behavior, a whole new set of values and concerns applies. There is much more to human sexuality than simple biological fact. If this were not the case, then we would all simply select mates and begin breeding indiscriminately at 15 or 16. No biological absolutes govern the use of sex or determine a morality by which normal can be distinguished from perverse. Each culture teaches its own values as absolutes, but they are absolutes only so far as that culture is concerned.

A quick survey of representative cultures shows that at one time or another practically every form of sexual expression has been both condemned as immoral or degrading and upheld as proper or normal. In rural Ireland, for example, female orgasm is practically unknown, and men rarely marry and have any sort of a regular sex life before their mid-30s. By contrast, the Greenland Eskimos consider wife sharing an important aspect of hospitality and set aside a special hut in their villages for unmarried adolescents to enjoy sex play and intercourse. To the ancient Greeks, homosexuality was the highest form of love civilized man could attain. In America, the status of homosexuals is the subject of bitter debate. Some psychiatrists consider homosexuality a mental illness, a point rejected by many homosexuals and also by many psychiatrists. Some people feel that homosexuals should be allowed to follow the life styles of their choice, while others would have them subject to continual arrest and harassment.

Biology can tell us quite a bit about the physical nature of man's sexuality. Continuing research will probably provide us with even more enlightening information. In no way, though, can biology provide any kind of absolute moral code that divides proper from improper sexual behavior.

There is much more to human sexuality than simple biological fact.

Controlling fertility

Humans have long tried to exert some control over their reproductive functions. Often this has been accomplished by rules forbidding certain kinds of sexual activity. The stricture against premarital intercourse probably arose as a way of preventing unwanted offspring. Some societies have taboos against sexual relations involving a woman who is nursing a child. Since in many primitive societies women nurse their children for 2 or 3 years, this taboo effectively spaces birth and keeps the woman from becoming burdened with more than one infant at a time.

In addition, humans have sought ways of separating the sexual and the reproductive aspects of intercourse by making it possible to have intercourse without pregnancy. In recent years, the number of available methods of birth control has increased greatly. Table 9–1 lists a number of methods and provides statistics on their rates of success. You can contrast these figures with the fact that under normal conditions of intercourse in the absence of birth control, 75 percent of the women would probably be pregnant after 6 months.

Selective celibacy or the rhythm method

One method of birth control is to engage in intercourse only at those times in the menstrual cycle when there is no egg available for fertilization. Since ovulation occurs on day 14, that is obviously a bad day for lovemaking, as is the day following, when the egg travels through the Fallopian tubes toward the uterus. In addition, sperm can live for up to 2 days after ejaculation, so days 12 and 13 are likewise out. Theoretically, then, the remaining 24 days should be safe. But ovulation can occur a day or two out of sequence and throw these calculations off. People practicing rhythm usually engage in sex from days 1 to 10, abstain from 11 through 20, and allow themselves to indulge from 21 to 28. A sharp increase in body temperature accompanies ovulation. Thus a woman practicing the rhythm method can keep track of her body temperature and determine her period of maximum fertility (Figure 9–25).

Besides requiring careful planning and practiced self-discipline, the rhythm method has one major drawback: It doesn't work. It is the least

Table 9–1 Effectiveness of several birth control methods. Effectiveness is measured in terms of 100 woman-years; that is, if 100 women used a given method for 1 year each, how many of them would become pregnant?

Method	*Pregnancies per 100 woman-years*
Rhythm	24
Withdrawal	18
Condom	14
Diaphragm	12
IUD	3–9
Pills	1
Sterilization	less than 1
Abortion	0

Adapted from Kenneth L. Jones et al., Sex. New York: Harper & Row, 1969, p. 71.

reliable method of birth control. The reason is the irregularity of the menstrual cycle, which is affected by so many different factors that ovulation can occur at the most surprising and inopportune times.

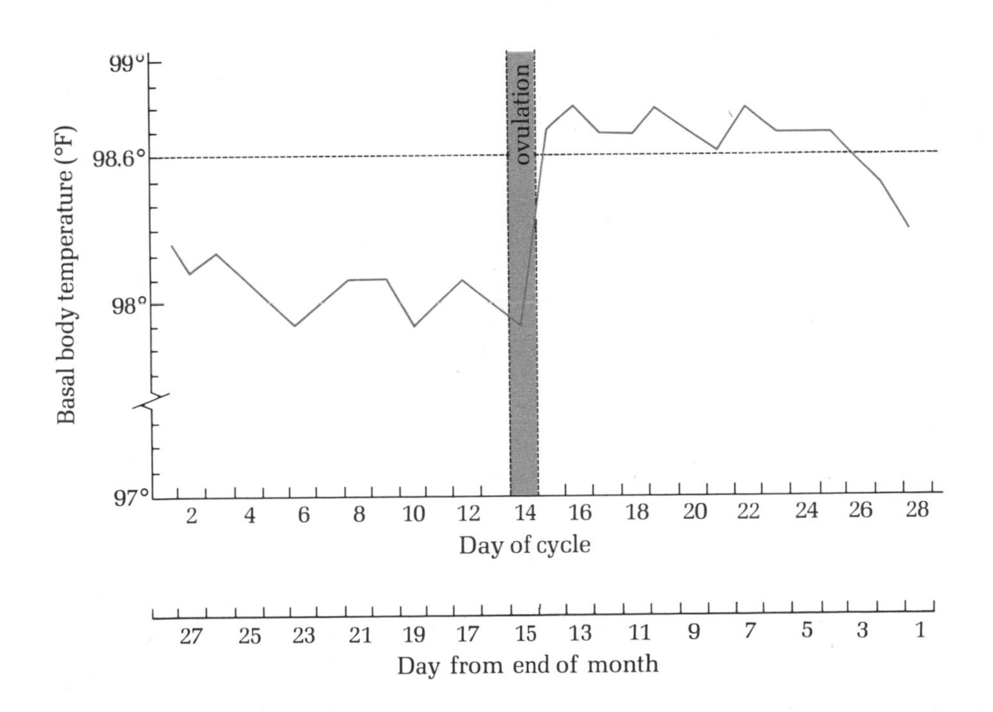

Figure 9–25 The fluctuation of basal body temperature during the normal menstrual cycle. Note the sharp increase at the time of ovulation.

Withdrawal

This method of birth control is sometimes called *coitus interruptus*, Latin for "interrupted intercourse," which is an accurate description. When the man feels his orgasm approaching, he pulls his penis out of the vagina and ejaculates outside the woman's body and away from the vaginal opening. Withdrawal requires considerable will power, and often leaves both partners feeling unsatisfied. Also, it doesn't work. A few drops of semen often escape the man's urethra well before he experiences the sensation accompanying ejaculation, and they contain more than enough sperm cells to start pregnancy. As you can see from Table 9–1, withdrawal is only a little more successful than rhythm.

The condom

A condom is simply a thin piece of rubber worn like a sheath over the penis. It captures the ejaculated semen and keeps sperm from entering the vagina. The penis and the condom should be withdrawn from the vagina shortly after ejaculation, because the condom loses its snug fit as the penis shrinks, and some of the semen may escape.

Used properly and carefully, the condom should be almost 100 percent effective, but accidents do happen. Occasionally a defective condom is used, which is about the same as using no condom at all. Also, if too little room is left at the tip of the condom to catch the semen, the explosive force of orgasm can split the condom open, rendering it worthless. Sometimes in the emotional intensity of intercourse, the condom can slip off unnoticed and not be there when needed. And, finally, using a condom can be distracting. It cannot be put on until the penis is fully erect, and fitting it may interrupt the sex act and dampen ardor.

The diaphragm

The diaphragm is a tight-fitting dome of rubber smeared with a spermicidal jelly or cream and placed over the cervix before intercourse. The diaphragm acts as a mechanical barrier by sealing off the cervix, and the cream or jelly adds extra protection by killing any sperm cells that touch it. Because women's vaginas differ greatly in size and elasticity, a diaphragm must be fitted by a doctor to make sure of the proper size. The diaphragm is more successful than the condom, but it can be torn, usually by careless handling when it is inserted. Also, in some rare cases, the contractions of the woman's orgasm may knock the diaphragm out of place.

The IUD

The IUD, the abbreviation for intrauterine device, is simply a small piece of flat, coiled, or curved plastic inserted into the uterus and left there. No one knows exactly how the IUD works, but apparently the presence of a foreign body prevents any fertilized egg from implanting in the uterine wall. The IUD has to be inserted and removed by a physician. Nearly all women experience some temporary discomfort or bleeding when the IUD is inserted, and, in some cases, the pain is so severe or long-lasting that the IUD must be removed. This happens particularly to women who have never borne a child. The variable failure rate given in Table 9–1 is due to the different shapes of IUDs; some are more successful than others. The principal advantage of the IUD is lack of concern. Unlike the condom or diaphragm, the IUD does not have to be put in or on or checked for fit at each and every sexual encounter.

The pill

When a zygote implants in the uterine wall, continued production of progesterone prevents subsequent ovulation. The idea behind the pill is to imitate the hormone balance of pregnancy and therefore to eliminate fertility. There are a number of different birth control pills, but most of them have the effect of raising the levels of progesterone and estrogen, inhibiting the pituitary's production of FSH, and therefore preventing ovulation. In general, birth control pills are taken from day 1 to day 20, and then omitted until day 28 to allow a normal menstrual period. Taken as directed, the pill is extraordinarily effective. The pill's occasional failures are due mostly to forgetfulness and missed pills.

The pill has aesthetic advantages over the condom or diaphragm, but it can have physical side effects. Women in early stages of pregnancy often suffer from weight gain, tenderness in the breasts, or nausea, and these symptoms can appear in women on the pill. There have been cases where serious side effects like blood clots in major blood vessels appear to have been caused by oral contraceptives. Statistically, though, taking the pill involves fewer risks than child bearing does.

Researchers are currently looking into the possibility of developing a pill for men. Because men produce sperm continuously, any hormonal preparation strong enough to suppress sperm production might very well have dangerous side effects. Hopefully, these side effects will be overcome, allowing both sexes to share equally the risks and responsibilities of fertility control.

Surgical sterilization

To avoid the hassles of taking pills or worrying about whether a diaphragm or condom is being worn properly, it is possible to be rendered sterile by surgery. This method of birth control is particularly popular with people who have had children and have decided against adding more to their families.

Both sexes can be surgically sterilized. In the case of a man, the operation is called a *vasectomy*, and it is simple enough to be performed in a doctor's office. Figure 9–26a shows the procedure. The patient is given a local anesthetic, and then two small slits are made in the scrotum to expose the vas deferens. Sections of both vas deferens are removed, and the ends are sealed shut by cauterizing. The man continues to produce sperm, but they remain in his testes and are reabsorbed into the body. In Europe, physicians are experimenting with a "gold valve" which is surgically implanted in the vas deferens. The valve can be adjusted to allow sperm to pass normally if fertilization is desired or to block them if contraception is preferred.

Vasectomy should not be confused with castration. A sterilized male suffers no loss in sexual potency or secondary sexual characteristics. He produces semen, but the fluid contains no sperm.

Making a woman sterile is a more involved surgical procedure (Figure 9–26b). A man's sexual organs are conveniently exposed to the surgeon, but reaching the woman's primary sex organs entails cutting into the abdomen. This operation, called a *tubal ligation*, is similar to a vasectomy. The Fallopian tubes, which carry the egg into the uterus, are cut

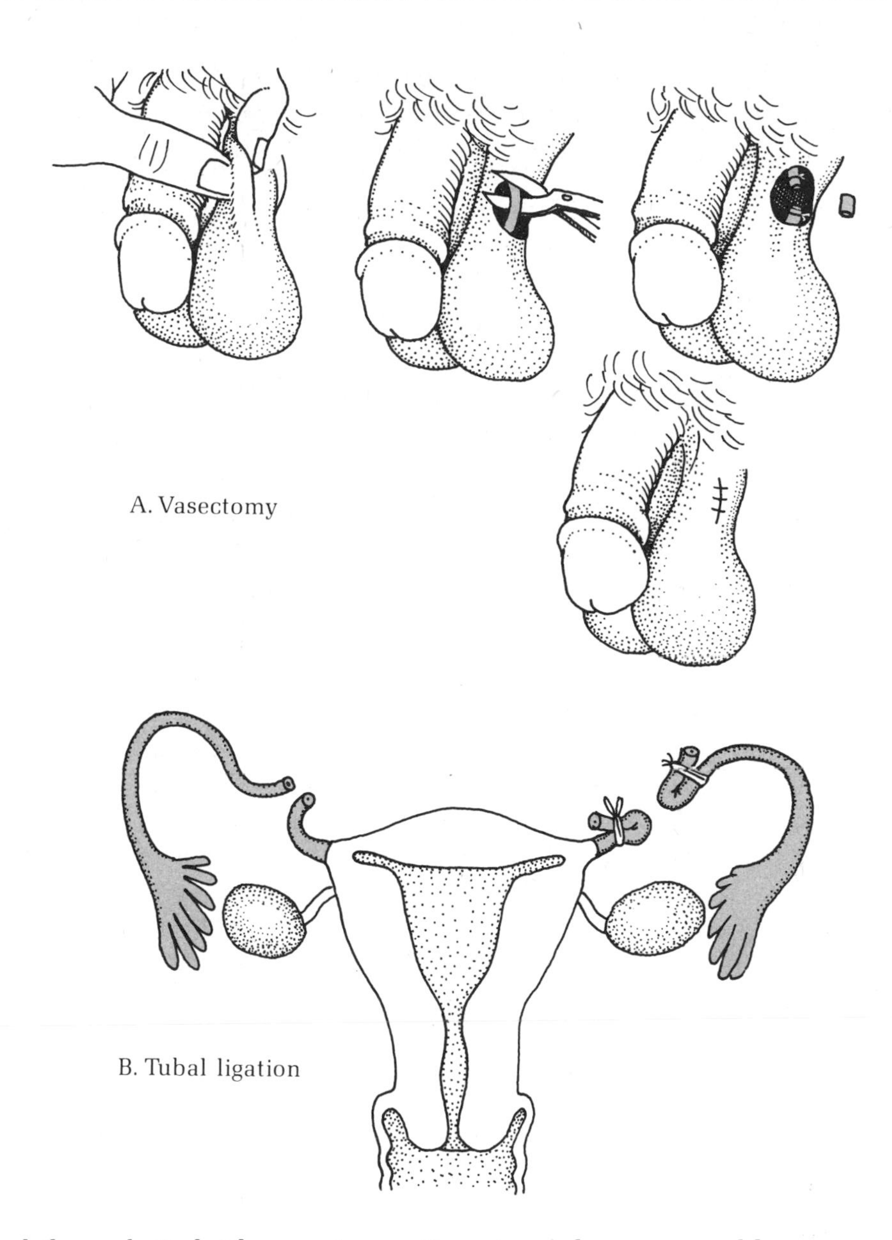

Figure 9–26 Sterilization through surgery. The most common methods of surgical prevention of conception involve disrupting the path of the gametes. In vasectomy (A) the vas deferens, which conducts sperm to the urethra, are cut, a portion is removed, and the ends are sealed. In tubal ligation (B) an analogous procedure is used: the Fallopian tubes are severed, and the ends are tied and sealed.

and the ends tied. The ovaries continue to produce eggs and hormones, but there is no way for sperm cells to get to the egg or for an egg to get into the uterus. Occasionally a vasectomy or a tubal ligation fails, when the ends of the severed tubes or ducts grow together again and the gametes can move through normally, but this happens only very rarely.

Who should be sterilized, the man or the woman? The answer depends on the couple. Tubal ligation involves more surgical risks than vasectomy, but few sterilized women suffer any psychological side effects from being made infertile. Some men, though, do suffer such side effects. They unconsciously equate sterilization with castration, and their sex lives—and their wives'—suffer accordingly.

Abortion

Apart from complete abstinence from sexual intercourse, abortion is the most effective method of birth control. Unlike the other methods of birth control, it aims not to stop pregnancies before they start but to end them once they have begun. However, the struggle over abortion in this country has been long and bitter. On the one hand are people who feel that abortion is the murder of a helpless victim and that to allow it is to accept the destruction of human life as a routine occurrence. On the other hand are people who feel that a woman should have the right to deal with her own body as she sees fit. Legal opinion has increasingly taken the latter position. As a result, abortion, which used to be a serious criminal offense, is now legal in most states.

One of two surgical procedures is used to abort the developing embryo. The suction method works best when the pregnancy has not advanced beyond the sixteenth week of pregnancy. The cervix is dilated and the contents of the uterus are pulled free by a suction tube. After the sixteenth week, suction is less successful. Instead, some of the fluid surrounding the embryo is replaced with a salt solution. The uterus expels the embryo within a day or two.

Abortion has been the most common method of birth control in Japan, where it has succeeded in lowering the birth rate markedly. In this country, it is unlikely that abortion will achieve such popularity. Instead, it will probably remain as an alternative to unwanted pregnancy when simpler methods of birth control either failed or were forgotten.

Summary

The simplest form of reproduction is the asexual division of microorganisms like the amoeba. Asexual reproduction is also found among some simple animals and among plants, both those that reproduce from runners and those that reproduce from spores.

Although asexual reproduction is simple and fast, sexual reproduction affords greater genetic variability. The asexual bacteria increase their variability by conjugation.

The reproductive method of plants is characterized by the alternation of gametophyte (haploid) and sporophyte (diploid) generations. In the sea lettuce, both forms look exactly the same. In the flowering plants, the gametophytes are dependent on the sporophyte. The pollen grains are the male gametophytes, and the female gametophytes develop within the ovules. Pollination involves a double fertilization, resulting in a diploid zygote and triploid endosperm. Relatively few plants pollinate themselves. Cross-pollinators spread the pollen by either releasing it into the wind or by attracting feeding animals to the flowers.

Animals have considerable sexual variety. Garden snails and sea hares are hermaphrodites. Jellyfish alternate between a free-swimming sexual form and a stationary asexual form. Aquatic forms use water to transport sperm to eggs. In land animals, the sperm is introduced directly into the female. The penis is specially adapted to this purpose.

Fertilization is the union of sperm and egg nuclei to produce the zygote. When the sperm finds an egg, the sperm's head penetrates the egg's cell membrane, bringing the two nuclei together. A wave of change alters the outer coating of the egg, generally preventing the entry of additional sperm.

Sexual differences comprise two classes of characteristics. The primary sexual characteristics are the gonads, the testes of the male and the ovaries of the female. Secondary characteristics comprise all sexual differences other than the gonads. Secondary characteristics can be either permanent or temporary, and they often serve some reproductive purpose.

The male gonads are the testes, which produce sperm and hormones. Semen comprises sperm and the secretions of three glands located along the pathway the sperm follow from the testes to the urethra. The penis erects because of the blood-holding ability of masses of spongy tissue within it. The testes are located outside the body to ensure the lower temperature needed to produce living sperm. Sterility is the lack of a sufficient amount of living sperm; impotence is the inability to get or maintain an erection. Castration produces sterility and some change in secondary characteristics, depending on the age of the man.

The female gonads are the ovaries. They are connected to the uterus by the Fallopian tubes. The uterus opens into the vagina through the cervix. The vulva cover and protect the clitoris and the exterior opening of the vagina. The menstrual cycle is controlled and coordinated by the varying production of different hormones in the pituitary, the ovaries, and the corpus luteum. Menstruation occurs in any cycle not resulting in pregnancy. Human and great ape females are exceptional among all species in that they are sexually receptive even when they are not ovulating.

Human intercourse comprises four stages. During excitement, the sexual organs become ready for intercourse. In the plateau phase, sexual tensions mount steadily, producing increasingly greater excitement. Orgasm is characterized by violent contractions and feelings of great pleasure. During resolution, the body returns to its preexcited state.

While biological research adds to our understanding of the physical side of sex, it cannot provide any basis for a sexual morality. Issues of morality and propriety are social and cultural, not biological.

There are a number of available birth control methods: rhythm, withdrawal, the condom, the diaphragm, the IUD, the pill, surgical sterilization, and abortion.

Questions to consider

1. Why can reproduction properly be called the "ultimate biological necessity"?
2. a. What is the meaning of asexual reproduction?
   b. What organisms reproduce in this fashion?
3. What advantage do bacteria gain from conjugation?
4. a. What is meant by "alternation of generations"?
   b. How does this phenomenon differ in the sea lettuce and in a flowering plant?
5. For genetic reasons, it's not surprising to find that cross-pollinators greatly outnumber self-pollinators. What are these genetic reasons?
6. Jellyfish alternate between sexual and asexual forms. How does this differ from the alternation of generations found in plants?
7. What happens when a sperm finds an egg?
8. A man can be distinguished from a woman by his testes, penis, beard, voice, height, and weight. Which of these differences are primary sexual characteristics, and which are secondary?

9. What pathway do the sperm follow from their site of production to their destination inside the vagina of a woman?
10. What are the principal female reproductive organs?
11. a. What are the major events of the menstrual cycle?
    b. How are these events coordinated and controlled?
12. a. What are the stages of intercourse?
    b. What happens during each one?
13. a. Why is rhythm a generally unsuccessful method of birth control?
    b. Withdrawal is about equally unsuccessful, but for a different reason. What is that reason?
    c. How does abortion differ from the other methods of birth control?

Glossary

alternation of generations the pattern of reproduction in which the life cycle alternates between prominent haploid and diploid forms

anther sac the small structure in a flowering plant that is the site of pollen production

asexual reproduction reproduction involving genetic material from a single parent organism

bulbourethral gland one of three types of glands in the male reproductive system whose secretions contribute to the semen

castration the removal of the testes

cervix the lower muscular portion of the uterus which joins the vagina

climax phase the third phase of sexual intercourse, characterized by intense physical response in both the male and female and ejaculation in the male

clitoris one of the female external sex organs, composed of erectile tissue and highly sensitive to stimulation

conjugation a simple form of reproduction in which single-celled organisms exchange genetic material before dividing by mitosis

corpus luteum the structure formed in the ovary after the release of the egg cell *cross-pollination* fertilization of one flowering plant by the pollen of another flowering plant

endosperm the layer of the seed in a flowering plant which provides nutrients for the developing embryo

epididymis the coiled tubelike structure in which sperm cells are stored after production in the testes

estrogen one of the major female sex hormones

estrus the limited period of sexual availability in many female mammals

excitement phase the first phase of sexual intercourse characterized by initial sexual stimulation

Fallopian tubes the two tubes connecting the ovaries and the uterus in mammals

fertilization the fusion of the male and female gametes to form a diploid zygote

follicle the sac-like structure in the ovary in which the egg cell develops

follicle-stimulating hormone (FSH) a female hormone produced by the pituitary gland which controls the development of egg cells in the follicles of the ovary

foreskin a flap of skin covering the glans of the penis

gametophyte the sexual, haploid phase of the plant life cycle

glans the fleshy tip of the penis

gonad one of the gamete-producing organs of animals

hermaphrodite a plant or animal form having both male and female sex organs

hormone a chemical substance produced in one organ of the body which can regulate the functions of other organs

hymen a membrane layer which partially closes the lower opening of the vagina

impotence the inability of a male to attain or sustain an erection

inner lips the interior folds of tissue surrounding the outer opening of the vagina *luteinizing hormone* a female hormone produced by the pituitary gland which regulates the release of the mature egg cell

masturbation sexual satisfaction through self-stimulation of the genitals

menstruation the flow of blood and uterine secretions which occurs regularly in females when a fertilized egg does not implant

orgasm the intense physical sensations characteristic of the climax phase of sexual intercourse

outer lips the external folds of tissue surrounding the vulva

ovary the female gonad; the site of egg cell production

oviducts the Fallopian tubes; the ducts connecting the ovaries and the uterus

ovule the female gametophyte in plants

penis the erectile male sexual organ

pistil the shaftlike organ in plants which is the site of egg cell production and fertilization

pituitary gland the gland at the base of the brain which produces many important hormones

plateau phase the second phase of sexual intercourse characterized by physiological changes preparatory to the climax phase

pollen grains the small particles carrying the male gametes of flowering plants

pollination the introduction of pollen into the female organs of a flowering plant

polyp a phase in the life cycle of some lower animals characterized by asexual reproduction

primary sexual characteristics structural and functional characteristics directly related to the gonads

progesterone a female hormone that regulates the preparation of the walls of the uterus for the implantation of a zygote

prostate one of the male glands that contributes substances to the semen

resolution phase the final stage of sexual intercourse in which the sex organs return to their normal size and shape *scrotum* the sac-like structure in which the testes are held away from the male torso

secondary sexual characteristics physical sexual characteristics not directly related to the gonads

self-pollination fertilization of a flowering plant by pollen grains of the same plant

semen the sperm-containing liquid secreted by the male reproductive organs

seminal vesicles one of the male glands which contributes to the seminal fluid

sex pilus the tubelike connection between conjugating single-celled organisms through which genetic information is exchanged

spore a small asexual reproductive particle in lower animals and plants

sporophyte the asexual, diploid phase of the life cycle of plants

sterility the inability of a male to produce living sperm cells

testis the male gonad which is the site of sperm production

testosterone a male sexual hormone that regulates the development of secondary sexual characteristics

triploid having a complement of chromosomes equal to three times the haploid number

tubal ligation a surgical method for sterilization in females in which the Fallopian tubes are closed

uterus the pear-shaped organ which protects and nourishes the developing embryo

vagina the muscular passage joining the uterus with the vulva

vas deferens the thin tube connecting the epididymis with the urethra

vasectomy a surgical method of sterilization in males in which the vas deferens is closed

vulva the external female sexual organs

10 Developing

Even a quick examination of the human body reveals that it comprises a great many distinct structures. Consider how different the clear, leathery cornea of the eye is from the soft and moist coating of the eyelid that covers it, the contracting and fibrous mass of the upper leg muscles is from the rigid thigh bone it is attached to, or the liquid blood is from the elastic arteries and veins it flows through. Remarkably, each of these structures, and the countless others found in the body, all originated from the single-celled zygote. In the 9 months separating fertilization and birth, that one cell gives rise to 2 billion body cells of vastly different types!

In this chapter, we'll first outline what happens between fertilization and birth. Next, we'll confront one of the toughest questions facing biology, a question only partly answered: How does a single cell know how to divide and grow in the precise fashion needed to produce the form of cellular order we call a human being? At the conclusion of the chapter, we'll look at another developmental process, aging.

In the 9 months separating fertilization and birth, one cell gives rise to 2 billion body cells of vastly different types.

Eggs of the Everglade apple snail. The undifferentiated tissues within these eggshells will give rise to complete and independent organisms by a process still not yet completely understood.

From fertilization to birth

The development of the zygote into a fully formed individual involves three processes. The first of these is, obviously, growth. Cell division gives rise to a great many more cells and an increase in the size, or mass, of the organism. Besides increasing in size, though, the cells also become different from one another. This second process is appropriately named *differentiation*. Differentiation implies that the cells take on a form that enables them to fill certain specialized roles. A bone cell is very different from a liver cell, and it is this difference that allows the one to act as a rigid structural unit of the body and the other to manufacture bile and other needed secretions. It's important to note, though, that differentiation does not necessarily make cells more complex. A red blood cell is highly specialized, fitted only for the task of carrying oxygen in the blood. But it is also a very simple cell, being essentially a bag of hemoglobin with no nucleus. It is far less complex than the single, unspecialized cell of a one-celled amoeba. The third process is *morphogenesis*, the development of a particular organ or structure with new form or shape. Morphogenesis results from a precise sequence of growth and differentiation. The appearance of the hand during development involves the growth of new cells and the differentiation of these cells into the types—bone, skin, blood vessels, muscles, and so on—needed to construct the hand. And, of course, each specific aspect of morphogenesis results in the development of the whole organism.

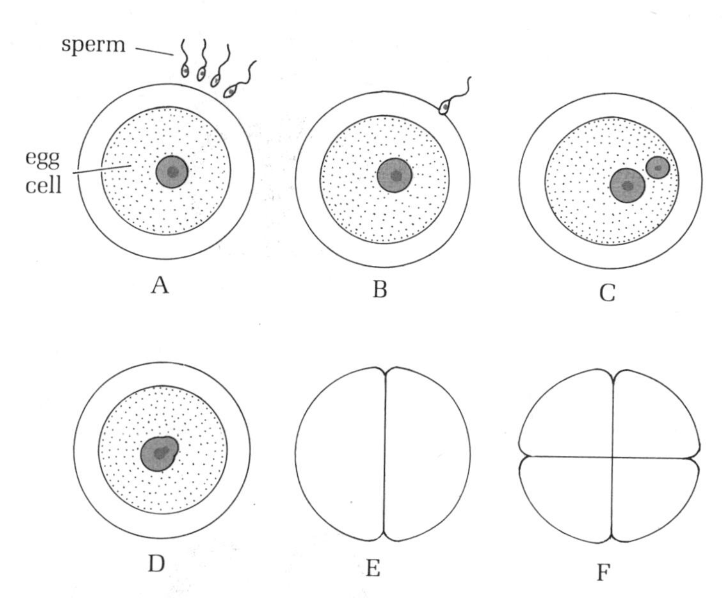

Figure 10–1 Fertilization and early cleavage of the egg cell. The haploid gametes—egg and sperm (A)—meet, and one of the sperm cells penetrates the outer membranes of the egg cell (B). The nuclear material of the two cells fuse, restoring the diploid number of chromosomes (C,D). The fertilized cell then begins to cleave mitotically, producing a ball of diploid cells (E,F).

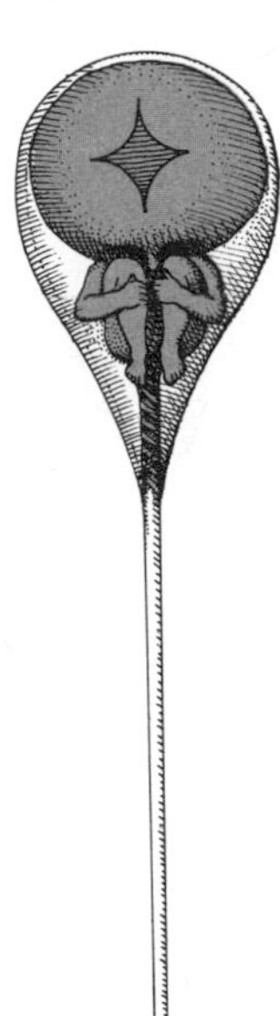

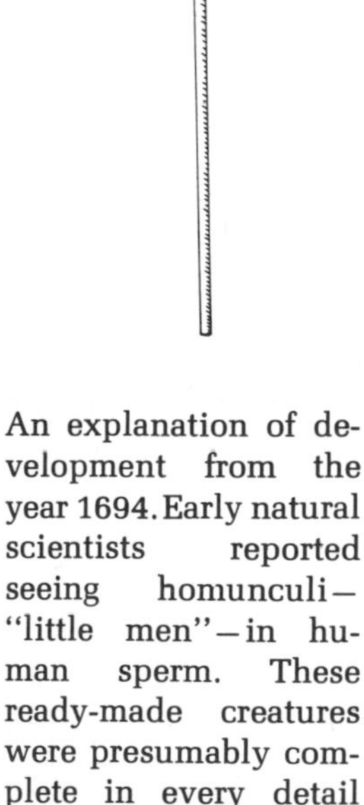

An explanation of development from the year 1694. Early natural scientists reported seeing homunculi—"little men"—in human sperm. These ready-made creatures were presumably complete in every detail and only had to grow, rather than differentiate, within the uterus during pregnancy.

Keeping the nature of these processes in mind, we will describe briefly what happens in time between the formation of the zygote and the introduction of a new individual at birth.

The first week

Once expelled from the ovary, an egg takes from 1 to 4 days to pass down the Fallopian tube to the uterus. However, the egg is receptive to sperm only for approximately 24 hours, and the sperm are active only for 48 hours. Thus, fertilization must take place during the first day of the journey. The sperm usually locate the egg in the upper reaches of the Fallopian tube, and fertilization occurs there. Once the egg and sperm nuclei fuse into one, the cell begins to divide mitotically (Figure 10–1).

There are some interesting variations on this theme. If *two* sperm encounter *two* separate eggs, *fraternal* twins will result. Because of their distinct genetic make-up, they are genetically unique individuals. If, however, a single egg fertilized by a single sperm begins its mitosis by a

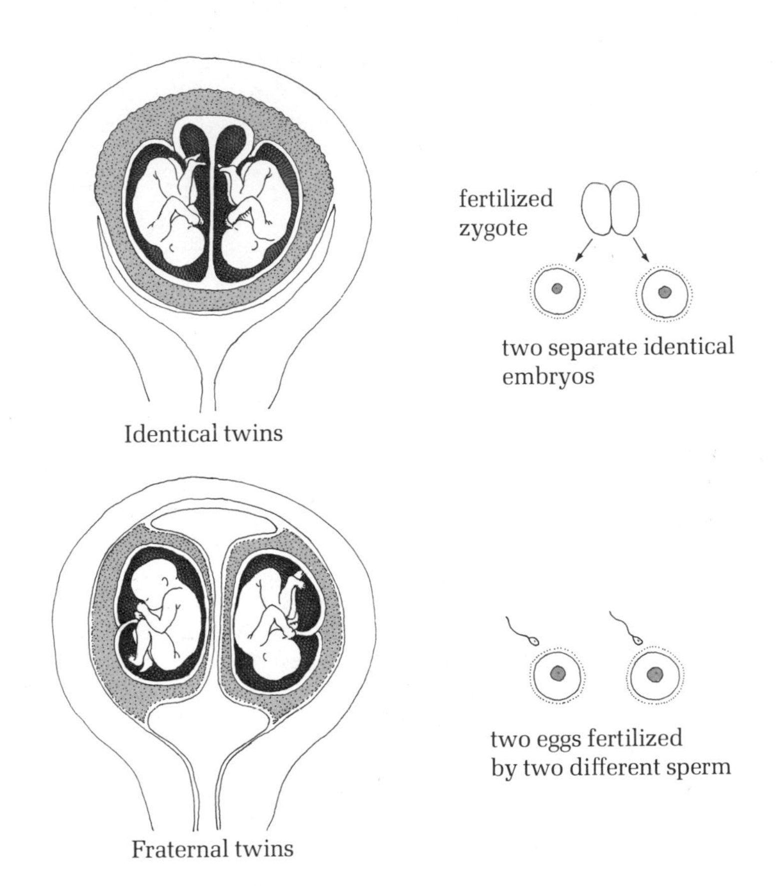

Figure 10–2 Identical and fraternal twins. Identical twins result from the separation of a single fertilized zygote into two identical embryos. Fraternal twins are actually two unique embryos, which happen to develop within the uterus at the same time.

complete division into two separate portions, *identical* twins will develop. As the name implies, they are genetically identical (Figure 10–2).

By the end of the second day, the zygote has divided twice, resulting in four cells. The multicellular zygote is referred to as the *embryo*. At the end of the third day, the embryo comprises 16 cells, and, with the fourth day, it is a compact ball of about 32 cells, a stage referred to as the *blastula*. There has been practically no growth so far; the 32-celled blastula is about the same size as the one-celled zygote. Although chromosomes have been duplicated so that each cell is diploid, the original cytoplasm of the zygote has simply been divided among the new cells.

About the time the blastula forms, the embryo has reached the uterus, whose walls have been thickened and supplied with extra blood to receive it. As cell division continues, the embryo attaches to the uterine

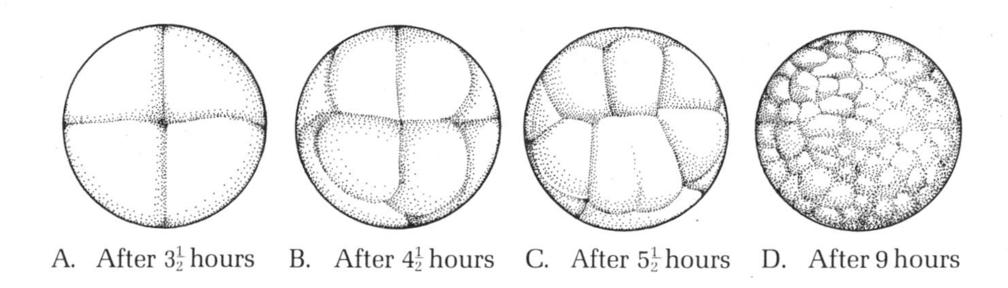

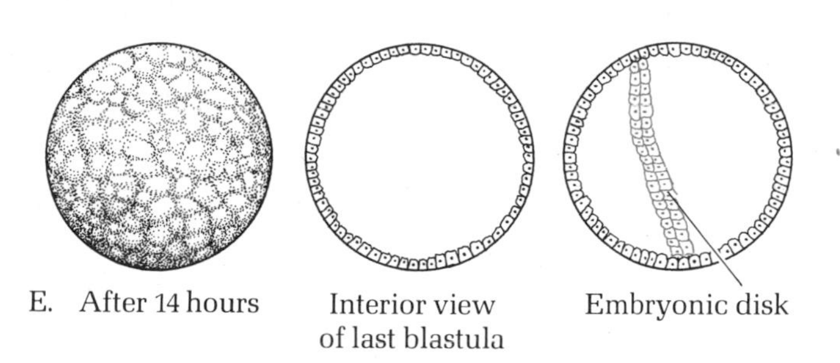

Figure 10–3 Cleavage and the formation of the embryonic disk. The frog has long been an excellent subject for the study of embryology. Within less than 24 hours, a frog embryo will have completed cleavage, and the embryonic disk will be formed.

wall and buries itself below the surface. If, by some error, the embryo implants in the Fallopian tube, an *ectopic* pregnancy results. An ectopic pregnancy must be corrected surgically to prevent the growing embryo from rupturing the Fallopian tube and causing fatal internal bleeding.

With the implanting of the embryo, the first of the many instances of differentiation can be seen. Some of the embryonic cells begin to divide faster than the rest, and form a coating of small cells over a central mass of larger cells. A cavity develops within the embryo. The cells then move and arrange themselves into a more complex form with two cavities. As you can see in Figure 10–3, the embryo consists of two groups of cells. The two-layered cluster of cells labeled the *embryonic disk* will continue to develop as the embryo proper, while the other cells are destined to become the membranes that support the embryo in the uterus.

Here we have a simple model of differentiation. A number of undifferentiated cells is first divided into two or more distinct groups, and then each group develops in a different fashion. This process occurs countless times: segregation of cells into groups, with each group following a different pathway of development.

Figure 10–4 A portion of the fetal circulatory system of the placenta. These minute branches of the fetal arteries and veins intertwine with those of the mother in the placenta. Through this delicate connection, materials are exchanged between mother and child.

The placenta and its role

Let's leave the embryo proper for a moment and look at the membranes that develop from the group of cells separated from the embryonic disk at the implanting of the embryo. These membranes support the embryo and are crucial to its survival and development. Of these, the *placenta* is surely the most important. The developing embryo, although contained within the body of the mother and totally dependent on her for food and oxygen, is separate from her and not a part of her body. The placenta serves as the organ of exchange between the mother and the embryo. The placenta is essentially a capillary bed from the mother running side by side with a capillary bed from the embryo. The embryo is connected to the placenta by the *umbilical cord*. The blood supplies of the two individuals never mix, but various materials can travel back and forth. Food and oxygen pass from the mother to the embryo, and waste materials pass from the embryo to the mother (Figure 10–4).

As the embryo grows, the placenta enlarges with it until it becomes a disk about 1 inch thick and 7 inches across. The side of the placenta embedded in the uterine wall is covered with fingerlike projections that increase the surface area available for exchange of materials. At full growth, the placenta contains roughly 140 square feet of exchange area, roughly equal to the floor space of the average living room.

The placenta is not the only support structure that develops from the embryo. The *amnion* surrounds the embryo, forming a watertight cavity filled with liquid. The amniotic fluid acts like a shock absorber to protect the embryo against the bumps of the mother's daily life. It also keeps the embryo suspended and floating, which is very important for proper development. If the embryo came to rest against the uterine wall for any lengthy period of time, the portion of the embryo actually touching it would very likely be malformed. The amnion will also soon be commonly used as a diagnostic tool. Experiments now being conducted reveal that cells taken from the amnion can reveal some genetic irregularities as well as the sex of the child in advance of birth.

Two membranous organs—the *yolk sac* and the *allantois*—found within the uterus are tag ends of human evolution. Birds, reptiles, and two species of primitive mammal, the duckbilled platypus and the spiny anteater, develop outside of the body of the mother in a shelled egg. The shelled egg is an amazing structure. No water can pass through the shell, but the shell does allow carbon dioxide and oxygen to pass back and forth. While in the egg, though, the developing embryo needs a source of food and a place to stow waste products. The yolk sac holds the food, and the allantois takes care of the wastes. In humans, the placenta fulfills these roles during uterine development of the embryo, and the yolk sac and the allantois are small and quickly become functionless.

The dividing of the cells

As the embryo implants and the cells that will develop into the special membranes are set aside, an important beginning step in development is taking place within the embryo proper. The embryonic disk comprises two layers of cells, one exterior and one interior. Some of the cells on the outside layer begin to migrate toward the inside of the embryo, a process known as *gastrulation*. When gastrulation is complete, the formerly two-layered embryo contains three layers.

Gastrulation is an important occurrence because each of these layers has a distinct developmental fate. The outermost layer of the gastrula is the *ectoderm*. The ectoderm will become the exterior layer of the skin and its related structures, such as hair and fingernails, and it is also the precursor of the brain and the nerves. The middle layer is the *mesoderm*. The

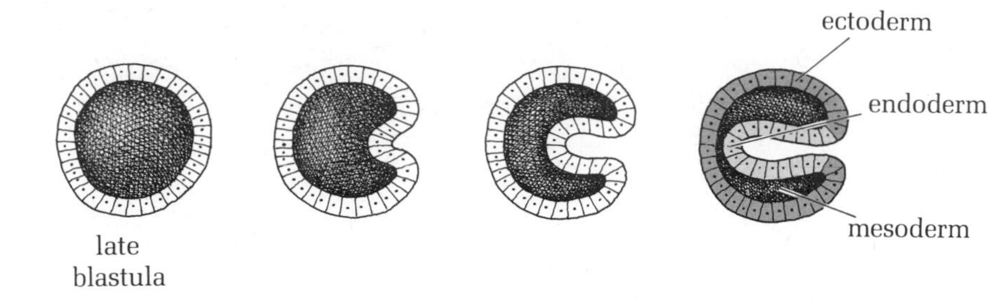

Figure 10–5 The process of gastrulation. In this first step toward differentiation, the cells of the blastula are reoriented to form three layers of cells, which will eventually give rise to different organ systems.

mesoderm makes up the great bulk of the adult form. It goes on to become the muscles, the urinary system, the outer layers of the internal organs, the inner layer of the skin, the circulatory system, and the bones. The innermost layer of the gastrula is the *endoderm*. The endoderm forms the glands, such as the liver and pancreas, and the inner linings of the respiratory and digestive tracts (Figure 10–5).

With gastrulation, the first step in differentiation of the embryo proper has begun. In the days remaining, these layers go through the further differentiation and morphogenesis that give the final form to the developing embryo.

The next 259 days

The time remaining to the embryo at the end of the first week is about 259 days. By no means does growth proceed evenly during this period. For the first 80 days, the embryo differentiates at a prodigious rate, but it grows relatively little. During the remainder of pregnancy, considerably less differentiation occurs, but the embryo grows rapidly.

By the twenty-fifth day, the rudiments of the first organs have appeared. The beginnings of the brain and spinal cord are clearly visible, and a tiny heart, at this point nothing more than a portion of blood vessel capable of constricting, has begun to beat. From the twenty-fifth to the fifty-sixth day, differentiation seems to run wild. At the beginning of the period, the embryo is only half an inch long and looks more like a tadpole than a human. On the fifty-sixth day, it has grown only another inch and weighs less than 1 ounce, but it has developed distinct eyes, ears, a mouth, limbs complete with fingers and toes, a brain and spine, many of the adult muscles, and such internal organs as the liver and kidneys.

By the sixty-third day, the external sex organs have begun to take shape, and the embryo can be seen to be either male or female. It's interesting to note that the male and female sex organs develop from precisely the same structures and that the organs of each sex are equivalent to each other. The male equivalent of the female clitoris is the penis. The male testes develop inside the body at the same site as the ovaries, and only later, about a month before birth, do they descend into their adult location in the scrotum.

At around 60 to 70 days, the embryo actually begins to move on its own—squinting, kicking, and gripping. But it is still too small for the mother to feel its movements. The embryo, though, has developed so much that it is given a new name: *fetus*.

By the hundredth day, differentiation is essentially over. Besides having all its parts, the fetus can perform such relatively complicated actions as swallowing, frowning, and sucking its thumb. Now the fetus begins to put on weight, and, because of its increasing size, the mother detects its presence. Usually the mother feels the first fetal movement during the eighteenth week of pregnancy, days 119 to 126. The fetus's heart beats strongly enough for it to be heard through a stethoscope placed against the mother's abdomen. These are the symptoms of *quickening*, the first strong signs of life.

At this time the fetus, although well formed, weighs only 11 ounces. Were the fetus expelled into the outside world at this time, or at any time up until the twenty-third week, it would not survive. The remaining weeks of pregnancy add weight to the fetus. If the fetus remains in the uterus full term and is not born prematurely—before the twenty-eighth week—its weight will increase nearly 10 times.

Bringing the child forth

The average date of birth is 266 days after fertilization, in the thirty-eighth week of pregnancy, but there is considerable variation. Most women give birth up to 14 days either side of the average 266. However, births falling anywhere from the thirty-fourth week, or 238 days, to the forty-third week, or 300 days, are not considered particularly unusual.

The birth process consists of two phases (Figure 10–6). The first is dilation of the cervix. Normally, the opening in the cervix is about as wide

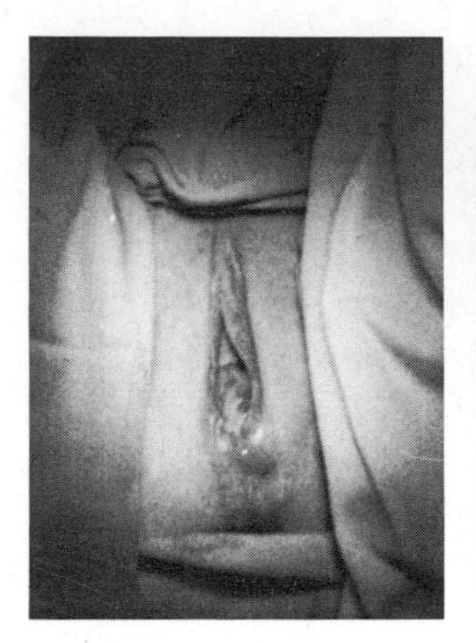
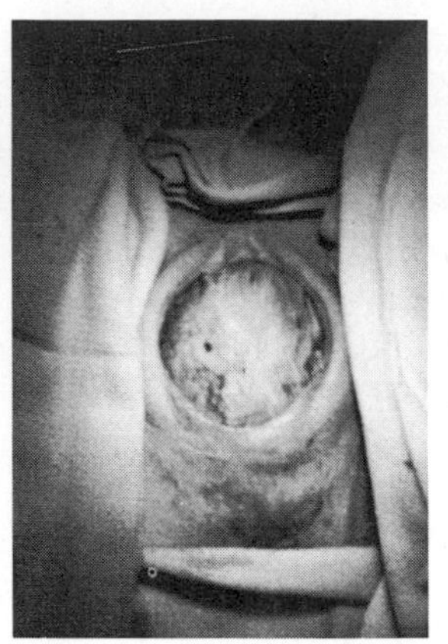
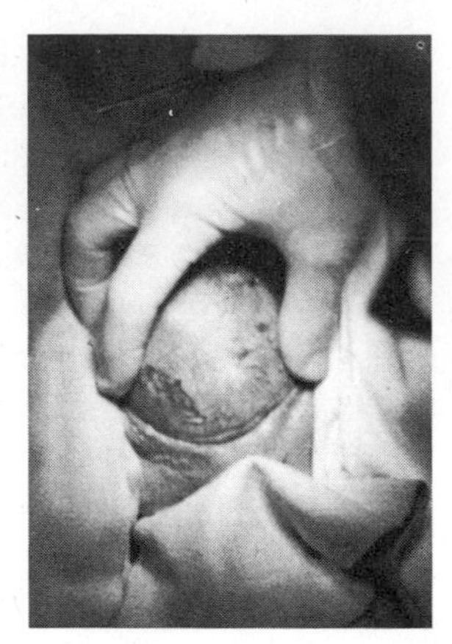
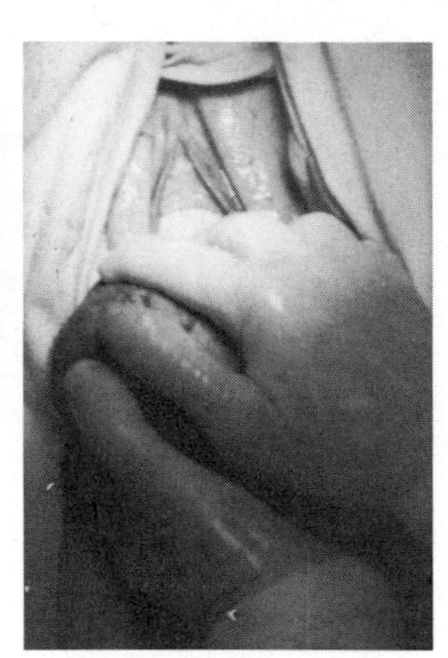

Figure 10–6 Four stages in the normal birth of a child. Rhythmic uterine contractions propel the movement of the child—from the first appearance of the child's head in the partially dilated vaginal opening to the beginning of the child's life as an independent organism.

as a pencil lead is thick, and it must dilate to a diameter of about 4 inches to allow the fetus to pass from the uterus into the vagina. While the cervix is dilating, the amnion ruptures, and the amniotic fluid ("bag of waters") escapes through the vagina. The second process, which accompanies dilation, is rhythmic contraction of the uterus. This regular contracting forces the fetus out of the woman's body, almost like squeezing toothpaste from the tube. The first contractions are relatively weak and irregular, but they become stronger as the cervix dilates. When the cervix is fully open, the contractions are frequent, intense, and long-lasting, coming about once a minute with only short spells between them. The very considerable force exerted by the uterus—so considerable that one doctor attending a birth put his finger in the wrong place and got it broken—squeezes the fetus through the dilated cervix and into the birth canal of the vagina. Usually the fetus comes head first and face down, but some babies are born with their feet or buttocks leading the way, in what is popularly called a "breech birth." After the fetus enters the vagina, continued contractions of the uterus push it out of the mother's body.

As soon as the baby is free of the mother's body, the umbilical cord is cut and tied off, severing the fetus's life-sustaining connection with the placenta. The fetus must now make the radical adjustment from the warm and secure aquatic environment of the uterus to the world outside. With the cord cut, the baby can no longer get oxygen from the placenta. The rising level of sodium bicarbonate in his blood triggers an automatic response that causes the infant to take his first breath, which usually sounds more like a cry of despair than a gasp of delight.

After the fetus has passed out of the mother, the placenta comes free from the wall of the uterus. Pushed by continuing uterine contractions, the placenta is expelled, sometimes in one piece, sometimes as a bloody flow. The expelled placenta is often referred to as the *afterbirth*.

How does development proceed?

Each one of the 2 billion cells of the newly born infant is a daughter cell of the original zygote, the result of a great many cell divisions. In our discussion of cell division in Chapter 8, we emphasized the fact that the two daughter cells resulting from a mitotic division are genetically identical. All of the cells in the infant have resulted from mitotic division. If all of these cells are genetically identical, how can it be that they have adopted so many different and distinct forms? Since we know that each of these cells carries the same genetic information, then somehow that genetic information must be selectively expressed in the proper order during the course of development. How are these selections made?

This question is one of the most significant issues in biology and the subject of a great many ingenious experiments. As the state of the science now stands, the question is only partly answered. But the answers that have been obtained so far give us both some idea of how development proceeds and an insight into what promises to be one of the most important and exciting research fields in the biology of the future.

Few of the experiments on development have been carried out on humans but, instead, on seemingly exotic creatures such as the sea urchin and the fruit fly and the more common frog. The reason for this choice of experimental material is not any inherent interest in these organisms

The rising level of sodium bicarbonate in his blood triggers an automatic response that causes the infant to take his first breath, which usually sounds more like a cry of despair than a gasp of delight.

themselves, but the serious ethical issues surrounding the use of human embryos for experiments. One can quite properly ask what right anyone has to manipulate a mass of cells that, given the right conditions, could become a human individual. Also, human embryos are nearly impossible to handle in a laboratory. They develop only in the chemical surroundings afforded by the uterus, and these conditions are too complicated to be set up and maintained in an experiment. As a result, scientists researching development use organisms, like the frog and sea urchin, which develop outside the body of the mother in a watery environment easily duplicated in the laboratory.

A model of development

Before we look at the influences on development, we would do well to set up a model of development, a plan which is purposely oversimplified but which makes a complex phenomenon easier to understand. Figure 10–7 shows the so-called landscape model of development. The undifferentiated cells of the early embryo are represented by ball bearings held in the cup at the top of the mountain. The developmental fate of each cell and its daughter cells is represented by the channel the balls follow as they roll downhill.

The important thing to note about the landscape model is the number of critical junctures, or points of no return. At the top of the mountain, where the embryo is still a blastula, any of its cells might become mesoderm or ectoderm or endoderm. One cell could be freely substituted for any other. But with gastrulation and the division of these cells into three groups, they follow distinct and separate pathways. No longer can the cells be substituted for each other. Within each of these segregated groups, though, substitution is possible. One mesodermal cell is pretty much the same as any other. But farther down the mesodermal pathway, the similarity ends, as some of the cells are shunted into the blood channel, and then others are shunted off to form muscle and bone. By a series of such

The important thing to note about the landscape model is the number of critical junctures, or points of no return.

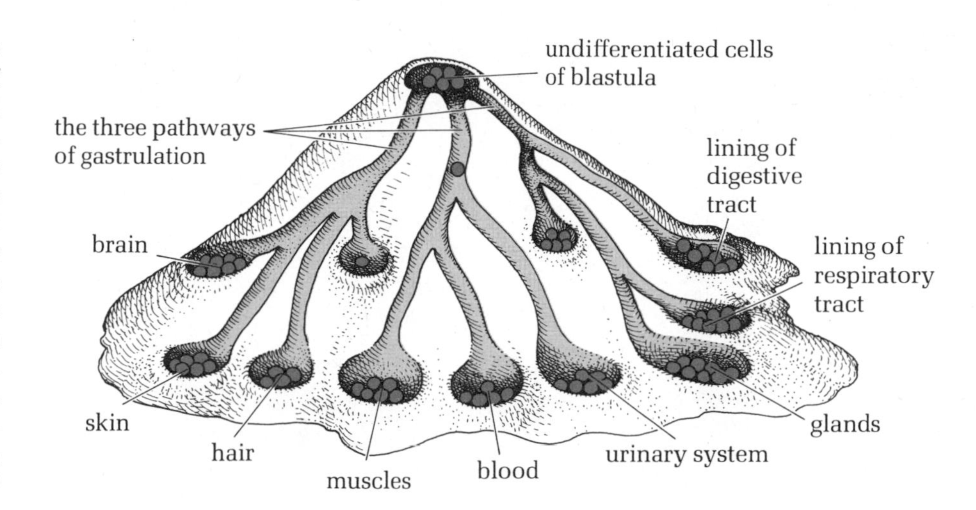

Figure 10–7 The landscape model of development proposed by C. H. Waddington. If we imagine the cells of the blastula at the top of an incline and the force of gravity as analogous to the force of differentiation, we can see how the path of a cell toward a particular type of development cannot be reversed after certain "forks in the road."

forks in each of the three major channels, the developmental possibilities of each cell are limited and the overall amount of differentiation increased. Once a ball—or a group of cells—enters a particular channel, it is committed to follow that pathway. There is no return.

A good example of this sort of development in humans is the formation of the external sex organs. Up through most of the second month of development, the tissues destined to become these organs are neither male nor female. But shortly before the end of the second month, some sort of a divide is crossed, and the organs take on the form of one sex or the other. Up until that time, the precusor structures could theoretically go either way. But at the critical juncture, some influence—presumably expressed in the twenty-third pair of chromosomes—pushes these structures to become one gender or the other.

The great value of the landscape model is that it gives our overall question about development a definite point of focus. What is the chemical nature of these critical junctures? How is it determined which channel a particular group of cells will follow?

What happens to the nucleus?

It's obvious that the nucleus plays an important role in differentiation because the chromosomes are contained within it. The important

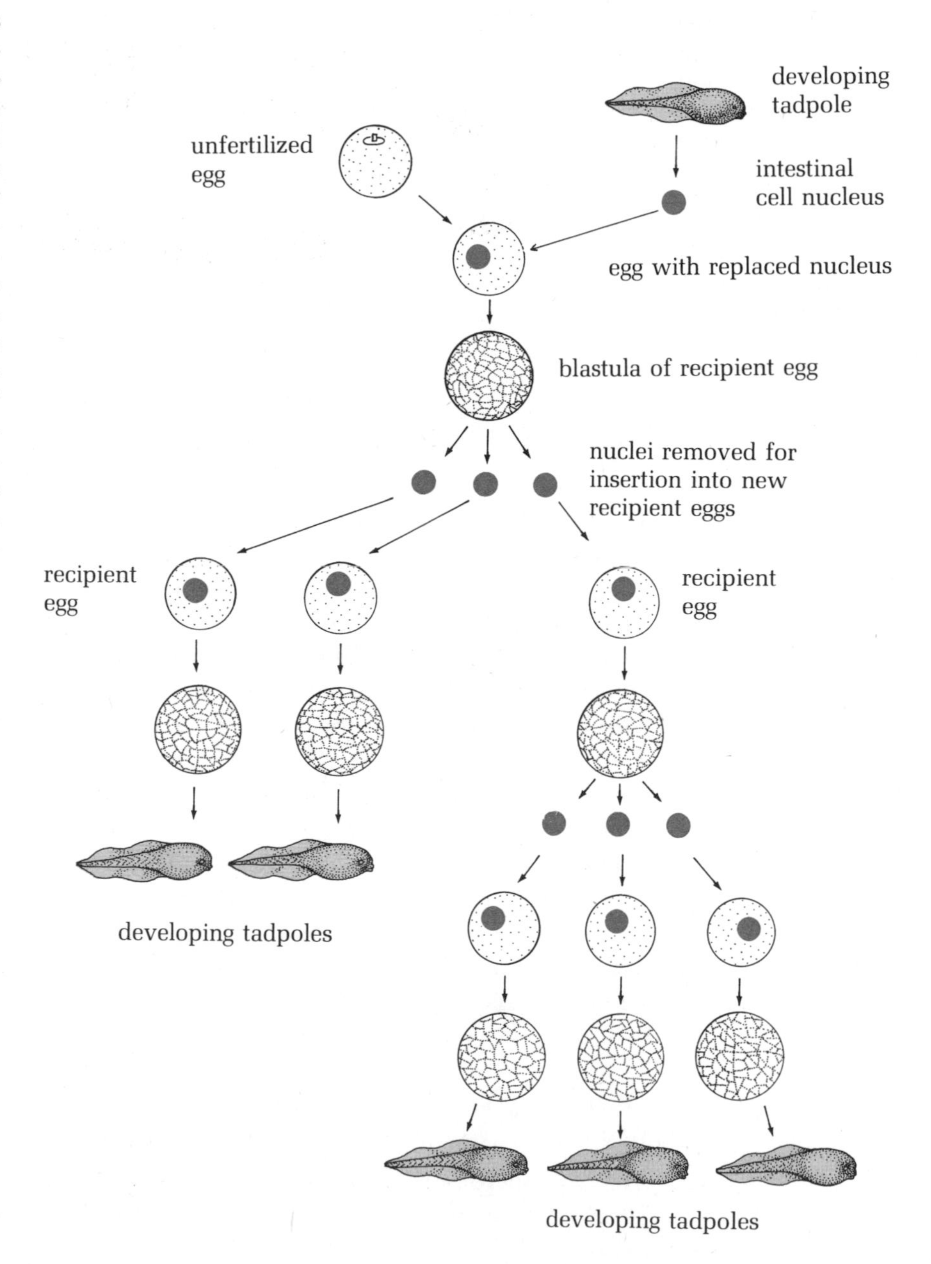

Figure 10–8 An ingenious experiment to test the role of the nucleus in development. A cell nucleus from a developing tadpole was inserted into an anucleated egg cell. Even though it had been removed from an organism that was far along the developmental path, the nucleus was still able to direct the differentiation and growth of the entire organism. Successive "transplants" from blastula to unfertilized eggs repeatedly demonstrated the significance of the nuclei's general regulation of development.

question is: Do the nuclei of the various cells change during development? In other words, does the nucleus of a brain cell contain only the information leading to the formation of the brain cell, or does it contain all of the genetic information of the entire organism?

A group of British scientists provided a clever experimental answer to this question. They took an unfertilized toad's egg and replaced its nucleus with one taken from an intestinal cell of a mature toad of the same species. The egg was then stimulated to begin development. It developed into a properly formed toad, which was genetically identical to the toad that contributed the nucleus. In another series of experiments, an egg provided with a transplanted nucleus was allowed to develop into a blastula. Nuclei were then removed from the blastula and transplanted to a second set of eggs. These eggs too developed into toads genetically identical to the original toad (Figure 10–8).

These experiments indicate that development causes no permanent changes in the nucleus itself. The chromosomes in each nucleus of every cell contain all of the information needed to produce the entire organism. But, obviously, not all that information is expressed in every cell. What determines what is expressed in any given cell?

Does the cytoplasm play a role?

The next obvious place to look for the determinants of the fate of a cell is in the cytoplasm, the region outside the nucleus. The subject for an important experiment on this question was the sea urchin, the spiny creature we first saw in Chapter 9. The yolk of the sea urchin egg clumps mostly at one end of the egg, called the *vegetal pole*. The other end of the egg is called the *animal pole*. A sea urchin egg was divided in half, so that one half was entirely vegetal and the other entirely animal. Then each egg half was fertilized by a sperm cell. Both halves developed abnormally. The vegetal egg developed almost entirely into gut, without the mouth, arms, or the fiberlike cilia typical of the normal sea urchin embryo. The animal egg contained no gut, but its cilia were both too large in size and too great in number. However, if a sea urchin egg is divided so that it contains equal portions of the vegetal and animal halves and is then fertilized, the resulting embryos develop normally (Figure 10–9).

What can be deduced from this experiment? Apparently the expression of the information contained within the nucleus is influenced by the

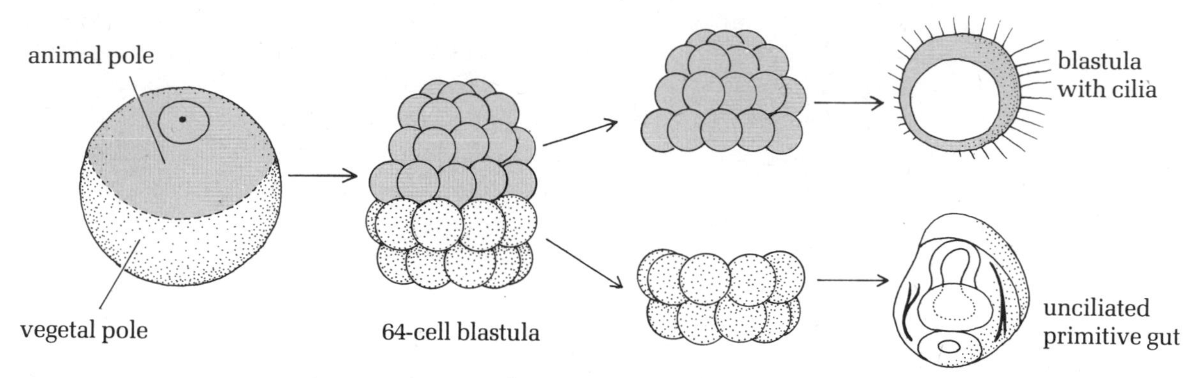

Figure 10–9 A demonstration of the cytoplasmic contribution to development. This simple experiment demonstrates the significance of both the animal and the vegetal portions of the egg to the normal development of the embryo.

cytoplasm in the early stages of development. Unless all the elements of the cytoplasm, both vegetal and animal in this case, are available in the right proportion, the zygote will not develop normally.

Do the cells affect each other?

Both the nucleus and the cytoplasm influence development. But, since so much of development has to do with increasing the number of cells, isn't it possible that the cells may somehow influence each other? Might the cells communicate back and forth and influence each other's development?

If a sea urchin blastula is agitated in a special solution, the blastula becomes unglued and the cells are separated. No physical or chemical harm is done to the cells; they remain intact and living. The only difference between these cells and those of a blastula is that these cells are no longer in close contact with each other. When the separated cells are returned to their normal saltwater environment though, all cell division

Might the cells communicate back and forth and influence each other's development?

ceases, and the cells quickly die. If the saltwater containing separated blastula cells is stirred so that the cells are moved about and collide with each other, they clump together, arrange themselves in the form of the blastula they once were, and resume development where they left off! Apparently, these cells are communicating with each other. This process of again becoming a blastula is too complex for it to be due to chance alone.

As an embryo develops, the various tissue layers are twisted and turned so that new relationships are formed between them. Do the cells of these layers communicate and shape each other's development? This question was answered in one of the classic experiments of developmental biology. A region called the *dorsal lip* of the blastula, which was known to develop into the spinal region, was removed from one salamander blastula and grafted onto another in a region that would eventually become part of the belly. The two species of salamander were different colors, so it was possible to see what was happening to the graft

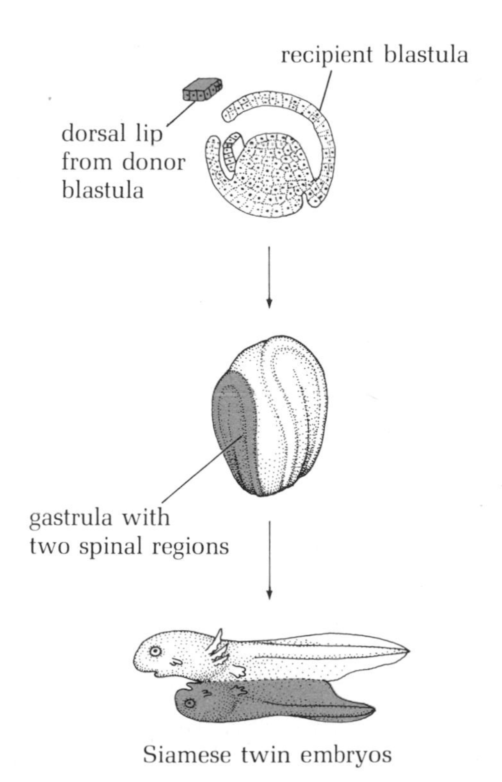

Figure 10–10 The role of the dorsal lip of the blastopore as a developmental organizer. With very little tissue the grafted dorsal lip could direct the development of an entirely new salamander embryo using the tissues of the host embryo.

and to the host. First, the graft developed into a spinal region, as would be expected, but it didn't stop there. It went on to become a second salamander embryo attached to the belly of the host like a Siamese twin. What was particularly remarkable about this second embryo was that, apart from the spinal region itself, all the tissues were derived from the host. The dorsal lip contained the elements needed to organize the cells of the host into a second embryo, an ability that led it to be called the organizer (Figure 10–10).

This experiment and others like it showed that the arrangement of tissues within the embryo does indeed influence the pattern of development. The ability of one tissue to influence another is known as *induction*, and, as the case of the Siamese-twin salamanders shows, it is a crucial aspect of development.

Induction provides a clue to the nature of the points of no return that we indicated on the landscape model. Tissues that work as organizers gain that ability only at some distinct point during development. If the cells of a salamander blastula before the appearance of the dorsal lip were moved about so that what would ultimately be the dorsal lip and what would be the belly region were reversed, then the "new" dorsal lip, not the "old," would become the organizer. But once the cells have become the dorsal lip, their developmental fate, and the fate of the cells they come into contact with, is sealed.

How does the organizer organize? What is the mechanism of induction? The organizer and the tissues it affects do not have to touch each other directly. A paper filter slipped between tissues does not stop induction. This leads us to believe that the inducing substance is a chemical that can pass through the paper. But so far no one has been able to isolate and identify an inducing substance.

A molecular theory of development

We have seen that the interactions between the nucleus and the cytoplasm in the cell and among various types of cells influence development in crucial ways. But what is happening at the basic level of the molecules that make up the cells?

In Chapter 8 we noted that genes are expressed in terms of specific proteins. The code of the chromosomal DNA is carried by mRNA to the

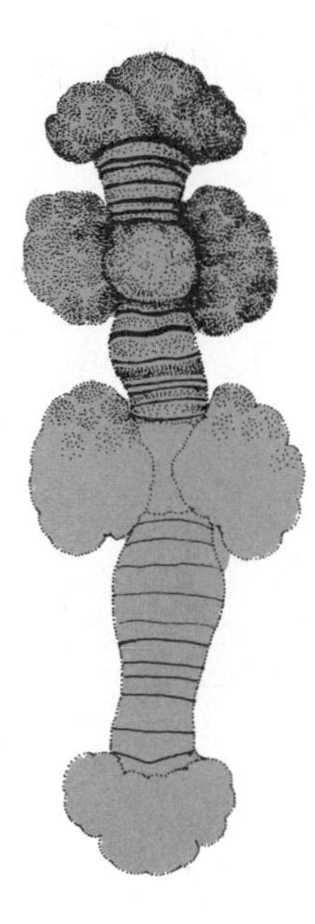

Figure 10–11 Chromosome puffs on the chromosomes of a fruit fly. It has been suggested that these bulges on the surface of the chromosome are actually the site of the transcription of the DNA code onto the mRNA. The photo shows giant chromosomes from the salivary glands of the midge, a common fly.

ribosomes, where the proteins are assembled. In chemical terms, differentiation means that different kinds of cells contain different structural proteins and different enzymes. How can cells contain different proteins if they all include precisely the same information coded into the DNA of their chromosomes? Why don't all cells have exactly the same set of proteins?

A tentative answer comes from the study of chromosomes in the salivary glands of certain insects, among them the fruit fly. These chromosomes have long been important to students of genetics because of their size. In these chromosomes, the DNA has been duplicated some 10,000 times even though the chromosomes have remained whole, making them many times larger than normal. They can be seen in great detail under the microscope. Researchers investigating development in an insect with these giant chromosomes observed what they called *chromosome puffs*, a

How can cells contain different proteins if they all include precisely the same information coded into the DNA of their chromosomes?

sort of bulge on the chromosome proper. Apparently the puffs are unraveled portions of the DNA of the chromosome that are being transcribed into mRNA (Figure 10–11).

This observation has led to a tentative hypothesis about the molecular side of development. You will recall that chromosomes contain proteins as well as DNA. The protein may act as a mask, preventing much of the DNA from being translated into mRNA and subsequently into proteins. Under the appropriate chemical conditions—presumably inducing substances from nearby cells—the proteins unmask portions of the chromosome, permitting the exposed DNA to be translated into mRNA. Unmasking various portions of the chromosomes would produce different sets of proteins, and therefore different cells.

This hypothesis is only tentative, with little experimental proof to back it up. It may well be years before any complete picture is available, simply because research into the chemistry of cells is extremely difficult. But this research may well prove to be of great importance, for if we learn how molecules are assembled and then organized into cells, we'll have come a long way toward answering that most fundamental of all biological questions: What is life?

Can differentiation be reversed?

The landscape model of development assumes that once a cell or a group of cells enters a particular channel it cannot retrace its steps. In other words, differentiation is irreversible. We have also seen that every cell contains all the genetic information of the organism and that the expression of various parts of genes is the basis for differences between cells. Are the unexpressed portions of the gene permanently turned off? Or is it possible to reverse differentiation by promoting the expression of masked genes?

The answer to these questions is a cautious yes. Why the caution? Because the answer depends to some extent on the organism one investigates. In man, as in most mammals, differentiation seems irreversible under most circumstances. If a human's hand is amputated, none of the body's cells will lose their differentiation and take up new roles as the missing bone or skin. The hand is gone forever.

In the salamander, though, amputation produces quite a different result. A cap of skin grows out over the wound. Then the tissues under the cap begin to lose their differentiation, or *dedifferentiate,* into cells like those found in the salamander embryo. These embryolike cells subsequently differentiate into the structures of the missing leg. Within a month of the amputation, an exact replica of the original limb will have grown from the wound. This is true regeneration, not simply outgrowth of the bone. If the bone is removed from the upper part of a salamander's leg, and the leg is then amputated through this boneless region, the limb will still regenerate, and it will contain bone. However, no one knows how this dedifferentiation occurs (Figure 10–12).

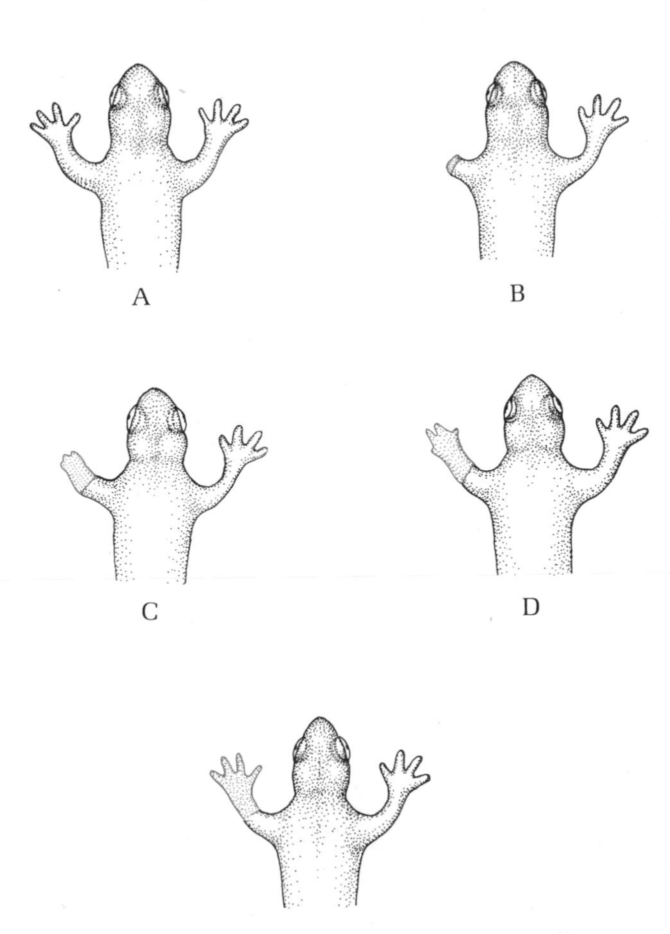

Figure 10–12 Regeneration in the salamander. The ability to completely replace lost tissues actually involves a form of dedifferentiation. Within a short period of time the cells at the point of amputation will "return" to an earlier state of the organism's development and repeat the growth of the limb.

In man, it is possible that dedifferentiation may occur now and again, but the effect of this is considerably less beneficial than it is with the salamander. Researchers investigating cancer—which is not one disease, but many—have noticed that the cells of certain kinds of cancers look very much like undifferentiated cells in the embryo. Could it be that certain kinds of cancer might result from a loss of differentiation? This does seem to be the mechanism of some skin cancers. It is known that the various layers of skin communicate chemically and that these communications maintain the differentiation of the cells and control the rate of cell division. Very possibly, skin cancer results from some disruption of these communications leading to a loss of differentiation and a greatly accelerated rate of cell division. It is a curious footnote on the way science proceeds to realize that research into the way a baby emerges from a simple ball of cells may give us an important tool against cancer.

Why do we grow old?

We first met the hydra in our discussion of budding as a form of asexual reproduction. One of the many remarkable things about the hydra is that it is continually replacing its own cells. A region near the base of the animal's tentacles is the site of constant cell division. As the cells divide, they migrate downward. At the opposite end of the animal, old cells are constantly sloughed off; that is, all the cells are constantly replaced. As a result, the hydra can live as long as it can avoid becoming a meal for a hungry fish. In fact, given the right circumstances, it is potentially immortal.

Not so humans. Our development continues after birth, reaching its culmination in full maturity at about the age of 21. After about 10 prime years, the body begins to deteriorate (Figure 10–13). Unlike the hydra, we go downhill. The process accelerates the closer one gets to 60, and the whole body works far less well than it used to. Hair turns gray and thin;

It is a curious footnote on the way science proceeds to realize that research into the way a baby emerges from a simple ball of cells may give us an important tool against cancer.

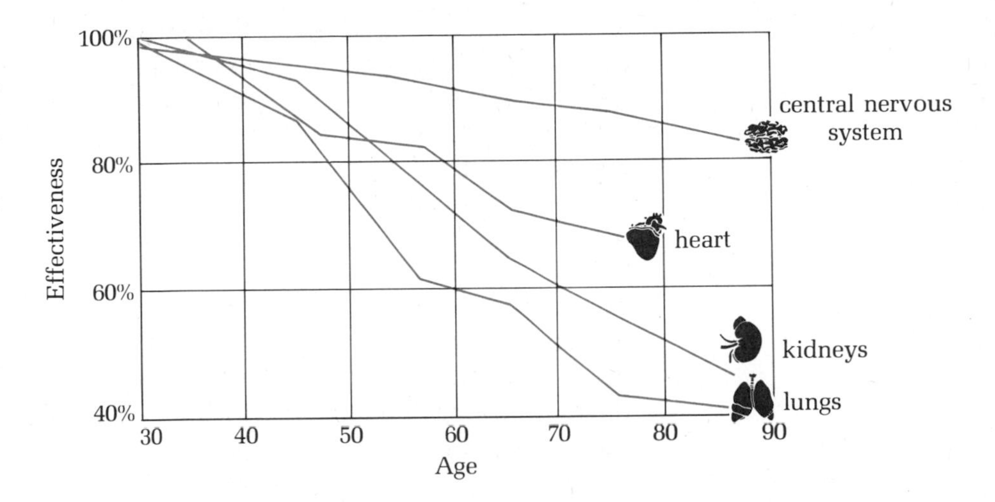

Figure 10-13 The decline of body functions over the course of the human life span.

the eyes lose much of their sharpness, as do the ears; the heart, lungs, and kidneys work much less efficiently; mental capacity often falls off; muscles lose their resiliency (tone) and strength, and once supple joints grow stiff. The body becomes subject to a variety of so-called degenerative diseases, such as cancer and heart disease, which involve or cause the breakdown of tissues and organs. We humans don't have the cellular turnover system of the hydra. But, what causes aging?

As is the case with all developmental questions, it is impossible to give a definitive answer. But we can point to what research into aging, essentially a new field of investigation, has turned up.

A biological calendar

There's a great possibility that the length of life may result from some sort of genetic mechanism. For one thing, there appears to be an upper limit to the life expectancy of any given species, almost as if the time of death were programmed into the individual. Houseflies live less than 2 months, while elephants occasionally live almost a full century. Dogs rarely live over 16 years, and, curiously, the time of death seems to be related to the size of the breed. Saint Bernards and Great Danes usually die between 10 and 12, while an occasional Chihuahua lives to be 20. In the case of man, 115 is the upper limit with rare exceptions. Long life seems to run in families, and it's been shown that identical twins—who have precisely the same genotype—have a greater similarity in their life spans than do fraternal twins—who have different genotypes. This indicates that long life, like blue eyes, might be inherited. But what sort of genetic mechanism could it be?

Scientists can take small samples of tissue and grow them on a nutrient broth to produce what is known as a tissue culture. For years biologists assumed that cells grown in culture were as immortal as the hydra. In the early 1960s, however, a cancer researcher found a limit to the number of times particular kinds of cells would divide. Human embryo cells divided about 50 times, and then stopped, while cells from the human lung doubled only 20 times. Remarkably, cells seem to know how often they have divided. If embryo cells are allowed to divide 20 times, and then put into a deep freeze to stop their division, they divide only 30 more times after they are defrosted.

Apparently there is some kind of a biological timing device built into cells. It might be one of two types. Possibly the upper limit of divisions is coded into the DNA, and when that number is reached, the cell stops dividing. It might also be that dividing ceases because of copying errors during mitotic division. Possibly there is a constant but small percentage of error during the transcription of DNA to mRNA. In the first divisions, these errors, and the mistakes they produce in protein, don't cause too much trouble. But, as the number of errors increases with each cell division, the cell's workings are increasingly impaired, until finally it can no longer function at all. Humans don't live long enough for all their cells to reach the last division, but it could be that aging results from progressive changes in the cells, perhaps because of accumulating errors.

The body's pacemakers

There is also evidence that changes in chemical secretions of the body, particularly hormones, may be a primary cause of aging. For example, estrogen, one of the female sex hormones, declines markedly during menopause, and this decline promotes the wrinkling skin and thinning bones of aging women. Similar changes also occur in the level of hormones manufactured in the brain. These hormones control many key chemical activities of the body (as we shall see in more detail in Chapter 12), and a decline in their level might well be the immediate cause of aging.

A loss of immunity

The blood contains special cells, called white cells, and globular proteins, called antibodies, whose job it is to recognize and destroy "foreign" invaders like bacteria and viruses and possibly even cancer cells. (We'll look at this immunity system in detail in Chapter 12.) Apparently, aging also involves some kind of a breakdown in the immunity system of the blood. The white cells and the antibodies no longer seem able to tell the good group from the bad. Bacteria and viruses invade the body easily and cause disease. Cancers that would have been stopped at an early age spread. And, at the same time, some of the antibodies begin to attack the body's own tissues rather than invaders from the outside. In other words, the body attacks itself and advances its own destruction.

Dealing with old age

Old age is not just a developmental phenomenon of interest to cell biologists. It's also an acute social problem made strangely worse by our present technology. In the rural America of a century ago, old age posed no particular problem. To begin with, there were fewer old people. Diseases like tuberculosis and influenza killed weaker people quickly, preventing many from ever reaching an advanced age. Those people who did live into their 60s and beyond generally could turn to their children for help, and live out their lives performing simple, but productive tasks on the farm. Contemporary medicine has greatly increased the life span, mostly because of antibiotic drugs that save many from disease. In the last century, more than 20 years have been added to the life expectancy of the average American. As a result, the number of old people has increased greatly. At the same time, though, the old have less of a place in society. Generally, a worker's economic usefulness is considered over by the time he is 60, yet he faces perhaps 10 more years of life deprived of

Old age is an acute social problem made strangely worse by our present technology.

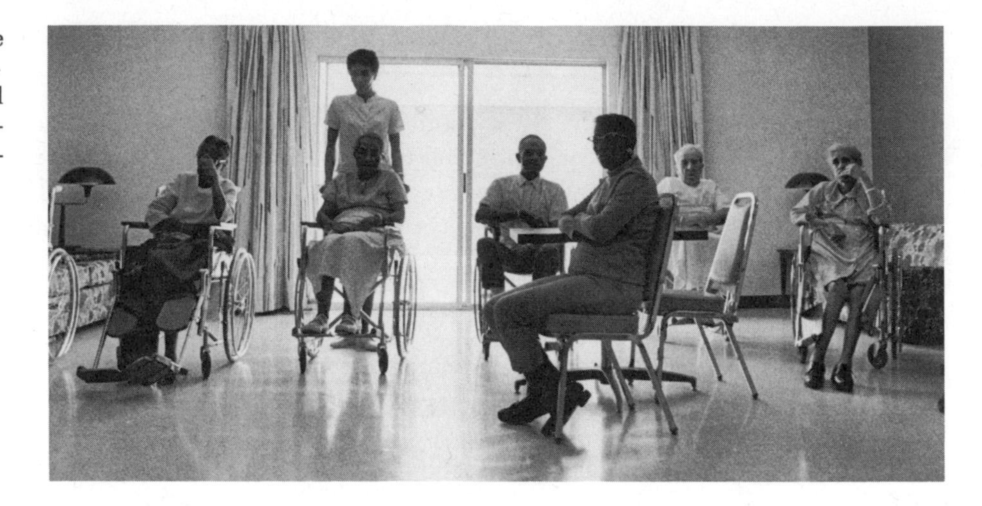

Figure 10–14 The problems of the aging—both physiological and emotional—are receiving long-overdue attention and concern.

enjoyable work, beset with increasing weakness, and possibly isolated in a nursing home (Figure 10–14). That prospect is depressing, and it's not surprising to find that nearly one-third of the suicide victims in this country are people over 60.

Research into development before birth and into aging itself offers some hope for the old—and for ourselves at some time 30 or 40 years into the future. If we can find out those most basic processes governing the growth and aging of our cells, we might be able to mitigate the effects of aging enough to make our last years something to look forward to and not something to dread.

Summary

The development of the individual from the zygote involves growth, differentiation, and morphogenesis. After fertilization, the dividing embryo descends into the uterus and implants in its walls. Some of the embryo cells develop into the supporting membranes. The placenta is the mechanism of exchange between the mother and the embryo, and the amnion encloses a protecting and supporting fluid. The gastrulation of the embryonic disk provides the basic cell layers—ectoderm, mesoderm, and endoderm—for differentiation. At the end of the first 3 months of pregnancy, most of the differentiation is complete. In the remaining 6 months, the fetus gains weight in preparation for birth. At birth, the cervix dilates, and strong contractions of the uterus and vagina push the fetus out of the mother's body.

The basic question of developmental research is: How is the genetic information expressed in a particular cell selected? Experiments involving the transplanting of nuclei show that differentiation entails no permanent change in the nucleus. It is now known that the cytoplasm influences the nucleus. And, as shown by the effects of transplanting the dorsal lip, we know that one tissue can influence the development of another by induction, presumably a transfer of chemicals from one tissue to the other. In molecular terms, differentiation means that only certain parts of the chromosomal DNA are transcribed onto mRNA. Possibly the protein in chromosomes acts as a sort of mask covering unneeded portions of the chromosomes.

In some cases, differentiation can be reversed. One example is the salamander's ability to regenerate limbs. In man, differentiation in reverse may be the source of certain cancers.

Aging appears to be at least partly genetic in nature. It also entails changes in such hormones as estrogen and a breakdown of the body's immunity system.

Questions to consider

1. a. What are the three processes of development?
   b. How does each one differ from the others?
2. a. What function does the placenta serve during pregnancy?
   b. What purposes are served by the yolk sac and the allantois in a chicken egg?
3. a. How many cell layers does gastrulation produce in the embryo?
   b. What are the names of these layers?
   c. In general terms, what structures will each of these layers develop into?
4. What are the major events of birth?
5. How does the landscape model make development more understandable?

6. In a classic experiment, a recently fertilized egg cell from a newt, an animal closely related to the salamander, was constricted with a tiny hair. The half of the zygote containing the nucleus began to divide normally. Later, the experimenter pushed one of the newly formed nuclei into the other half of the egg, and cut the egg in half by pulling the hair tight. Both egg halves—the half with about 20 nuclei and the half with only the one nucleus—developed into normal embryos.
   a. What does this experiment indicate about the role of the nucleus in development?
   b. Had cell division in any way altered the original zygote nucleus?
   c. What is the reason for your answer?
7. In the experiment in which a dorsal lip from one blastula was transplanted onto a host blastula, all of the tissues of the grafted embryo except for the spinal region came from the host.
   a. Why is this a significant fact?
   b. How does this observation contribute to our understanding of induction?
8. All cells contain all the genetic information of the organism, but only part of this information is actually expressed in any given cell. What is the present, tentative explanation for this phenomenon?
9. A salamander can regenerate an amputated limb by dedifferentiation within the stump, leading to the growth of new tissues. Does such a process or one similar to it occur in man?
10. a. Why does there seem to be some genetic basis to aging and the length of life?
   b. There are two ways that this genetic mechanism can be explained. What are they?
   c. What other factors have been identified in aging?

Glossary

afterbirth the placental tissues expelled from the uterus after the birth of a child

allantois the membranes in bird and reptile eggs which receive wastes from the developing embryo *amnion* the sac of fluids which protects and cushions the developing embryo

animal pole the region of the egg cell which includes the nucleus and which contains little yolk

blastula an early form of embryonic development composed of a sphere of a single layer of undifferentiated cells

chromosome puff a distinct bulge on the surface of a chromosome

dedifferentiation a loss of differentiation in which cells become less specialized

differentiation the development of specialized cells and tissues within the embryo

dorsal lip the region of the blastula which gives rise to the nervous system of the embryo and which appears to organize the differentiation of the blastula

ectoderm the outer layer of the embryo after gastrulation which gives rise to the brain and skin

ectopic pregnancy the implantation of the zygote outside of the uterus, usually in the Fallopian tubes

embryo the multicellular form of a developing organism resulting from the cell division of the zygote

embryonic disk the layers of cells in the early embryo which give rise to the new organism

endoderm the inner layer of the embryo after gastrulation which gives rise to the linings of the internal cavities of the new organism

fetus the most advanced stage of development in which the embryo has taken on the structure and appearance of the new organism

fraternal twin one of two children born of the same pregnancy which developed from separate zygotes

gastrulation the folding of the embryo to result in the formation of three layers of cells

identical twin one of two children born of the same pregnancy which developed from one zygote which divided to form two separate cells

induction the process in which one group of cells influences the differentiation of another group of cells in the embryo *mesoderm* the middle layer of the embryo after gastrulation which gives rise to the internal organs of the organism

morphogenesis the process by which tissues organize and develop to form complete, mature organs and structures

placenta the layers of membranes that anchor the implanted embryo in the uterus and that supply it with nutrients

quickening the first indications of movement and life in the developing fetus

umbilical cord the ropelike connection between the placenta and the embryo

vegetal pole the region of the egg cell that includes most of the yolk

yolk sac the portion of the egg cell that stores food for the developing embryo

11 Moving

Movement is characteristic of each and every form of life. To survive, an organism must be able either to move or to make use of the movements of the environment surrounding it. A motionless organism living in a motionless world would soon smother in its own wastes or starve once it consumed the nearby food supply. Bird and reptile embryos, which develop within hard-shelled eggs, deal with this problem by means of the yolk sac and the allantois, but they can stay in the egg only temporarily. Once the allantois is full and the yolk gone, the animal must emerge or die. Stationary organisms like flowering plants and corals depend to some extent on air and water currents to bring food and carry off wastes, but even these stationary organisms have ways of moving in response to their environment. Organisms that aren't anchored to any one spot move about to seek food, escape enemies, and locate partners for reproduction.

In this chapter, we'll take a look at the many ways living things have of moving in their many environments, and we'll also briefly examine man's technology of motion.

A motionless organism living in a motionless world would soon smother in its own wastes or starve once it consumed the nearby food supply.

Movement of the single cells

Some one-celled organisms move by shifting their weight within their cell membranes, almost like a man in a sack. The amoeba moves in such a specialized fashion that its style of locomotion is called *amoeboid motion*. The cell membrane of the amoeba pushes out in temporary projections called *pseudopodia* (literally, "false feet"). The clear, semiliquid cytoplasm then flows into the pseudopodia, moving the organism along. Other one-celled organisms use fibers capable of contracting. They attach one end of these contractile fibers to some fixed object that acts like an anchor. When the fibers contract, the organism is held securely to the anchor, but when the fiber relaxes, the organism can drift about in its watery environment. We'll see soon that in multicellular animals, contractile fibers similar to these are basic to the mechanisms of motion.

A great number of one-celled organisms use external, hairlike appendages to move themselves about. Figure 11–1 shows organisms with *flagella* and *cilia*. Flagella and cilia are the same in internal structure but different in size and number. Flagella are long, and they generally appear singly or in pairs. Cilia, on the other hand, are much shorter and come in such great number that a ciliated microorganism almost seems to be covered with fur. Flagella and cilia move the organism by beating against the water in much the same way that oars pull a boat along.

Flagella and cilia are found not only among microorganisms, but also in certain specialized cells of many-celled animals. Compare, for exam-

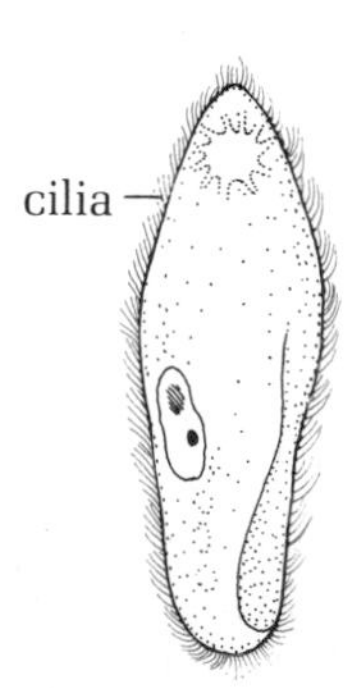

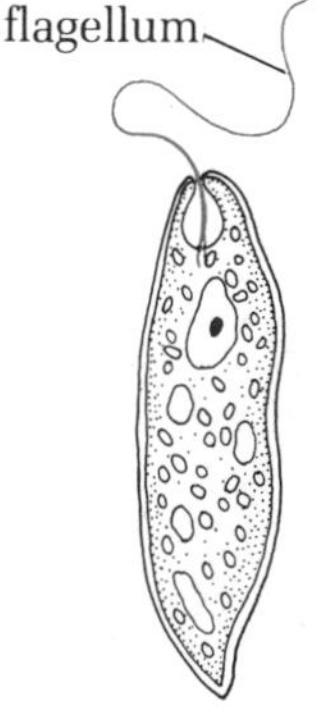

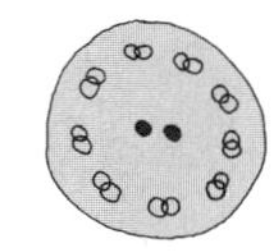

Figure 11–1 Cilia and flagella. The *Paramecium* moves by the action of its many cilia. The *Euglena*, on the other hand, is propelled by the whiplike action of its single flagellum. Both cilia and flagella have the same internal structure.

ple, Figures 9–14 and 6–7. The tail that powers a sperm cell on its journey to seek out the egg is a kind of flagellum. The ciliated cells lining the trachea don't themselves move, but the beating of their cilia pushes foreign matter trapped in mucus out of the respiratory tract. Cilia are also found in the human reproductive tract, in the Fallopian tubes, for example. Can you explain their presence there?

The plants: moving by growing

Rooted into the ground as they are, land plants hardly seem able to move. Plants, like the embryo in the shelled egg, are using up the nutrients and water in the soil about them. How do they find new sources of these needed items? By growth of the roots. All the time that a plant is alive, the roots grow, seeking out water and nutrients suspended in the soil. The roots of an apple tree, for example, can grow one-eighth to three-eighths of an inch each day under ideal conditions, and a rye plant left to grow for 4 months has somewhere in the vicinity of 7,000 miles of roots! In plants, growth accomplishes many of the functions served by movement in the animals.

This general statement applies as well to the typical phenomenon in which the tips and leaves of a plant seem to move in response to light. Figure 11–2 shows what happens when a plant's sole light source is moved from one side to the other. The light, required for the plant's photosynthesis, acts as a stimulus to the plant, and the plant's leaves seem to bend to find it. Such a directed movement in response to a stimulus is known as a *tropism*, and, in the case of light, it's called a phototropism. Phototropism is not true movement, but growth. Growing plants produce a class of hormones known as *auxins* that, among other things, promote growth and elongation of the shoot (the plant's new growth). Auxins are produced in the tip of the plant and flow downward toward the base of the stem. When a plant receives light equally from all sides, auxins are found in all shoot cells in equal amounts, and the shoot

A rye plant left to grow for 4 months has somewhere in the vicinity of 7,000 miles of roots!

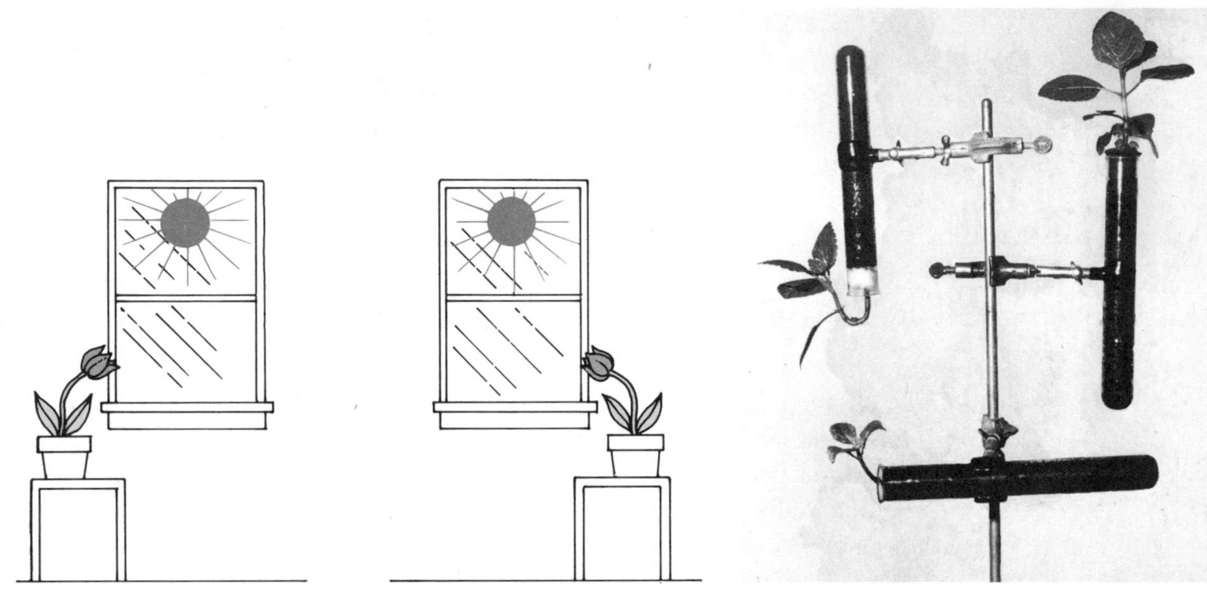

Figure 11–2 A. An example of a positive phototropism. A plant will continually maintain an orientation toward its light source. B. Plants show other types of tropisms. Geotropism, the response to gravity, is shown here. Plants tend to know up from down and usually begin their growth against the pull of gravity.

grows straight and tall. If the light strikes the plant mainly or solely from one side, the auxins are destroyed on the light side. Under the influence of higher auxins, the shoot cells on the dark side grow longer than those on the light side, bending the whole stem in the direction of the light. If the light source is shifted from one side to the other, dark side and light side are reversed, as is the flow of auxins. The formerly shorter cells grow longer, bending the stem toward the new light source.

There are a few plants capable of what is true movement. A good example is the so-called sensitive plant, technically named *Mimosa pudica*. If you touch a leaf of this mimosa, the leaf and its neighbors fold and droop almost immediately (Figure 11–3). How does it react so quickly? The reaction to touch is not caused by a change in hormone flow, but by water leaving the cells of the leaves and moving into other parts of the plant. Plant cells are rigid because they are filled with water. If the water drains out of the cell, the cell becomes limp, much like an empty water balloon.

Figure 11–3 The sensitive plant, *Mimosa pudica*. The response of these delicate leaves to touch is almost instantaneous, and involves a transfer of water away from the site of contact.

A water-flow mechanism similar to the mimosa's powers the workings of the Venus's-flytrap. The flytrap is one of a small group of carnivorous plants, ones that consume insects as a source of nitrogen. Most of these plants catch their prey with either a gummy pad that works like a piece of flypaper or with a no-exit trap that the insect unwittingly enters and then cannot escape. The Venus's-flytrap, though, works just like a bear trap. Each trap is actually a pair of highly specialized leaves supplied with trigger hairs. When an insect brushes against these hairs, cells in the hinge of the trap suddenly lose their water—and therefore their rigidity—and the trap snaps shut. Then, specialized glands secrete juices that digest the captured insect.

The water bag of the sea anemone

Unlike plants, many multicelled animals possess special tissues whose sole job is supplying the motor force behind movement. These

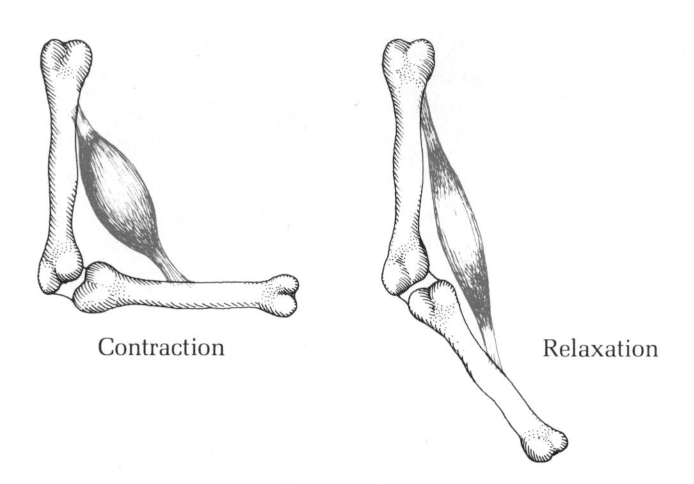

Figure 11–4 The contraction of a human muscle. The muscle can either contract and thicken or relax and lengthen.

tissues are the *muscles*. Essentially, a muscle is a bundle of fibrous cells capable of contracting. In the relaxed state, a muscle is thin, but it can contract into a short and thick mass, as illustrated in Figure 11–4. Muscles have no ability to lengthen, only to contract. The range of an animal's movements is dictated by the arrangement of muscles in its body.

Take, for example, the sea anemone, which spends most of its life attached to a rock and waiting for food to come by. Structurally, the anemone is quite simple. Its hollow, cylindrical body is usually filled with water. A mouth fringed with stinging tentacles provides an entryway into the body cavity for both water and food. At the other end of the anemone is a closed disk anchored securely to the sea bottom. The body wall of the anemone contains two sets of specialized cells that, although not the same as our muscles, serve the same purpose. The long muscles run in bundles through the length of the anemone's body. The circular muscles form a thin, continuous sheet around the anemone, except for a couple of thicker rings about the mouth that open and close it.

As long as the anemone keeps its mouth shut and its body cavity full of water, it can move in a great variety of ways by contracting some muscles and leaving others relaxed. Sometimes the anemone is tall and thin, but at others it is short and squat. When it's tall, the long muscles are all relaxed, and the circular muscles are contracted. By contracting some long muscles and relaxing others, the anemone can bend and twist, say, to catch one of the small fish it feeds on. The prey is taken into the central cavity of the body, and the indigestible remains are ejected by means of pulsing, peristaltic rhythms like those of the human digestive tract (see Chapter 7).

If the sea anemone is strongly disturbed, say, by a large fish or a scuba diver, the animal squeezes all the water out of its body cavity and collapses into a wrinkled "pancake" on the rock. Once the anemone has pulled this escape trick, it is helpless for a while. The muscles are of no use until the body cavity is again filled with water, and the animal has to wait for the slow, steady beating of ciliated cells near its mouth to sweep in enough water before it can again move normally.

Variations on a theme: muscles and skeletons

Obviously, a sea anemone is considerably less complex than a human being, and it is capable of far fewer kinds of motion or movement than we are. However, the basic principles governing the movement of all animal forms are shown in the example of the anemone. Two are particularly important.

The first is that muscles work in antagonistic pairs. That is, one muscle contracts in one direction or plane, while its mate contracts in the opposite direction. In the anemone, the long muscles contract vertically and the circular muscles horizontally. However, such muscle pairs don't pull against each other. When one of the pair contracts, the other relaxes.

The second principle exemplified by the anemone is that the muscles have to contract against something to produce movement in the organism. All animals have some sort of a rigid frame—loosely speaking, a *skeleton*—that does the double task of determining the structure of the animal's body and of giving the muscles something to contract against. In the case of the anemone, that skeleton is the water in the body cavity. Water contained within a sealed container can be compressed—forced into a smaller volume—only with great force, far greater than the anemone can exert, so the anemone's bag of water remains rigid enough to do the job of a skeleton. When the anemone expels its water and shrinks into that inconspicuous pancake, it can no longer move because it has very literally thrown away its skeleton. A variety of simple organisms use such a water bag, or *hydrostatic skeleton* (Figure 11–5).

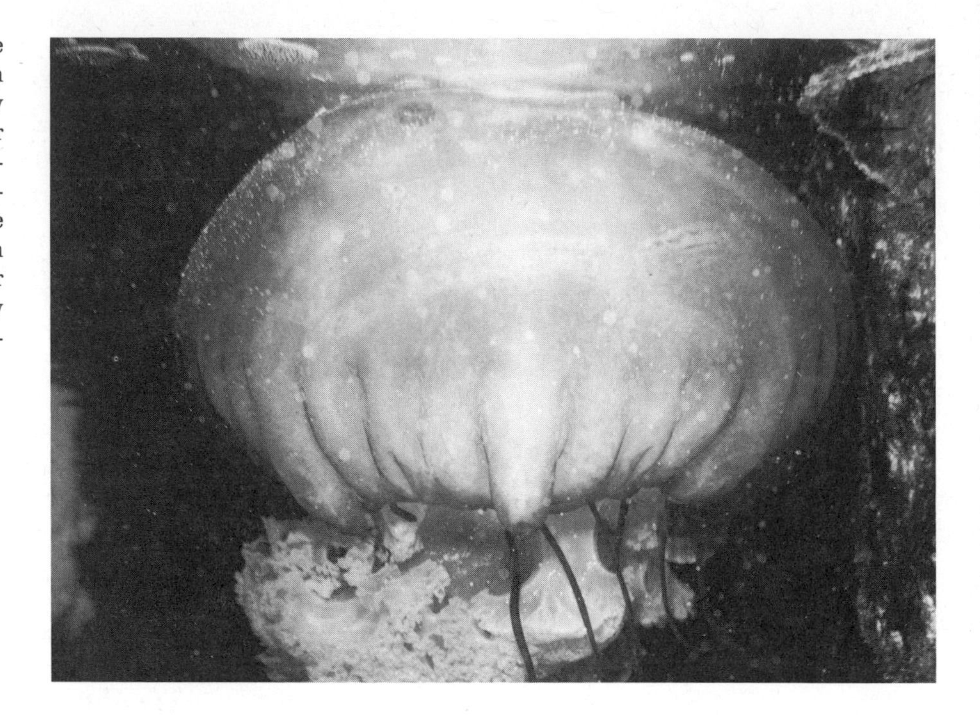

Figure 11–5 The hydrostatic skeleton serves as a temporary structural support for many aquatic organisms, such as the jellyfish shown here. The degree of compression of the retained water determines the rigidity of the skeletal structure.

But more complex creatures are provided with permanent and rigid frames. What are they, and what effect do they have on the ways that animals move?

Skeletons on the outside

Accustomed as we humans are to thinking of skeletons as being on the inside of the body, it is a bit surprising to realize that the most numerous group of organisms on the earth have exterior skeletons, or *exoskeletons*. This group is the *arthropods*, a classification that includes insects, spiders, mites, and such aquatic forms as lobsters, crabs, crayfish, and shrimp. Exoskeletons also appear in other forms besides the arthropods, such as the *mollusks*—clams, mussels, scallops, snails, and abalones. But, among the arthropods, particularly the insects, the exoskeleton is found in its most advanced form (Figure 11–6).

Figure 11–7 diagrams the characteristic movement of a crayfish swimming backward with powerful strokes of its tail. Look closely at the out-

Figure 11–6 Exoskeletons. An important structural form common to the beetle, the dragonfly, and the fiddler crab is the exoskeleton. This rigid skeleton must be shed at specific stages in the organism's growth and replaced with a larger one. The jointed limbs of the arthropods permit movement within what would otherwise be an inflexible structure.

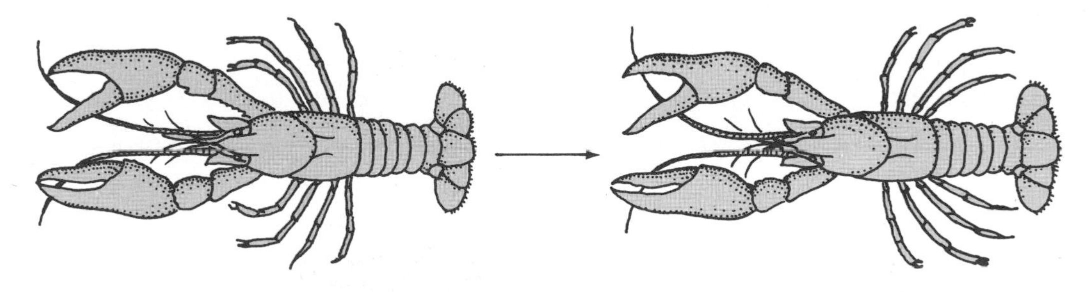

Figure 11–7 The crayfish is able to use its jointed swimmerets and its tail to move quickly through shallow water.

side detail of the tail and at its interior structure. The tail is covered with a series of overlapping, jointed plates. The muscles that power the tail span a joint, and the ends of the muscles are attached to the exoskeleton on either side of the joint. Contraction of one set of muscles pulls the tail from its normal horizontal position to a point under the abdomen, sending the crayfish off in a spurt. These muscles then relax, and the other member of the pair contracts, returning the tail to its original position where it is ready for another swimming stroke. All the muscles in a crayfish or a grasshopper or any other arthropod work in this basic way: spanning flexible and bendable areas of the skeleton in pairs.

The hard, armorlike exoskeleton of an arthropod doesn't grow with the animal. Instead, the animal sheds, or *molts*, its old exoskeleton and grows a new, larger one. Many insects have evolved intricate life cycles that entail complete changes of form (or *metamorphosis*), an adaptation made possible by molting. Flies, ants, wasps, bees, beetles, weevils, moths, and butterflies all pass through four stages: egg, larva, pupa, and adult. When the egg hatches, the *larva* emerges, a wormlike creature with small, inconspicuous legs and no wings or eyes. Larvae eat prodigiously and grow very fast. After several molts, the larva enters into the resting stage of a *pupa;* in some species, the pupa is protected by a special covering called a *cocoon*. While the organism rests, great physical changes

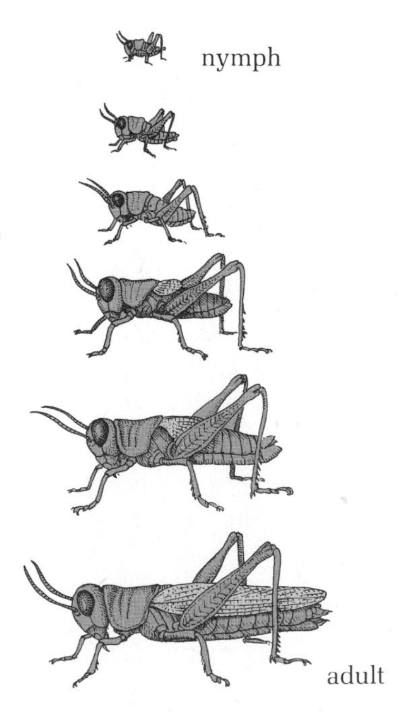

Figure 11–8 The development of the grasshopper. In this simple pattern, the growing insect sheds its exoskeleton at several points, each time emerging slightly larger and slightly more similar to the final adult form.

occur. Certain larval organs break down and are absorbed into the insect's body, while such adult organ structures as wings and eyes develop. When these changes are complete, sometimes after as long as several years, the adult emerges. However, not all insects go through the larval and pupal stages. The grasshopper has a simpler life cycle (Figure 11–8). The immature grasshopper looks like a miniature adult, and is called a *nymph*. Every time the nymph sheds its exoskeleton, it appears larger and more complex. After a specific number of molts, the grasshopper reaches full adulthood.

Skeletons on the inside

Most fish and all amphibians, reptiles, birds, and mammals have interior skeletons, or *endoskeletons*, made of bone. Since bones feel hard to the touch, we often tend to think of them as something more like the steel frame of a building than as a kind of living tissue. Figure 11–9 shows a cross section of a human bone. The exterior of the bone is covered with a membrane that contains many small vessels carrying blood to and from

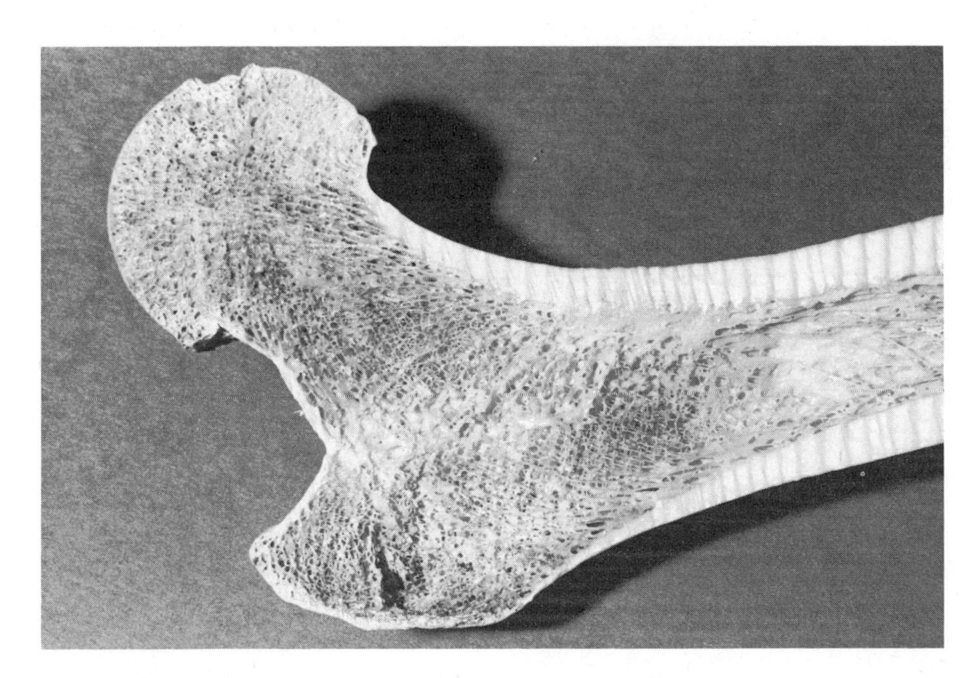

Figure 11–9 A cross section of human bone. This strong skeletal element is composed of a durable calcium outer shell and a fine supporting web of calcium salts.

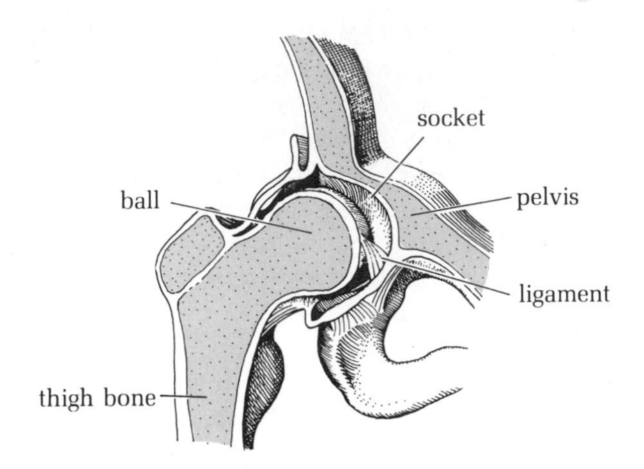

Figure 11–10 The human hip joint, an example of a ball-and-socket joint.

the living cells of the bone. The hard part of the bone is a thin shell of calcium salts. Within this hard shell are the living bone cells and a supporting web of calcium salts secreted by these cells. The central section of the bone is a hollow cylinder. Because of certain laws of physics, a cylinder or tube fashioned from a given amount of material is stronger than a solid rod made from the same amount of material. Also, the tubular structure cuts down the weight of the bone, allowing the animal to move it with a smaller expenditure of energy. But how can a thing so rigid be moved?

As was true with exoskeletons, the answer is the joints. Figure 11–10 shows the structure of the human hip joint joining the long bone of the thigh to the pelvis. Because we stand upright, the leg bone has to bear considerable weight at the same time that it moves through a wide variety of positions. The hip joint is a marvelously simple answer to a difficult problem. The upper end of the thigh bone is a smooth ball, while the receiving section of the pelvis is a socket the ball fits into. The two bones don't actually touch. If they did, friction would wear them away. Instead, a special capsule filled with a lubricating fluid is positioned between them. Strong elastic bands called *ligaments* hold the joint together but allow it to move freely in practically every direction.

The muscles in an animal with an exoskeleton work by spanning a flexible portion of the outer shell. The muscles of an endoskeleton work similarly, by spanning the moveable area of a joint. Figure 11–11 shows the muscles and bones of the upper arm. The two muscles of the antago-

Figure 11–11 A. The alignment of the muscles and bones of the human arm. B. The muscle-bone alignment of the human arm can be compared with the structure of a steam shovel.

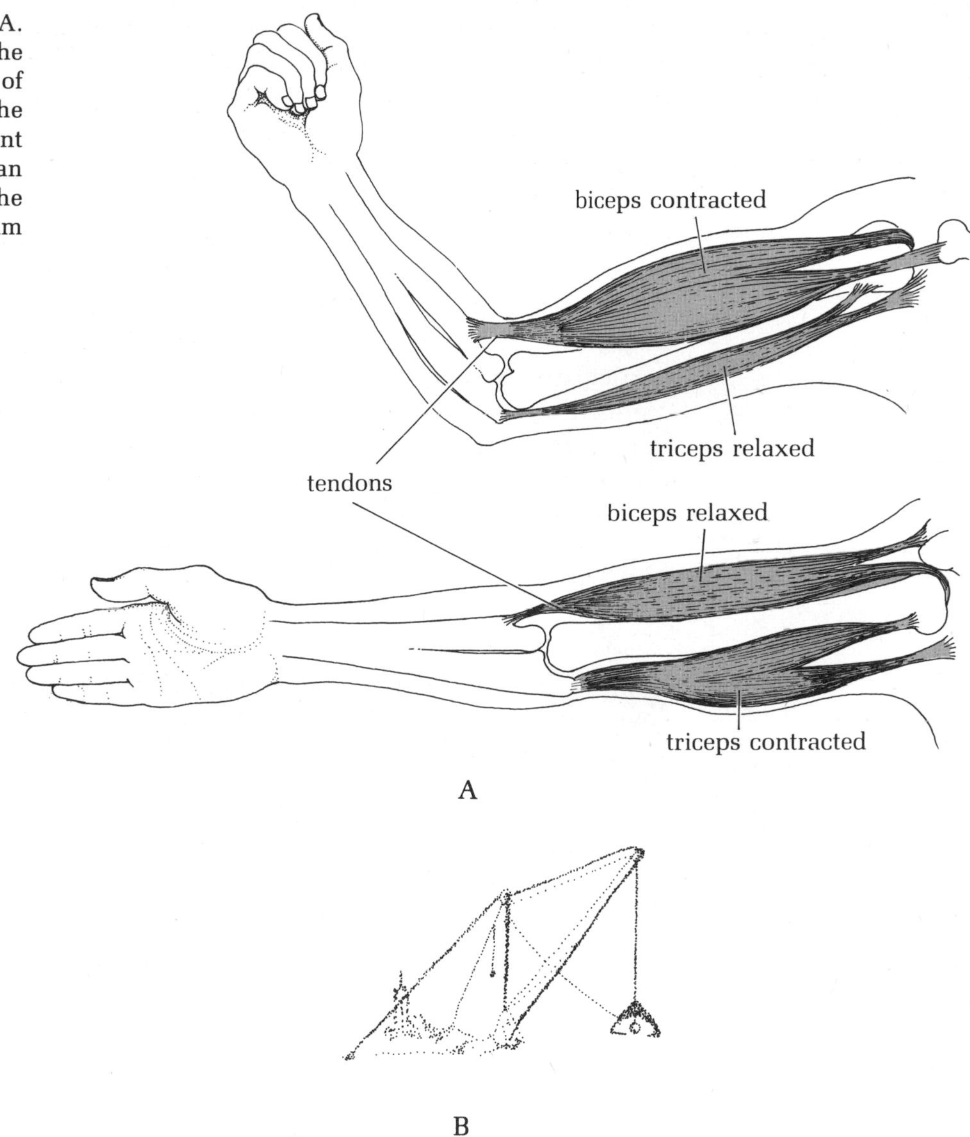

nistic pair are found one on each side of the arm: the biceps in the front, and the triceps in the back. The ends of both muscles taper into *tendons*, tough bands of fibrous tissue, that are anchored to the bone. One tendon is attached to the bone of the upper arm, while the other spans the elbow and connects to one or the other of the two bones of the forearm. When the triceps relaxes and the biceps contracts—shortening into the hard bulge of "make a muscle"—then the forearm and the hand are drawn up toward the shoulder. Reversing the process—relaxing the biceps and contracting the triceps—returns the forearm to its original position. In other words, the human arm uses the same system of levers as the steam shovel.

An outgrown exoskeleton can be thrown away for a new one, but an endoskeleton has to grow with the organism and provide working connections for the muscles all the while. In the embryo, some of the bones, particularly those of the limbs, appear first as *cartilage*, a tough, elastic material much like bone but lacking its calcium deposits. You can feel the resilience and toughness of cartilage in the tip of your nose or in the ear. As the animal grows after birth, bone cells penetrate the cartilage and transform it, so that there is progressively less cartilage and more bone.

Some bones are preformed not by cartilage but by a tough membrane. This happens to the human skull, which is not one bone but several. The

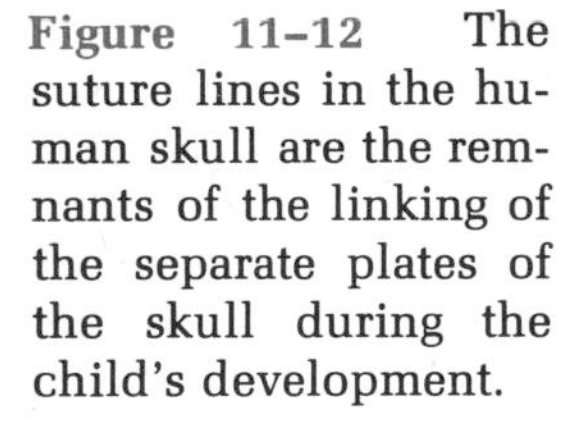

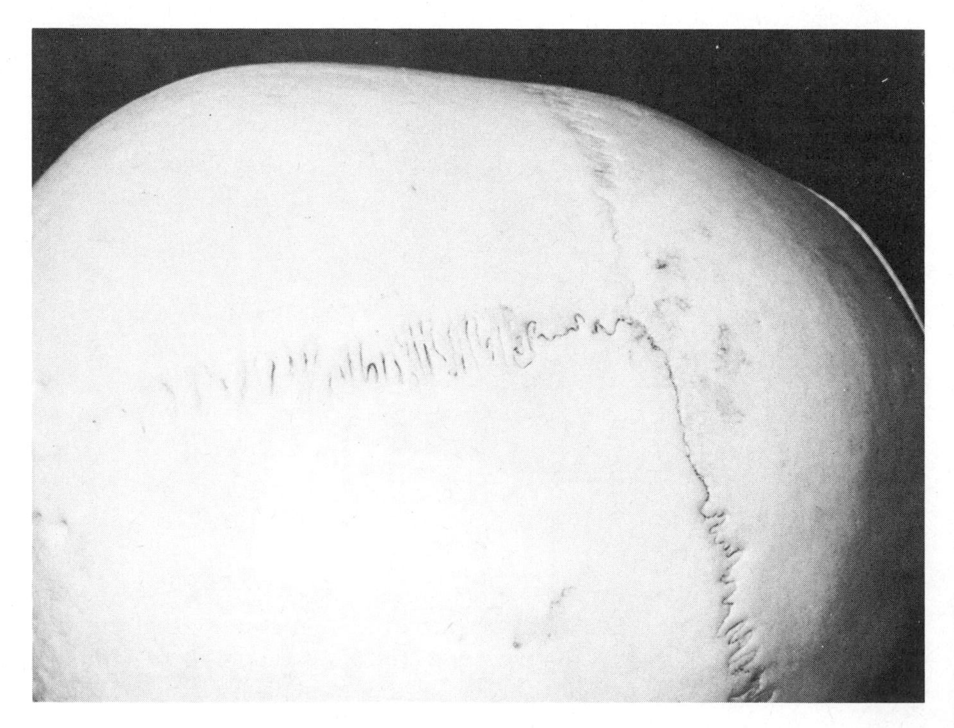

Figure 11–12 The suture lines in the human skull are the remnants of the linking of the separate plates of the skull during the child's development.

brain grows very rapidly after birth, and the skull must provide room for this growth at the same time that it provides protection for the vital nerve centers in the brain. At birth, the infant's brain is surrounded by bony plates spaced some distance apart and by a tough coating of skin and membrane filling the area between the plates. As the child develops, the plates grow until they meet in the jagged, interlocking pattern of the suture line shown in Figure 11–12.

The ways vertebrates move

Obviously, movement is not simply a matter of flexing the biceps. It's one thing to understand how antagonistic muscle pairs move a joint back and forth, and quite another to understand the complexity of movements like swimming, running, or flying. How do vertebrates get around?

The basic swimming stroke

As odd as it may sound at first, the swimming motion of a fish is the basic pattern of movement in vertebrates. Variations on the fish's swimming stroke have provided the mode of movement for other vertebrates.

In most fish, the backbone, or *vertebral column*, is the largest part of the skeleton. The fish's muscles are arranged in blocks down its sides and attached to the backbone or to the ribs (Figure 11–13a). The corresponding muscles on opposite sides of the body act as antagonistic pairs, so that when a certain part on one side of the body contracts, the opposite side relaxes. Muscle contractions begin at the head on one side of the body and pass down to the tail in a wave. As soon as the first wave has traveled the length of the fish, a second wave begins on the opposite side and also runs from head to tail. These alternating waves of contraction and relaxation swish the fish's tail back and forth and push the fish forward (Figure 11–13b).

A fish that spends most of its time swimming free faces the same problems as an aviator—avoiding spins and nose dives, remembering which way is up, and keeping track of water currents while moving with them or against them. The membranous, paddlelike organs called *fins* act

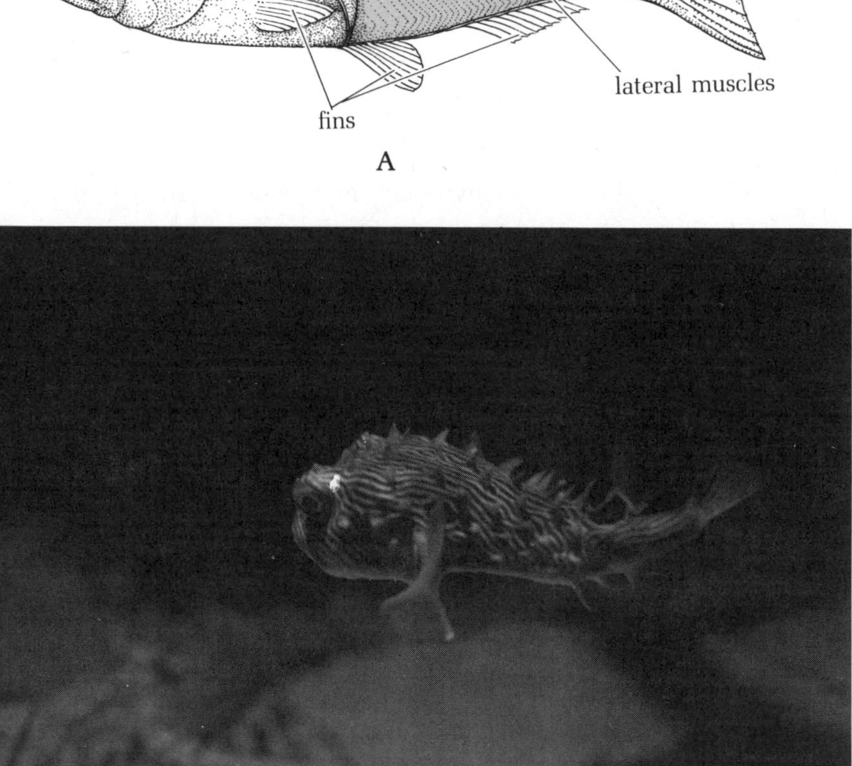

Figure 11–13 A. The muscles and fins of the fish are designed to assure swift, streamlined motion through water. B. The movement of the fish's tail propels it forward. This motion is controlled by the muscles along the side of the fish's body, which alternatively contract and expand. C. The marine blowfish, an example of a fish with a specialized fin structure.

Figure 11–14 Rainbow trout in a shallow mountain stream.

as fine adjustments to the powerful strokes of the tail. By varying the position of the fins or the speed of their movement, the fish can rise, drop, or turn to either side (Figure 11–13c). A fish can even move in order to hold still. Figure 11–14 shows a trout holding a steady position in the swift current of a mountain stream. The fish keeps itself headed straight upstream with its small fins, and matches the power of its swimming stroke to the speed of the current. As a result of this delicate balancing, the fish remains in the same place.

Some fish spend most of their lives on the bottom, and they face different problems. The bottom-dweller is far less subject to the power of

Many fish have adapted to the special structural needs of life at the bottom of bodies of water. The catfish shown here is a common scavenger who can seek his food from sunken refuse and dead animal matter.

A bottom-dwelling fish doesn't have too much of a problem remaining upright because of the flotation provided by the water, but to a land animal stability is crucial.

the current, but living on the solid surface of the bottom means that the fish must be able to stand up without falling over and that it has to pick itself up in order to move. The channel catfish is a bottom-dwelling species common to warm freshwater streams and lakes of the United States. The catfish actually supports itself on the tips of the four fins on its underside. The catfish is a powerful swimmer as well as a bottom feeder, and its fins do double duty as swimming stabilizers and bottom supports.

Moving on four legs

In Chapter 4, we saw that the first land animals evolved from lungfish that came up onto the land for extended periods of time. These lungfish had the long, strong fins of bottom-dwellers, and they used these fins to push themselves about on the land just as if it were the ocean floor. Time and natural selection produced the first simple four-legged animals.

Modern salamanders, among other species, have retained this basic structure. Figure 11–15 shows how the salamander's legs jut from its body, almost like jointed fins. A bottom-dwelling fish doesn't have too

Figure 11–15 The salamander has a low center of gravity; that is, most of its weight is close to the ground. Combined with the added advantage of four legs, this provides for nimble and quick movement.

much of a problem remaining upright because of the flotation provided by the water, but to a land animal stability is crucial. The salamander's legs provide a very stable base of support. It takes considerable force to overturn a rock slab lying flat on the ground, but the same slab standing on its side can be toppled easily. The reason is that the first slab has a lower center of gravity than the second one. The salamander's legs keep the animal barely clear of the ground, giving the animal a low center of gravity.

A salamander moves on land much as a fish moves in water. The short limbs have little muscle, and most of their movement comes from the body muscles. Salamanders and such aquatic reptiles as crocodiles and alligators walk and swim with the same set of muscles. When swimming, they keep their legs tucked up close to their bodies. When they are walking, their legs provide stable anchors that keep them moving forward.

Short legs jutting out from the side of the body work well for an animal that spends much of its time in the water, but they are comparatively inefficient on land. A crocodile, for example, can reach relatively great speed on land, but only over short distances. Getting increased speed over great distances means giving up some stability. The legs are swung under the body and bent so that they act as shock absorbers. This structure raises the body off the ground and makes the animal considerably less stable, but instead of simply being able to scurry quickly, it can now run and jump.

Changes in the bone structure and musculature of the limbs of this modified four-leg plan allow for a wide range of movements. What are the possibilities of four-limbed movement, and what structures allow them to be used?

Stability in most land animals is directly related to the ability to support the organism's center of gravity. It takes considerable force to overturn a rock slab lying flat on the ground, but the same slab standing on its side can be toppled easily.

Flat feet and tiptoe walkers

The legs of a bear are much like ours. The feet are broad, flat surfaces in contact with the ground, and the ankle joint is placed low. Like man, a bear can stand upright by putting its center of gravity over its hind feet. The bear's ability to stand upright frees its forelegs to reach into a beehive for honey or to threaten an enemy. The forelegs are heavily muscled, and the bear can perform a number of involved and complex movements with them. Back down on all fours, the bear is stable because of its short legs and flat feet. However, the bear's structure sacrifices speed for complexity of movement and stability. A bear can run at top speed only for a short distance. The limbs are too heavy to be moved at high speed for a long period of time, and the shortness of the bear's limbs keeps its strides short. In fact, because a bear's hind legs are shorter than its forelegs, it can go faster uphill than down.

Greater speed is gained by the legs characteristic of dogs and their relatives, the wolves, foxes, and jackals. The driving muscles that power the dog's strides are located in the upper ends of the legs. The remainder of the limb is a thin, light stilt. Such inboard muscles save energy because the animal isn't burdened with swinging the dead weight of large muscles located at the far end of the limb. The dog's ankle is located high off the ground, increasing the overall length of the leg. This structure increases the dog's speed because of the general principle of physics that the longer the arm of a lever, the faster the movement of its tip (Figure 11–16). Dogs are also equipped with a flexible spin that bends with the hind legs, giving more power and greater length to the stride.

A variation of the dog's limbs is found in the large running animals like horses and antelope. These animals use very little spine motion in their running, and make up the difference with legs that are proportionately longer than a dog's or a wolf's. The lever length of a horse's leg is

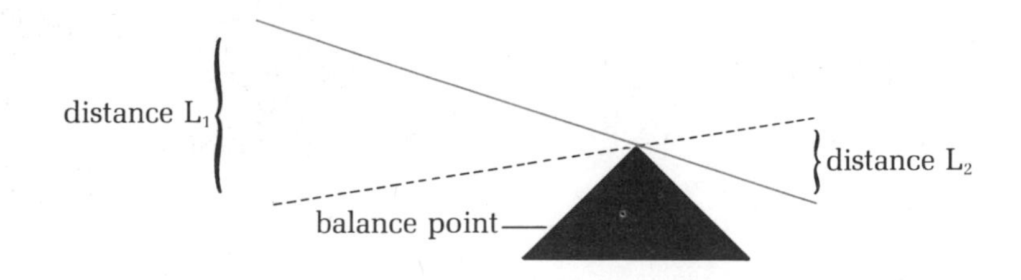

Figure 11–16 Distance and speed in a lever. If you were to hop onto the short end of this seesaw, the short end would drop in the same amount of time that it would take for the longer arm to shoot upward through a much longer distance. Thus, the longer arm is moving faster than the shorter arm.

Figure 11–17 Athletic skill and specialized body types. Jim Ryun, the famous distance runner, has the long, thin body structure essential for his type of athletic skill. Al Feuerbach, a former member of the United States Olympic team, has a heavy, muscular frame which contributes to his skill as a shot putter.

increased by the fact that it actually runs on the tip of only one toe. The other toes, reduced to mere remnants, contribute to the length of the limb below the knee.

As good as long, thin legs are for running, they are useful for little else. Dogs and horses can't do much with their legs other than run. The bear versus dog comparison reveals a general principle about vertebrate limbs. Heavily muscled and shorter limbs are good for complex movements and for strength, but are of limited value in running. Long, thin limbs are good for running, but they trade power for speed. This general rule helps to explain not only differences between two species but also differences within one species. For example, track and field athletes represent a variety of body types, suited to various events. A shot-putter is built more like a bear than an antelope. The power and strength needed to push a 16-pound ball over 60 feet requires heavily muscled arms and legs. Distance runners are quite another breed. They tend to have long, thin legs, which are low in weight but long in stride (Figure 11–17). In running, the heavy muscles of a shot-putter are a liability.

Two legs from four

Man was by no means the first creature to stand upright on the hind feet, thus freeing the forelegs for purposes other than locomotion. Such a body plan was found among predatory dinosaurs. The dinosaur's body was balanced on the pivot of the hip, so that the elongated tail provided a stabilizing counterweight to the outsized head. The dinosaur's forelimbs held the prey tight, while the huge and powerful jaws tore it to pieces.

Some smaller animals have small forelimbs and highly developed hind limbs built for hopping. This particular form of movement is especially effective on loose sand, and many hopping mammals are desert dwellers. Like the dinosaur, desert rats and kangaroos use their tails as counterbalances. The forelimbs are used much like a pair of hands. Hopping animals that lack a heavy tail, like rabbits and frogs, catch themselves on their forelimbs at the end of each jump. Lacking the counterbalance, they cannot become totally two-legged.

Primates moved toward two-legged walking when they began using their forelimbs to swing through the trees. With evolution, the hand became more and more adept at grasping, and was increasingly able to perform a wide range of movements. Primates that have moved out of the trees to the ground, like the plains chimpanzee, the mountain gorilla, and the baboon, generally walk on all fours. However, they support themselves on the knuckles of the hand, not the palm, and the hands retain their range of movement. These ground-level primates use upright posture at least part of the time, but, with man, standing and walking on two legs became a permanent feature.

In the apes, the structure of the hip and pelvic bones puts the weight of the body in front of the legs. Human hip bones are bent under, putting the weight of the body directly atop the legs and affording the point of balance needed for walking on two legs. The human foot is a marvelous evolutionary adaptation to walking on two legs alone. The big toe of the foot of a great ape is opposed to the other toes, so that it functions much like a thumb. This makes the foot a good instrument for grasping and holding onto branches. In humans, the joints of all the toes are brought into a straight line, forming the ball of the foot. We are able to walk forward by shifting our weight from the heel to the ball in an unbroken stride. The ape can't do this, and he must shuffle much like a human trying to walk on

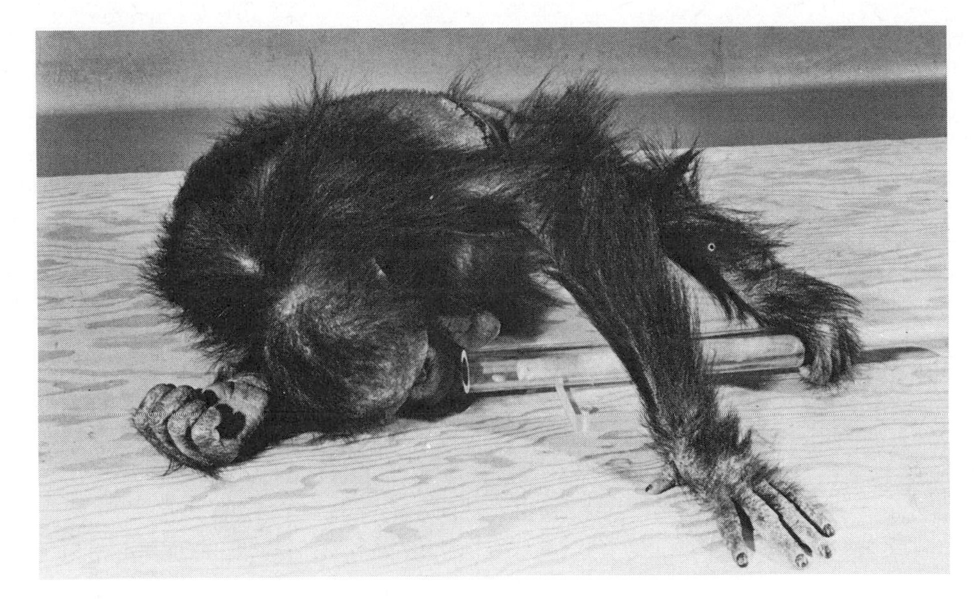

Figure 11–18 The resourceful orang-utan can use his hind legs in a manner analogous to our use of our hands. Using a wooden pole, he is able to push the small piece of candy inside the plastic tube out toward his mouth.

his hands. But we can't peel a banana with our toes, a feat that a chimpanzee would hardly consider extraordinary. (See Figure 11–18.)

The marvel of flight

Among the higher animals only birds and bats, a group of mammals, have the ability to fly. Some other species, like the misnomered flying fish and flying squirrel, glide but don't actually fly. The wings of flying animals are a special kind of forelimb, an adaptation of the general bone structure in the forelimbs of all animals with endoskeletons. A bat's wings, which are actually only the "hand" of a forelimb, are covered with a light, tough membrane. A bird's wings are covered with bulky but lightweight feathers. An insect's wings are not forelimbs, but a separate light-

All wings found on living things, as well as the wings on airplanes, work according to the same rules of physics.

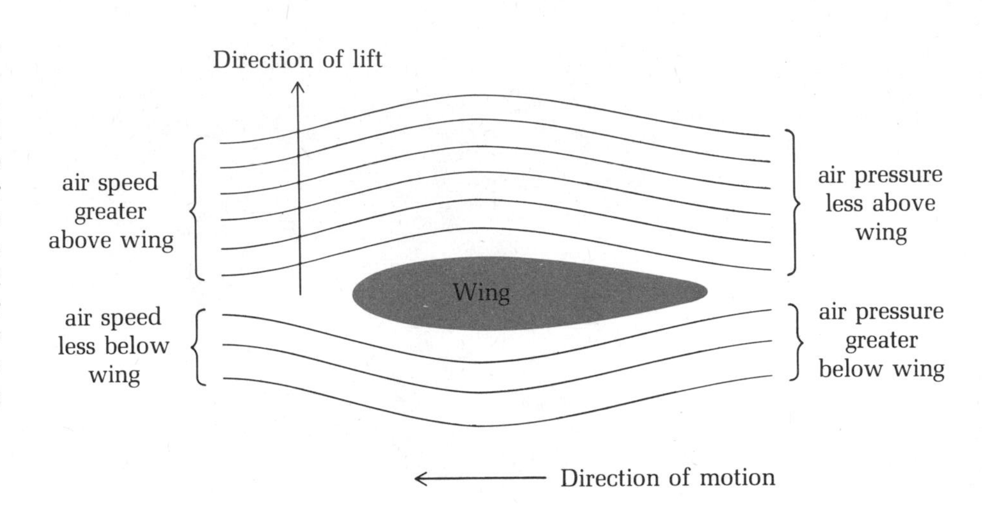

Figure 11–19 Bernoulli's principle explains the "lift" experienced by a wing-shaped structure in flight. The air speed, being greater along the upper surface of the wing, assures lower air pressure at that point. The higher pressure along the lower portion of the wing lifts the wing structure upward.

weight structure similar to the exoskeleton. Despite these differences, all wings found on living things, as well as the wings on airplanes, work according to the same rules of physics.

That rule is *Bernoulli's principle*, named after the scientist who formulated it. Simply stated, the principle says that the pressure of air decreases at those points where its speed increases. Figure 11–19 shows the shape of a wing and how air flows around it. The air traveling over the longer distance of the upper surface of the wing moves faster, and therefore has a lower pressure, than air passing under the wing. Because of the second law of thermodynamics (see Chapter 5), the higher-pressure air under the wing tries to flow toward the lower-pressure air on top, thereby exerting a lifting force on the wing. You can test this effect yourself by taking a strip of paper 10 inches long, holding one end of it against your lower lip, and blowing steadily across it. The strip will rise and straighten itself in the direction of the airstream.

A gliding animal like the flying squirrel is actually falling to earth the long way around. It begins its flight by climbing to some high point and jumping off. The folds of skin between the squirrel's hind and fore limbs act like a parachute, slowing the squirrel's fall and allowing it to travel safely to some lower point or to the ground. If a gliding animal can find an updraft and stay with it, it can remain aloft as long as the draft rises faster than its body falls. Although capable of flight, vultures and condors often glide for hours on end to save energy in their search for the rotting animal carcasses they feed on.

Figure 11–20 The graceful osprey in flight.

Birds, bats, and insects are not dependent on a slow, controlled fall to keep air flowing over their wings (Figure 11–20). The flapping of their wings pushes them forward in the air, somewhat analogous to the way that an oar pushes a boat across the surface of the water. Weight is obviously an important factor in flying, and through natural selection birds have evolved bodies that economize on weight. Birds have hollow bones that are so light that the entire skeleton of the oceanic frigate, a species so large that its wings span 7 feet, weighs but 4 ounces. The biggest bone in a bird is the breastbone, which serves as an anchor for the massive muscles that power the wings. The exoskeletons of flying insects are also lightweight, but the muscular arrangements are, of course, different. The muscle pairs in a bird are attached by one end to the wing and by the other to the breastbone. The wings are the moving part. In the case of insects, the wings are mounted rigidly into the walls of the thorax, that section of the body roughly equivalent to the human chest. The muscle pairs move the walls of the thorax and with them the wings.

Sea gulls and hawks sometimes hover at the edge of a cliff by cupping their wings just enough to match their rate of fall to the strength of the wind blowing up the face of the cliff. This kind of hovering is much like gliding. However, only one kind of flying animal can hover under its own power—the tiny hummingbird, whose wing movements create a continuous lift steady enough to allow it to search a flower for a meal of nectar.

The technology of motion

It is the peculiar characteristic of humans to go beyond their physical limitations by making special tools. This has certainly been true of movement. Very possibly the development of ways to transport people and their belongings began with nomadic herdsmen and hunters many thousands of years ago when they found that certain kinds of domesticated animals could be persuaded—or forced—to carry a load. In the desert countries of Asia and Africa, the camel did this job, while the llama performed a similar function in the high mountains of South America. Some cultures had the animal pull special carrying devices. The travois of the American Indian was pulled originally by dogs and later by horses. The Eskimo dog sled falls into the same class. And, of course, animals carried not only man's belongings but man himself. The horse and the camel have perhaps served in this capacity more than any other animals.

These methods of locomotion are all exploitative: that is, they simply transfer the work from one animal—man—to another—say, a horse or a dog. But with the wheel, a new way to move was actually invented, a method that provided the basis for a great many moving machines. Some small animals do roll to get themselves around, but the whole animal does the rolling. No animal makes use of a rolling mechanism separate from the main body, nor has any animal ever been found that was equipped with a wheel mounted on an axle. However, making a wheel roll does involve an expenditure of energy, and the original source of that energy was animal. A simple cart pulled along by a man, an ox, or a dog was very likely the first wheeled machine. Since that time, humans have developed

Humans have developed a great many kinds of carts and incorporated nonbiological sources of energy to power them, but the basic idea has remained the same.

The invention of the wheel was one of the major advances in the history of civilization.

a great many kinds of carts and incorporated nonbiological sources of energy to power them, but the basic idea has remained the same. All our land vehicles are pushed along on wheels by some energy source such as a steam or diesel engine.

Human flying machines sometimes imitate, and sometimes differ from, flying organisms. The balloon is a passive form of motion. Lighter than air, it rises like a drop of oil to the surface of water and is pushed along by air currents. The balloon's pilot can control his flight to some extent by dropping weight or by releasing gas, but the source of his motion comes from the air around him. Although heavier than air, a glider is also dependent on air movement. But the glider's long wings provide so much lift that if the pilot can locate enough updrafts he can remain airborne for many hours.

Many of the first attempts at inventing a heavier-than-air craft capable of flying under its own power made the mistake of imitating slavishly the flight of birds and insects. The early inventors tried to fabricate wings that supplied both forward power and lift. The Wright brothers' successful plane and all planes since then have separated power and lift. Propellers or jet engines push the plane forward, and the wings supply the

Over the many generations of self-powered motion, organisms have developed ways of recognizing the world about them and changing their motion to suit it.

lift. This separation of lift and power has allowed man to fly faster and to carry more weight than any bird. But no man-made flying machine can approach the kind of control that allows a hawk to power-dive from an altitude of several hundred feet, snatch up a mouse in its talons, and return to the air all in one motion.

The side effects of technology

Our machines of motion and speed are true marvels of human genius. In some ways, though, the myriad machines that have followed the wheel have caused us acute troubles.

Some of these problems arise from the mere physical presence of these machines. The automobile is the principal mode of local transportation in America, and what was once seen as a blessing takes on more of the trappings of a curse each year. As we saw in Chapter 6, the automobile is the principal source of air pollution. Also, our cities have been cut and divided by the roadways needed by cars. And no matter how fast new freeways are built, the number of cars always stays just far enough ahead that the twice-daily traffic jams grow steadily worse. To save cities from cars, mass transit will have to become more of a reality than an idea in the near future.

Another class of problems arises from the fact that these machines go faster than our natural abilities can handle. Over the many generations of self-powered motion, organisms have developed ways of recognizing the world about them and changing their motion to suit it. Man can judge the environment and act on that information very fast. Consider a man running full speed down a narrow trail. He rounds a bend and sees a rhinoceros standing in the middle of the trail 30 feet ahead. In about a quarter of a second, the runner can recognize the rhino as a danger, decide to hop into the bush for hiding, and start hoping. Even at the top human running speed of a little over 20 mph, that's more than enough time to avoid a collision. But at 60 mph in an automobile, that quarter of a second means that the car travels 22 feet before the driver even begins to apply the brakes. Imagine what happens in a jet aircraft going twice the speed of sound. Our reaction time doesn't match the speeds we can achieve. Instrumentation of the sort that fliers depend on is of great value, but the fact that we can travel at about 1,500 mph when we are biologically equipped to handle speeds only one seventy-fifth that fast means that we are working at the very edge of our physical abilities.

Summary

Movement in response to the environment is a characteristic of life. Microorganisms move by means of pseudopodia, contractile fibers, flagella, or cilia. In plants, growth rather than movement is the rule. Phototropism in plants is caused by a change in the flow of auxins in the shoot, resulting in increased growth in cells on the dark side. Water flowing out of cells provides movement for the quick-acting plants like *Mimosa pudica* and the Venus's-flytrap.

Animals move by virtue of muscles arranged in antagonistic pairs. When one contracts, the other relaxes. Muscles pull against the rigid framework of one or another kind of skeleton. Simple organisms like sea anemones have hydrostatic skeletons. Mollusks and arthropods have exoskeletons, while endoskeletons of bone characterize most fishes and all amphibians, reptiles, birds, and mammals. In both exo- and endoskeletons, muscles provide movement by spanning flexible joints. Arthropods molt their exoskeletons periodically to allow for growth, an important aspect of their life cycles. In many vertebrates, the endoskeleton begins as cartilage, which is gradually replaced as the organism grows.

The waving of a swimming fish's body is the basic vertebrate movement. The simplest four-legged animals are essentially equipped with jointed fins that provide good stability but low speed. Swinging the legs under the body provides more mobility at the cost of some stability. Animals with flat feet and short limbs can move in a variety of ways, but

Our reaction time doesn't match the speeds we can achieve.

they have none of the speed of those long-legged species who run on the tips of their toes. They, in turn, can do little other than run. Man can walk upright because of the structure of the pelvis and the foot.

Some species, such as predatory dinosaurs, hopping desert mammals, and primates, have evolved special hind limbs for locomotion, and use the forelimbs for other tasks. The birds and bats have specialized wings that give them flight, an ability they share with many insects. Despite differences in structure, all flying animals achieve flight through the physical rule of Bernoulli's principle.

Man has developed specialized tools of motion powered either by animal or technological energy. The wheel and axle has been the key to all man-made movement on land. Aircraft use wings for lift and engines for power. This technology of motion has caused special problems because of the effects of these machines and man's physical limitations.

Questions to consider

1. a. How does an amoeba move?
   b. What is the name given to this style of motion?
2. a. What is a phototropism?
   b. Is phototropism growth or movement? Why?
3. What mechanism provides the quick-shutting action of the Venus's-flytrap?
4. a. A sea anemone extended full length draws down so that it is short and squat. Which muscles relaxed and which contracted to permit this motion?
   b. When a sea anemone expels all the water in its body cavity, it can no longer move. Why is this the case?
5. a. What kinds of organisms have exoskeletons?
   b. How are muscles arranged in an exoskeleton?
   c. How does an arthropod exoskeleton grow with the animal?
6. a. Why are many vertebrate bones hollow?
   b. What are the structures and workings of the human hip joint?
7. a. What is the basic mode of vertebrate movement?
   b. The legs of a salamander can be said to be fish's fins adapted to land. Why is this true?

c. Why is an animal that stands on long, stiltlike legs less stable than one with angled legs jutting from its side?
d. In what ways are the legs of a horse adapted for high-speed sprinting?

8. What features of the human anatomy allow for permanent upright posture?

9. a. What is the physical principle behind flight?
b. How is this principle incorporated into the structure of a typical wing?

10. a. Is the wheel an imitation of nature?
b. How does the flight system of an airplane differ from that of a bird?

Glossary

amoeboid motion the pattern of movement characteristic of some single-celled organisms like the amoeba, in which the cytoplasm circulates within the cell membrane to propel the cell forward

arthropod one of the group of animals characterized by an open circulatory system, jointed legs, and an exoskeleton

auxin a type of plant hormone that regulates the rate and direction of plant growth

Bernoulli's principle the principle that states that the pressure of a fluid (air or water) decreases when its velocity is increased

cartilage the resilient and flexible protein tissues that provide support and connections in the skeletal system

cilia the short projections that provide locomotion in many single-celled organisms; the linings of tissues of some higher organisms

cocoon the protective capsule formed by some insect pupae

endoskeleton the internal skeletal structure characteristic of mammals, birds, and reptiles

exoskeleton the exterior skeletal structure characteristic of mollusks, crustaceans, and insects *fin* the flat, paddlelike extension used for locomotion in fish and other aquatic organisms

flagella the long, whiplike projections used for locomotion in some single-celled organisms

hydrostatic skeleton the structural system found in some aquatic organisms, characterized by the temporary flooding of internal cavities

larva the developmental stage in the life cycle of most insects which follows the egg phase and which is characterized by a lack of resemblance to the adult form

ligament the tough, protein, hingelike tissue that supports bones at the joint

metamorphosis the developmental pattern characterized by sharp changes in form and structure between stages

mollusk one of the group of animals characterized by soft body tissue sometimes protected by a hard shell

molt the process in which an animal sheds its exoskeleton or some other external coating

muscle the specialized tissue found in most higher animals which contracts and relaxes to move the organism

nymph the late developmental stage in the life cycle of insects characterized by an immature form which most closely resembles the adult

pseudopodia the projections of cytoplasmic material which move amoeboid single-celled organisms

pupa the developmental stage in the life cycle of insects characterized by a resting stage

skeleton the firm structural support system found in many animals

tendon the strong protein tissues that connect muscles to bones

tropism directed growth, equivalent to movement, by a plant in response to a stimulus from the environment

vertebral column the hollow shaft of bone and nerve tissues found in the skeletal systems of most higher animals

12 Communicating: Internal

One of the remarkable characteristics of all organisms is their ability to maintain themselves despite their environments. We humans, for example, have amazing chemical and physical consistency. At 100°F above zero or 20°F below, the body stays at 98.6°F. In the steamy humidity of a rain forest and in the bone-dryness of a desert, our water content remains the same. Even though we eat only two or three times a day, the concentration of glucose in the blood hovers about the same point all day long. We are bombarded continuously by disease-causing microorganisms, but we are more often healthy than sick. Caught in a situation of danger—confronting a mugger bent on robbery or a watchdog trained to defend its turf—the body is able to mobilize all its strength to fight the threat or to flee from it, and then, after the danger is past, to return to its normal state. This ability to remain in a steady state in the face of environmental change is called *homeostasis* (meaning, same position).

Homeostasis can be likened to the workings of a furnace thermostat regulating the temperature of a room. The thermostat controls the burning of the furnace to keep the room at some desired temperature, say 70°. If the temperature falls below 70°, the thermostat starts the furnace burning and keeps it going until the room warms to 70°. Then the thermostat shuts the furnace off until the temperature falls below 70°, and heat is again needed. No matter how much the temperature outside varies, the thermostat will keep the room comfortably constant.

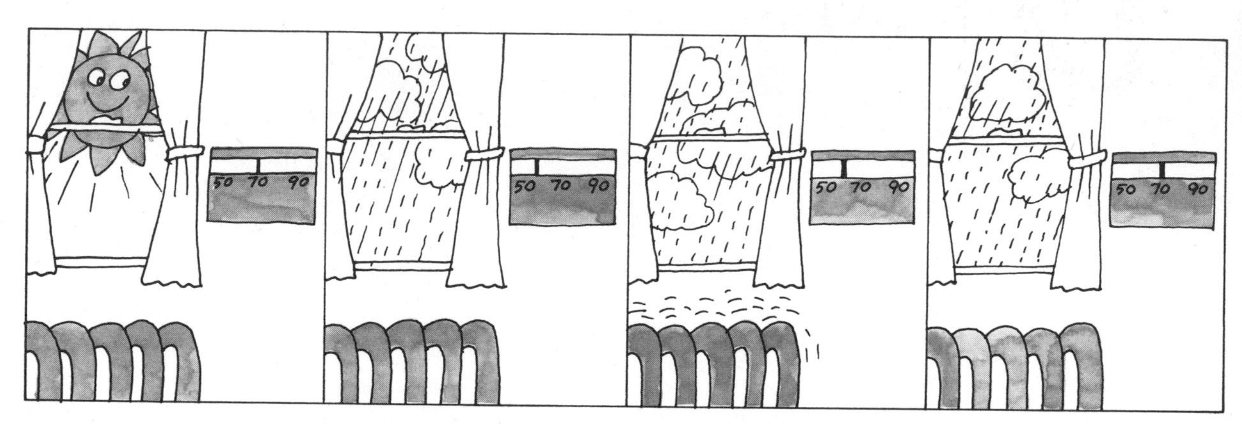

The thermostat operates by constantly comparing an external stimulus (temperature) with a fixed reference point, in this case, 70° F. When the temperature drops below this point, the heating equipment is adjusted to increase the temperature to the desired reference point.

But homeostasis is more complicated than the workings of a thermostat. The environment changes in many ways other than temperature, and we have to be able to respond to each kind of change appropriately. In other words, we need many kinds of thermostats. Also, organisms need to change their internal states to allow for periodic activities, such as flowering in plants and the menstrual cycle in women, and these changes have to be made to occur at the right time and in the proper way. And all the various parts of the body have to be made to work together, so that the left hand very literally knows what the right is doing.

Every organ—indeed, every cell of the body—is involved in some regulatory task or other. But certain systems play homeostatic roles almost exclusively. In Chapter 7, we examined how the kidney functions to maintain the proper amounts of water and ionic salts. In this chapter, we'll look at three more of these special systems: the immunity response, the hormones, and the nervous system.

Immunity: telling self from nonself

Every day we come into contact with literally millions of microorganisms—particularly viruses, bacteria, and fungi—that can induce disease. We meet them in the air we breathe, the water and food we consume, the people we rub elbows with. How does the body defend itself against invasion? The skin and the entryways into the body are equipped with a variety of mechanical and chemical defenses that stop microorganisms before they can gain a foothold. The ciliated cells of the trachea, for example, move foreign matter trapped in mucus out of the respiratory system. The caustic hydrochloric acid of the stomach kills practically all the microorganisms carried into the digestive system. The secretions of the tear glands in the eye, which bathe the surface of the eye continuously, contain an enzyme that destroys bacteria. The sebaceous glands associated with the hair follicles of the skin release an oil highly toxic to fungi. This is the reason that athlete's foot, caused by a fungus, appears on the soles of the feet and between the toes, which are devoid of hair.

But what happens to those bacteria, viruses, and fungi that get past these defenses and do actually invade the body?

The white knights of the blood

Most of the cellular portion of the blood is made up of the red blood cells. But the blood also contains *white blood cells,* outnumbered by the red cells about 1,000 to 1, but crucial to the body's defense against invasion. Unlike the red blood cells, which are swept along by the blood like twigs in a stream, the white blood cells can travel against the current and through the walls of the capillaries to the site of an infection. There they make war on the invader, oftentimes being destroyed in the process. The pus that accumulates in an infected wound is made up mostly of dead white cells (Figure 12–1).

How do the white blood cells stop the invader? First, they can engulf and dissolve foreign microorganisms. The white blood cells throw out pseudopodia (see Chapter 11), just as if they were amoebas, to trap the microorganism and draw it within the cell, where it is dissolved with the aid of enzymes. Second, the white blood cells manufacture *antibodies.* An antibody is a globular protein that reacts with an invading organism or substance and renders it inactive. Anything that causes antibodies to be produced is referred to as an *antigen.* The mechanical and chemical defenses of the skin and the engulfing action of the white blood cells

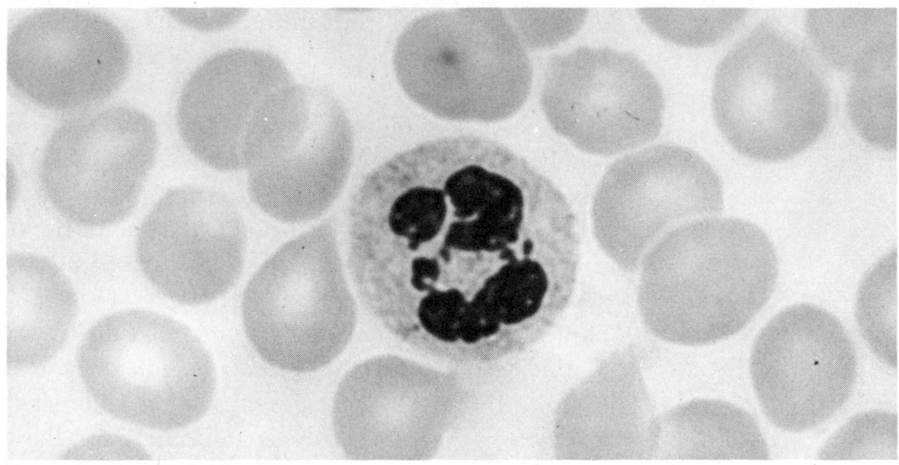

Courtesy Carolina Biological Supply Company

Figure 12–1 White blood cells are the specialized structures in the bloodstream that attack infecting materials. They thus serve as one of the primary lines of defense against disease, along with the cilia of the respiratory tract and the skin.

work against all invading microorganisms. But a particular antibody works against a specific antigen. An adult human is estimated to have 10 million different antibodies. Antibodies work much like enzymes by combining with the antigen, destroying it outright or chemically transforming it in such a way that it can be engulfed by a white blood cell (Figure 12–2).

Remembering an invader

We are not born with a full array of antibodies ready to take on every possible antigen. Instead, antibodies develop in response to the introduction of an antigen. The first time an antigen invades the body, it is somehow analyzed, perhaps by the white blood cells, and chemical instructions for manufacture of the proper antibody are given. It can take anywhere from 1 to 30 days for the antibodies to appear in the blood, but usually they hit a peak concentration in 4 or 5 days. Subsequently, the concentration of antibodies drops markedly, sometimes to the point where they cannot even be detected. However, the body somehow remembers the antigen and its antibody response. The second time the antigen is introduced, antibodies against it appear in a matter of hours, and the infection is stopped or limited quickly. This ability of the body to remember an antigen and respond to it is known as the *immunity response.*

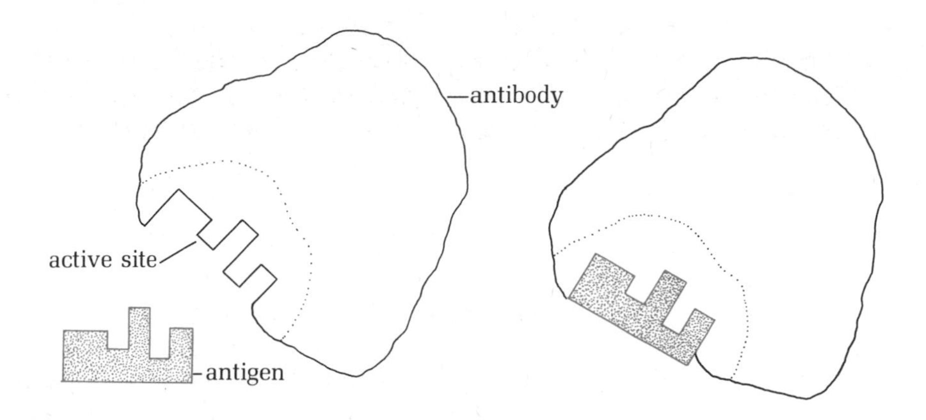

Figure 12–2 The specificity of antibodies. In general, antibodies are extremely specific in their match with antigens. A lock-and-key mechanism, similar to the function of an enzyme, provides for the "identification" of a specific antigen by its appropriate antibody.

How long can the body remember? In other words, how long does immunity last? It depends on the particular antigen. A first bout with the so-called childhood diseases—principally, German measles, red measles, mumps, and chicken pox—usually produces lifelong immunity. Occasionally, though, people do catch these diseases a second time, indicating that they have lost their immunity. In the case of other diseases, immunity lasts only a few months.

One doesn't actually have to be visibly sick from a particular invader to develop immunity against it. Experiments in this country have shown that many people are immune to spinal meningitis even though they have never actually had the disease. Apparently they were exposed to the disease, probably during childhood, and manufactured antibodies fast enough to stop the infection early, producing immunity thereafter.

Borrowing immunity

For countless centuries, smallpox was the scourge of Europe, a violent disease that killed many of its victims and left the rest disfigured. Then an eighteenth-century English doctor noticed, as many others had, that people who had contracted the similar but much milder cowpox were immune to smallpox. Working on a hunch, the physician, Edward Jenner, took a sample of pus from a sore of a woman dairy worker with cowpox and inoculated a young boy with it. Subsequently, he inoculated the boy

Jenner had hit on a lucky coincidence: the virus that causes cowpox is similar enough to the one responsible for smallpox that the antibodies specific to it also act against the smallpox antigen.

with pus from a smallpox victim. Despite the exposure to the killer, the boy remained healthy. Jenner had hit on a lucky coincidence. The virus that causes cowpox is similar enough to the one responsible for smallpox that the antibodies specific to it also act against the smallpox antigen. Introducing the cowpox virus into the boy provided him with the antibodies needed to stop the later smallpox invasion. Jenner's experiment, one of the cornerstones of modern preventive medicine, was the first case of *vaccination* (from Latin *vacca,* meaning cow).

The seemingly endless series of shots given nowadays to children in their early years are all meant to stimulate the body's production of antibodies and confer immunity. Modern vaccines are of two types, and they confer two different kinds of immunity. Most vaccines consist of the antigens themselves, oftentimes killed or weakened to lower the chance that they might themselves cause disease. These vaccines cause the body to produce its own antibodies, thus making itself immune. This sort of immunity is called *active immunity*. Active immunity is the sort conferred by the childhood vaccines against diphtheria, whooping cough, and tetanus (lockjaw). Some vaccines, though, contain not antigens but antibodies. Instead of inducing the body to manufacture its own antibodies, the antibodies are presented ready-made. This sort of immunity, called *passive immunity,* results from the vaccine against cholera, among others. Passive immunity is short-term, and generally doesn't last over 6 months.

Passive immunity occurs in nature as well as in the doctor's office. When an infant is born, it is removed from the sterile environment of the womb and meets a great array of potential antigens. Since this is the infant's first contact with these antigens, he has no immunity of his own. Why then don't all newborns succumb to disease almost immediately? During the last month of pregnancy, the mother's antibodies cross the placenta into the fetus. At birth, the infant has the same immunities as his mother. The child's passive immunity is further bolstered by drinking the antibody-rich *colostrum,* the fluid produced by the mother's breasts before true milk begins to flow. Thus, in the crucial first months of life, the infant is protected by his mother's antibodies. But this immunity lasts only about 6 months, the reason that necessary vaccinations should be begun at that time.

Keeping tabs on the self

In finding disease-causing microorganisms and other antigens, the antibodies are displaying the ability to tell the cells that belong to the organism from those that don't. This ability extends to more than microorganisms. The immunity system can actually distinguish the cells and substances of the self from all other cells and substances, and it reacts to these aliens as it would to any antigen.

Precisely how the antibodies accomplish this feat remains a great and intriguing mystery. It is known that an organism "learns" to tell self from nonself during development. Obviously, for example, the salamander gastrulas involved in the dorsal lip transplant described in Chapter 10 could not yet tell self from nonself. Otherwise the transplant would have been rejected. A human is immune to nonself while still in the uterus, but in the mouse immunity doesn't appear until after birth. Also, there are certain kinds of cells in the human body that are not distinguished as self —sperm cells, the thyroid gland, nerve tissue, and the lens of the eye. If a man's sperm cells are injected into his own bloodstream, antibodies attack and destroy them. These types of cells never enter the bloodstream under normal circumstances, and apparently the immunity system has never learned whose they are.

The ability to tell self from nonself represents a powerful homeostatic mechanism that provides great protection. But it wreaks havoc with medical attempts to transplant organs. So far as the immunity system is concerned, a transplanted organ is foreign matter to be destroyed. In the case of kidney transplants, for example, it is sometimes possible to match the tissue types of donor and recipient close enough for the new organ to take properly. The most successful organ transplants to date involve those

As far as the immunity system is concerned, a transplanted organ is foreign matter to be destroyed.

between identical twins. Rejection of the transplanted tissues is minimized because of the genetic and antibody similarities between the two individuals. Transplant patients receive drugs that suppress the workings of the immunity system. Unfortunately, these drugs also have the effect of inhibiting the antibody response to microorganisms, raising the risk of disease.

When immunity goes awry

From a chemical point of view, the immunity response is incredibly complex. The organism has to recognize an antigen, distinguish it as friend or foe, and make the appropriate response. Because of this complexity, the immunity system occasionally runs off the track, seeing friend as foe or responding in a damaging way.

Sometimes an organism actually becomes immune to its own cells, mistaking self for nonself. This phenomenon is called *autoimmunity*. A person suffering from the disease hemolytic anemia is immune to his own red blood cells. Antibodies attack the red blood cells and destroy them just as if they were invading bacteria. Autoimmunity is also thought to be the cause of rheumatoid arthritis, a crippling disease involving severe inflammation of the joints. In this case, the disease involves immunity to collagen, the major structural protein of the cartilage and connective tissues.

Allergies result from an overresponse to an antigen. Pollen, for example, is a weak antigen; that is, it is an antigen to which most people respond mildly or not at all. But in an individual with hay fever, pollen induces a major immune response located primarily in the mucous membranes lining the upper reaches of the respiratory tract. The pollen triggers certain cells to release histamine, a chemical that causes the itching, swollen eyes and dripping nose associated with hay fever. A great many weak antigens may cause allergies: bee and wasp stings, animal hair and dander, egg white, milk, even carbon paper and money. Hay fever and other allergies generally run in families, indicating that the conditions are inherited.

The antibodies donated by the mother to the fetus in her uterus work to the obvious benefit of the fetus, except in one instance. Blood contains a chemical component known as the *Rh factor*, named after the rhesus monkeys in which it was first discovered. Most people—about 85 percent—have the Rh factor and are called Rh positive, but some are lacking it and are called Rh negative. If the Rh factor is introduced into an Rh-negative person, antibodies are formed against it. The presence or absence of the Rh factor is determined genetically. An Rh-negative woman mated with an Rh-positive man may produce an Rh-positive fetus. In the first such pregnancy, this causes no problem, because the blood supplies of mother and fetus don't mix. At birth, though, when the placenta is pulled loose from the uterine wall, some of the fetus's Rh-positive blood may enter the woman's blood, causing antibodies to be produced against it. The next time the woman is pregnant with an Rh-positive fetus, these anti-Rh antibodies travel across the placenta and attack the fetus's red blood cells, killing the unborn child, leaving it blind and deaf, or causing brain damage. One way to stop the reaction is to transfuse the fetus in the uterus, but this is a very involved surgical procedure. A more recent treatment is a substance called Rhogam, which destroys the Rh antibodies of the mother. The blood of an Rh-negative woman who has given birth to her first Rh-positive child is tested for Rh antibodies. If they are found,

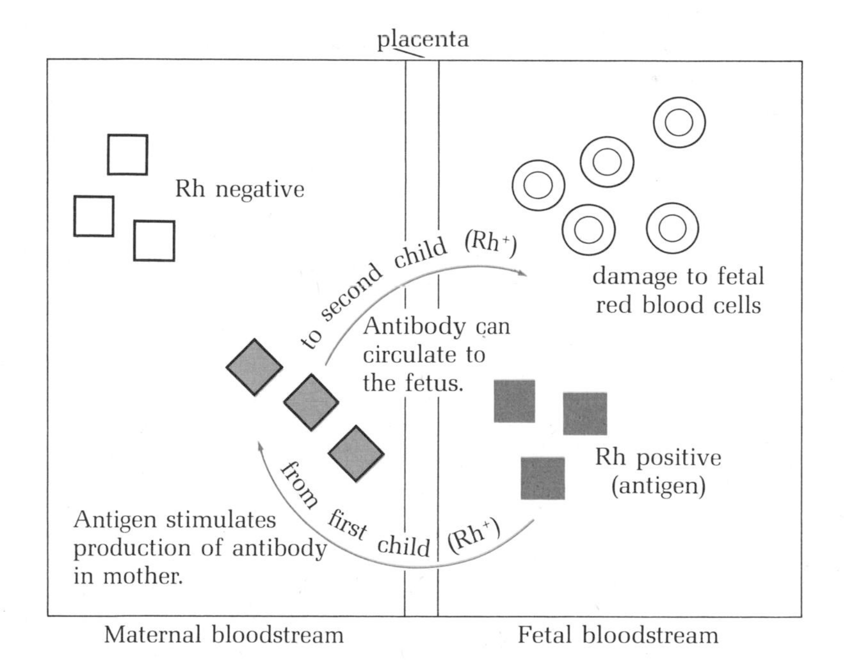

Figure 12–3 The potential danger from differences in Rh factor between mother and child. If an Rh-negative mother is carrying an Rh-positive child, the possible mixing of the two bloodstreams at birth could result in the production of antibodies against the Rh factor in the mother's blood. If she then becomes pregnant with an Rh-positive child, her anti-Rh antibodies will cross the placental barrier and damage the fetal blood cells.

the woman is given Rhogam, which suppresses the antibodies in her bloodstream and eliminates the risk of an Rh reaction in any subsequent pregnancy (Figure 12–3).

The chemical messengers

In Chapter 9, we saw the important role of hormones in developing and maintaining sexual characteristics. A hormone, you will recall, is a chemical substance produced in one part of an organism and then transported to another area, where it exerts its effect. The testosterone produced in the testes and the estrogen and progesterone in the ovaries affect a great many different parts of the body. Hormones in great variety are found in many kinds of organisms, ranging from man to the simple invertebrates and the plants. Hormones are essentially chemical messengers, ways for one part of an organism to tell another to respond to some external or internal change. The messages they bear are an important aspect of homeostasis.

Total control: plant hormones

We saw in Chapter 11 that phototropism in plants is directed by a class of hormones known as auxins. Hormones like the auxins are the major regulatory mechanisms in plants. They regulate the internal environment, allow the plant to respond to external stimuli, and direct such periodic activities as flowering and the dropping of leaves and fruit.

The auxins are important growth regulators in plants, and they do more than just orient the plant to light. Auxins are responsible for the upward growth of shoots and the downward growth of roots. While auxins promote the growth of shoot cells, their effect on roots depends on the amount of auxins. In small proportions, they promote growth, but large amounts inhibit it. When a seedling is planted right side up, the auxins are distributed equally throughout, and the plant grows evenly. If, however, the seedling is laid on its side, auxins accumulate on the bottom side of the shoot and the root. Growth on the bottom of the root is inhibited, but growth on the top side is promoted. As a result, the root turns down-

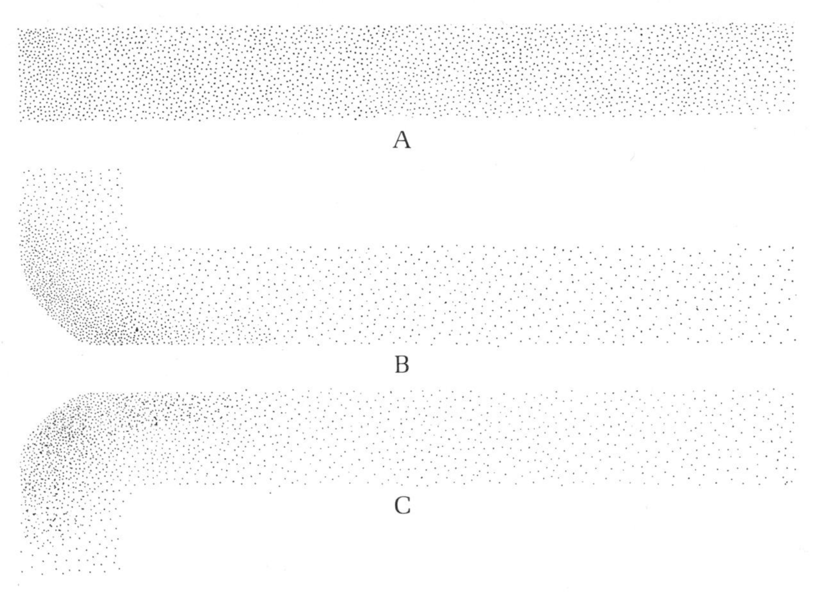

Figure 12–4 The photo shows a bud on a birch twig. Plant hormones regulate both the location and the rate of growth. The illustration shows auxins functioning through the stimulation and regulation of the growth of plant cells. The proportional concentration of a plant hormone can stimulate growth in one region of a plant tissue and depress growth in another portion of the tissue; this differential rate of growth is analogous to movement in animals.

ward. As for the shoot, its bottom cells grow faster than the top, bending the shoot upward (Figure 12–4). The auxins also play a role in the ripening of fruit.

One plant hormone was first isolated because of the misfortune of some ambitious fruit growers. In the early part of this century, the ripening of fruit picked while still somewhat green was hastened by storing the fruit in a closed room with a kerosene stove. The growers assumed that the extra heat was the reason for the accelerated ripening. Some of them installed more modern equipment that used fuel sources other than kerosene. Surprisingly the fruit ripened much more slowly. As it turned out, what hurried the ripening along wasn't the heat, but the gas ethylene given off by the burning kerosene. Ethylene is produced naturally in ripening fruit under the influence of the auxins. The ethylene given off by the burning kerosene was a hormonal shot in the arm, so to speak, for the fruit.

A close relative of hormones with effects similar to those of the auxins is the *gibberellins*. They too promote cell elongation and division. They are also known to play a role in the flowering of mustard and cabbage plants. The *cytokinins*, like the gibberellins and auxins, are growth stimulators. They are found in rapidly growing parts of the plant like the fruit and the seeds. Cytokinins also help the plant repair itself when damaged. The bleeding sap that covers up the wound left when a plant is pruned, for example, contains a goodly amount of cytokinins, which promote the growth of tissue layers to cover damaged areas.

The dropping of a leaf or fruit from a plant is called *abscission*. Abscission results from the growth of a special layer of cells that marks the line where the leaf or flower will fall off (Figure 12–5). Abscission is promoted by a hormone named *abscisic acid*. Abscisic acid is produced continuously by the plant, but it is counteracted by the auxins. At the time when abscission should occur—in October or November, in the case of a maple tree—the amount of auxins falls off, and the abscisic acid's effects are seen as the leaves turn color and fall to the ground. Citrus growers often spray their ripening groves with auxins to keep the fruit on the trees until picking time.

These plant hormones provide the organism with a way of coordinating its activities and of responding to such environmental factors as light and gravity. And there are some important things to note about the

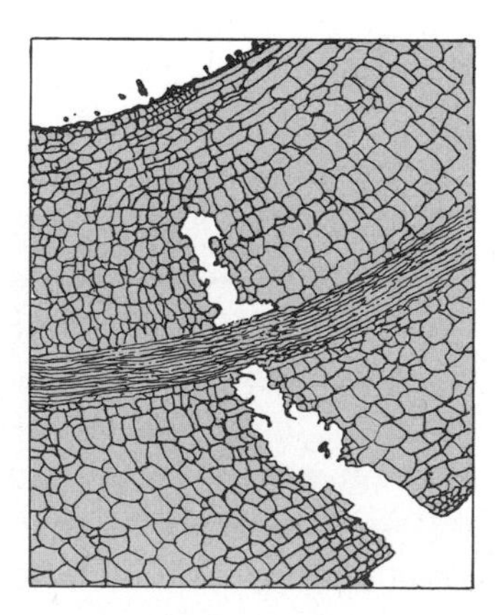

Figure 12–5 Abscission in a plant stem. The production of the hormone abscisic acid is a chemical way of limiting the growth of the plant.

In man, the hormones make sure that the body is not just an assembly of parts but a working unit.

workings of hormones that apply to man and other animals as well as to plants. For one thing, a hormone's effects depend not only on the hormone but also on the target tissue. A given amount of auxins, for example, will promote growth in the stem and inhibit growth in the root. Also, hormones often work against each other, as shown in the case of the counterbalancing of auxins and abscisic acid. And, finally, hormones usually work their effects in small doses. Auxins in overabundance are lethal, and many commonly used agricultural herbicides (plant killers) are synthetic auxins. So far as the plant is concerned, a little of the hormone is a good thing, but an overdose is deadly.

Making the parts a whole: human hormones

In man, the hormones play the primary role of controlling and regulating many of the multitudinous functions of the body. They make sure that the body is not just an assembly of parts but a working unit. The human hormones are produced in a group of organs referred to as the *endocrine glands,* also called the *ductless glands* (Figure 12–6). The liver, which is not an endocrine gland, is connected to the duodenum by a duct. The endocrines, by contrast, release their secretions directly into the bloodstream without the aid of a duct. Table 12–1 lists the various parts of the human endocrine system, the principal hormones produced by each one, and the effects these hormones have.

The size of the pituitary gland, roughly that of a kidney bean, belies its importance to the endocrine system. We met this gland previously (see Chapter 9), when we saw its role in the menstrual cycle. But some of the pituitary hormones also trigger the workings of other endocrine glands. These interendocrine messengers are called the *tropic hormones,* and the glands they influence are the thyroid, the gonads, and the adrenals. The tropic hormones promote the production of hormones by the target organs.

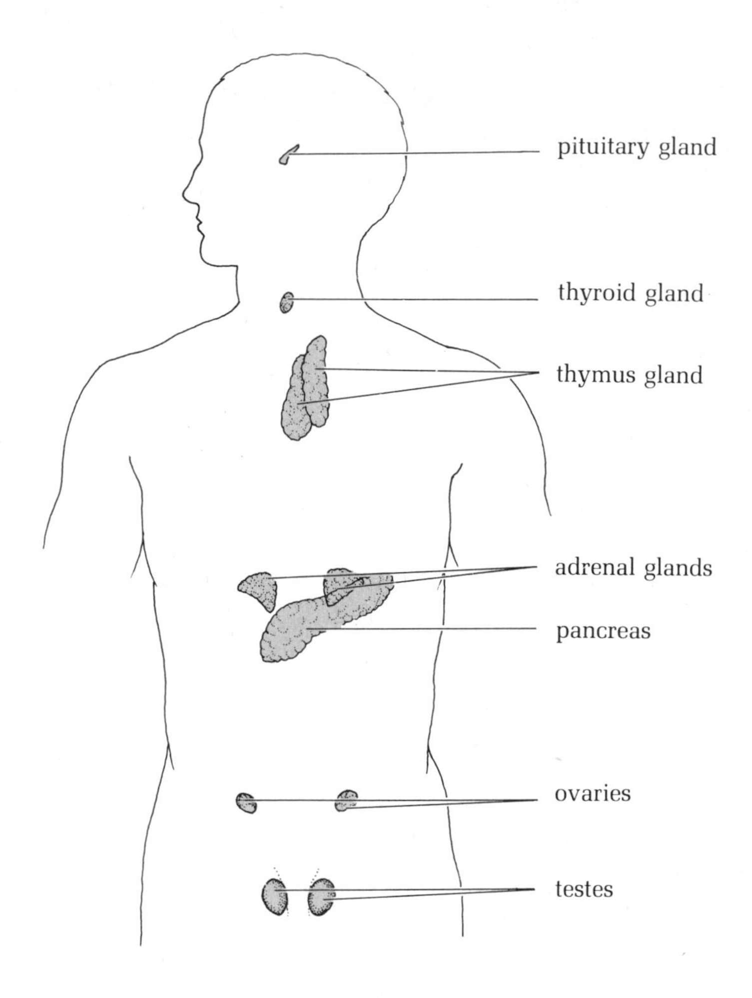

Figure 12–6 The human endocrine system.

Under the influence of thyrotropic hormone (TTH) released by the pituitary, the *thyroid gland* produces thyroxine. Thyroxine controls the rate of cellular respiration, also called the *basal metabolic rate*. Thyroxine acts as a sort of throttle, controlling the rate of the body's basic chemical machinery.

As we have seen, the gonads produce the hormones that develop and maintain secondary sexual characteristics—testosterone in the male, estrogen and progesterone in the female. The gonads are influenced by a number of pituitary hormones.

The *adrenals*, relatively small glands fitted like caps on the top of the kidneys, are perhaps the most important chemical factories in the

Table 12–1 The human endocrine system

Gland	*Hormone*	*Effect*
Pituitary	Adrenocorticotropic hormone (ACTH)	Stimulates adrenal cortex
	Thyrotropic hormone (TTH)	Stimulates thyroid
	Luteinizing hormone (LH)	Stimulates testes and ovaries
	Follicle-stimulating hormone (FSH)	Stimulates follicle of ovary in ovulation
	Growth hormone (somatotropin)	Promotes growth in bone and muscle
	Prolactin	Stimulates breasts to produce milk
	Vasopressin	Controls excretion of water in kidneys
	Oxytocin	Stimulates contractions of the uterus at birth
Ovary	Estrogen	Promotes female sexual characteristics
	Progestrone	Stimulates thickening of uterus lining
Testis	Testosterone	Promotes male sexual characteristics
Thyroid	Thyroxine	Controls basal metabolism
Adrenal	About 50 cortical hormones, e.g., cortisone, aldosterone	Controls many basic chemical mechanisms
	Adrenaline	Increases sugar in blood, raises heartbeat, dilates arteries
Pancreas	Insulin	Lowers level of sugar in blood

body (Figure 12–7). The outer layer, or *cortex*, of the adrenal glands produces at least 50 hormones, and by no means are the effects of all of them known. These *cortical hormones* are produced under the influence of the pituitary's adrenocorticotropic hormone, commonly and mercifully abbreviated to ACTH. The cortical hormones play an important role in regulating the body's chemical handling of carbohydrates and in main-

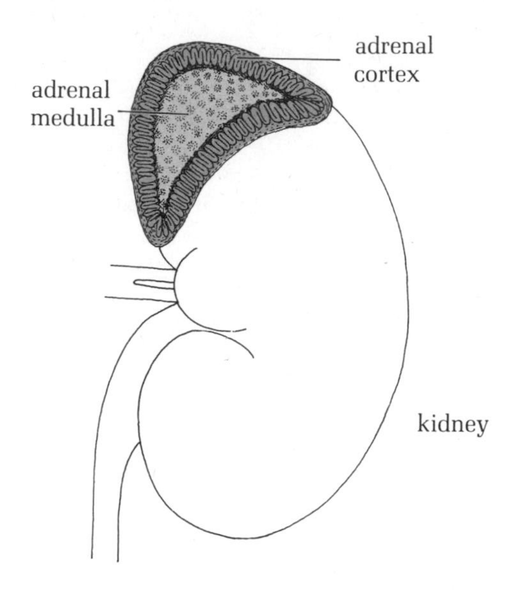

Figure 12–7 The adrenal gland and its position in relation to the kidney.

taining the proper level of salt and water in the body. Male hormones are produced in the adrenal cortex of both men and women. In women, these male hormones apparently have something to do with promoting sexual desire. A woman whose ovaries are removed surgically generally suffers no change in her sexual interest, but if her adrenals are removed, physical desire often falls off completely. The interior, or *medulla*, of the adrenal functions independently of the cortex. The medulla's best-known product is *adrenaline*. Adrenaline is the hormone behind the fight-or-flight response of the body. It increases the level of blood sugar—making energy readily available—speeds the heart, and dilates some blood vessels—increasing the amount of oxygen carried to the muscles. The release of adrenaline is triggered by the nervous system, which we will look at shortly.

A system of checks and balances

The hormones of man, and those of other vertebrate animals, perform a variety of important tasks. They orchestrate and synchronize periodic processes like growth and menstruation. For example, a rise in testosterone or estrogen brings on the development of sexual characteristics at puberty and maintains them afterward. Hormones keep the body's chemistry in order, regulating everything from the basal metabolic rate to the concentration of sugar and ionic salts in the blood. And they provide an important part of our response to the environment. Adrenaline, for example, prepares the body for unusual physical demands.

Hormones are powerful chemicals that have to remain in the right balance for the body to work properly. Too much or too little of a hormone can devastate the normal chemical workings of the organism. An oversupply of thyroxine maintains an abnormally high metabolic rate, rendering the individual underweight and overactive, while too little of the same hormone produces obesity and listlessness. If the pancreas lags in its production of insulin, which promotes the transport of sugar from blood to cells, diabetes results. Then the blood contains an overabundance of sugar, but the cells are sugar-starved. A tumor on the adrenal gland may cause overproduction of male hormones and the development of the male characteristics they promote. The bearded ladies displayed as sideshow freaks in the circus are often the victims of such tumors. The proper amount of any given hormone is often quite small. In a woman's whole lifetime, estrogen production totals less than a single teaspoonful.

But how is the amount of hormone produced and released kept in balance with the body's needs? The secretions produced in response to the tropic hormones affect not only the various organs and tissues of the body but also the pituitary. For example, TTH stimulates the thyroid's production of thyroxine, and thyroxine in turn inhibits the pituitary's release of TTH. When the body's thyroxine level is low, TTH stimulates production of thyroxine. As the concentration of thyroxine in the blood increases, it reduces the amount of TTH released by the pituitary until the thyroxine hits its optimum level and no more TTH is needed. This kind of control mechanism—in which a substance that promotes the release of a certain product is inhibited by that product—is called *negative feedback* (Figure 12–8).

Negative feedback characterizes as well the control of hormones not directly regulated by the pituitary. High blood sugar stimulates the pancreas to release insulin. As the blood sugar level falls because of the chemical workings of the insulin, the pancreas receives less stimulation and releases less insulin. In time, the insulin and blood sugar come into balance.

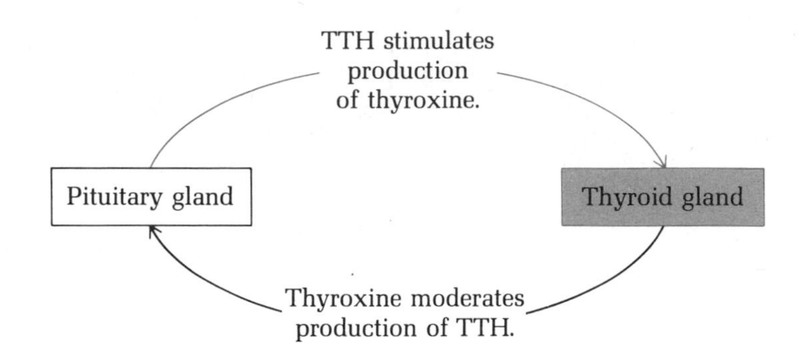

Figure 12–8 An example of negative feedback. The production of thyroxine by the thyroid gland regulates the production of thyrotropic hormone (TTH) by the pituitary gland and vice versa.

Electrochemical communications: the nerves

Hormones provide a well-balanced and coordinated homeostatic system. However, hormonal responses generally lack speed. Phototropism in a plant generally takes at least an hour or more, even in fast-growing shoots. Plants lead slow-paced lives, and the relatively low speed of a change triggered by a hormone poses no particular problem. Such is not the case with the multicelled animals, whose survival often depends on the ability to respond quickly to some change inside themselves or in the environment. In the more advanced animals, the burden of quick response falls on the *nervous system*, which can send messages through the body in a matter of thousandths of a second. The nervous system acts as a homeostatic network of communications between the various parts of the body and with the outside world (Figure 12–9).

The parts of the neuron

The basic unit of the nervous system is the nerve cell, or *neuron*, shown in Figure 12–10. The most prominent feature of the nerve cell is its extensive surface area. Although the cell body proper, which contains the nucleus, is of average cellular size, long protrusions branch out in opposite directions.

The many separate branches on the top side of the cell are the *dendrites*. They pick up coded information from other nerve cells or from

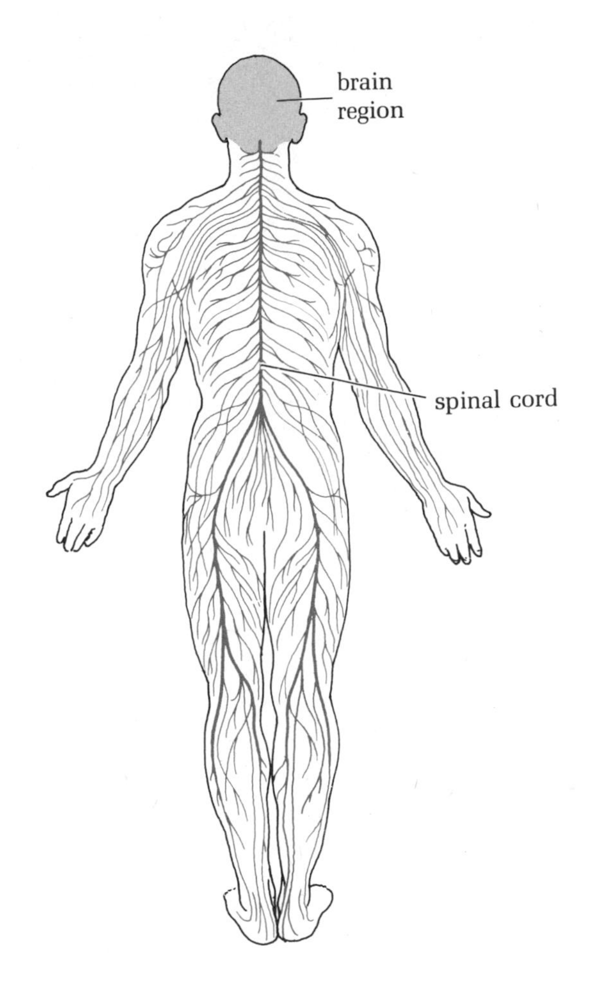

Figure 12–9 The human nervous system.

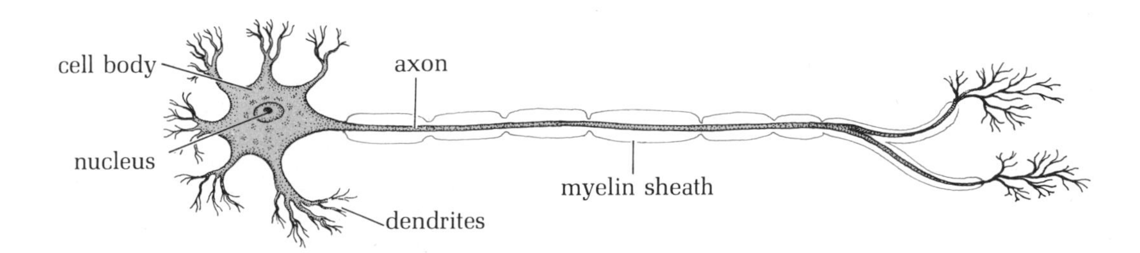

Figure 12–10 The structure of the neuron.

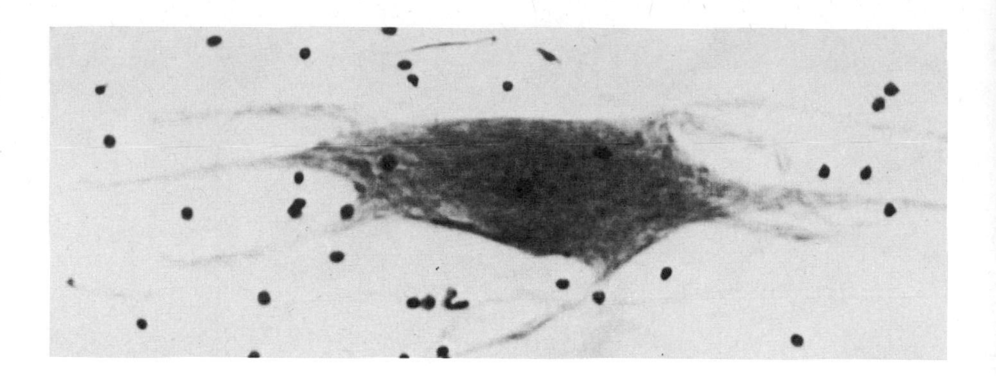

A photomicrograph of a human nerve cell, showing the cell body and dendrites.

specialized receptors. At the opposite end of the cell body is a single, long extension known as the *axon*, which may vary in length from a fraction of an inch to as long as 15 feet, depending on the size of the animal in question. The longest axons in the human body are about 4 feet long. The axon carries information from the nerve cell to other cells. Although dendrites come in bunches, a neuron usually has only one axon.

Essentially, a nerve is an axon or a bundle of axons. For example, the nerve passing through the elbow that causes the painful "funny bone" reaction is a bundle of axons connecting the hand and lower arm with the spine. The cell bodies of some of these axons are located in the spine, and they carry messages to the hand. The cell bodies of the remaining axons are in the hand and lower arm, and these axons carry information in the opposite direction. Each axon is wrapped in a sheath of *myelin*, a fatty material that insulates the axon from its neighbors.

A nerve cell's ability to repair itself depends on its location. If outlying nerves, such as those in the hands and feet, are severed, they can grow back together again slowly. This is the reason that feeling returns to a severely cut hand only after several months. But nerve cells in the brain or spine rarely if ever repair themselves, making brain or spinal damage irreversible.

The nerve impulse and its code

Given the disproportionate size of the neuron's surface, it's not surprising to find that the *nerve impulse* is a phenomenon of that surface, particularly of the membrane of the axon (Figure 12–11). This membrane

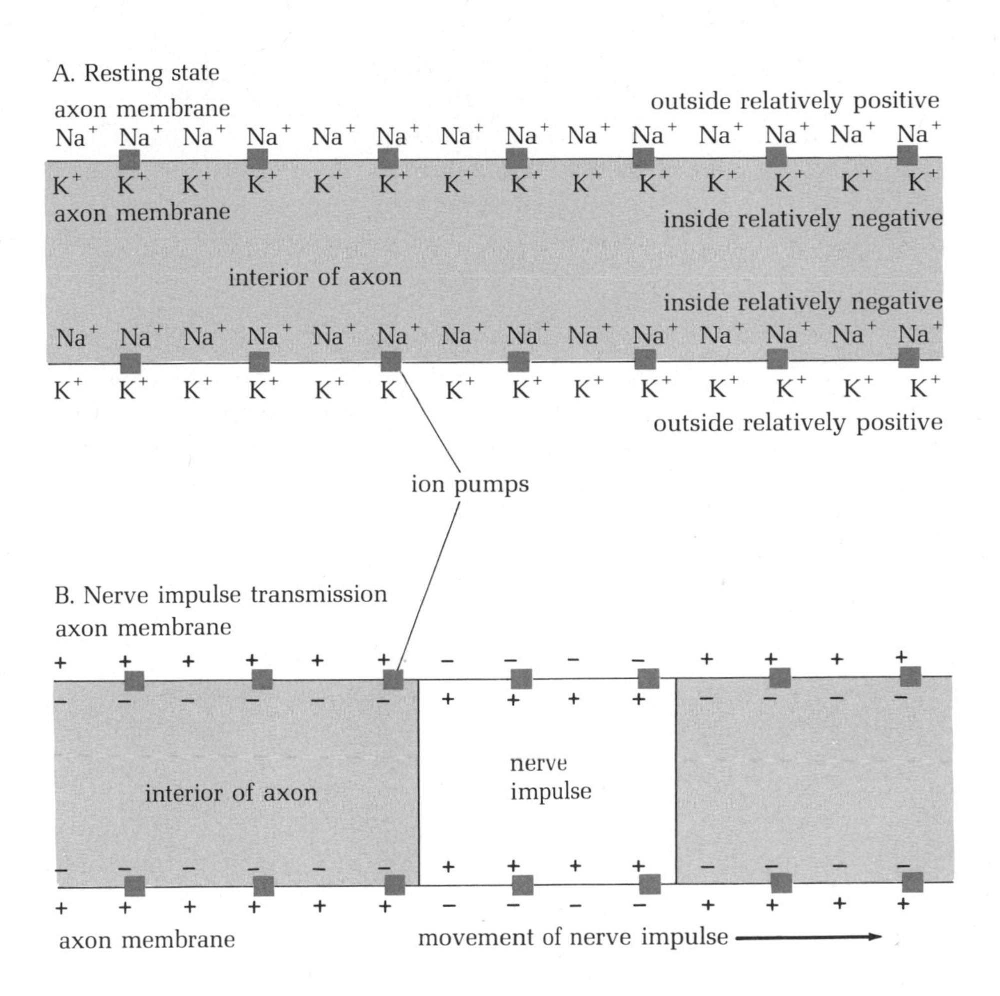

Figure 12–11 The transmission of a nerve impulse along the axon of a neuron. A. In the resting state, when no impulse is being transmitted, the interior of the axon contains more potassium ions (K^+) than the external fluid, which contains more sodium (Na^+) ions. Relative to the interior, the exterior is more electrically positive. B. As a nerve impulse is transmitted along the axon, however, this relative charge changes. Ion pumps, represented here as small squares, permit some K^+ ions and Na^+ ions to be exchanged across the membrane, thus temporarily reversing the charge. This "front" moves along the axon until it comes to the synapse; the electrical message crosses this gap chemically, and the process is repeated along the next axon.

contains *ion pumps*, poorly understood chemical mechanisms that move sodium ions (Na^+) out of the axon and bring potassium ions (K^+) in. Inside the axon, the concentration of K^+ is about 20 times greater than in the watery solution surrounding it, and this watery solution contains some 10 times more Na^+ than the axon. As a result, the axon's exterior is electrically more positive than the interior. A nerve impulse involves a change in this alignment. Stimulation of the dendrites shuts down the ion pump in a portion of the axon membrane, allowing K^+ out and Na^+ in. A small electrical current actually flows from the outside of the cell to the inside.

The current shuts down the ion pumps in the adjoining sections of membrane, setting up an electrical flow there as well. This flow shuts down the pumps in the next section, and so on, until the pulse has traveled the entire length of the axon. As soon as the electrical current has flowed through a portion of membrane, the ion pumps begin working normally again, so the nerve impulse can only travel in one direction.

How can information be coded in an electrical current flowing along the membrane of a cell? The answer is that the code works in terms of frequency: different rates of electrical flow signal different sensations. A relatively slight sensation, like a delicate touch on the hand, will send one pulse of current through the neuron and down the length of its axon. But a stronger sensation, like a crackback block thrown by an All-Pro tackle, sets off many more pulses. Generally, the frequency of the pulses doubles for every tenfold increase in sensation.

The speed with which the impulse travels depends on the size of the axon: the thicker the axon, the faster the speed. In the largest human axons, the impulse hits 400 mph, but for most human neurons the speed varies between 3 and 300 mph.

Bridging the gap

If you look closely at Figure 12–12, showing the end of the axon of one neuron in close association with the dendrites of another, you will see that the two cells don't actually touch. If the nerve impulse actually flows along the membrane of the axon, how does it jump the gap, the *synapse*, between two neurons?

When the impulse reaches the end of the axon, the neuron releases one of four different kinds of chemicals known as *transmitter substances*.

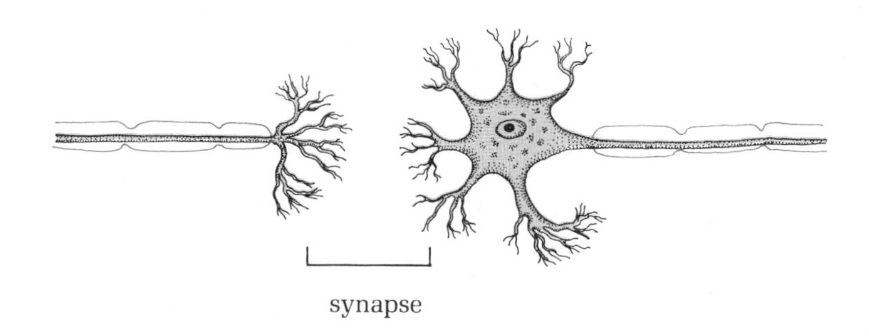

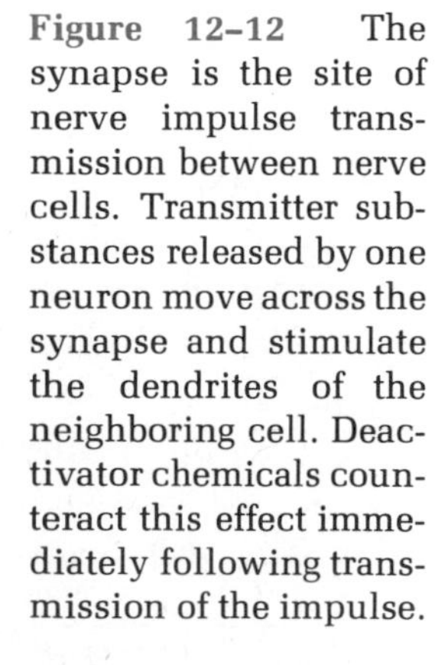
Figure 12–12 The synapse is the site of nerve impulse transmission between nerve cells. Transmitter substances released by one neuron move across the synapse and stimulate the dendrites of the neighboring cell. Deactivator chemicals counteract this effect immediately following transmission of the impulse.

There has to be some kind of brake or dampening device to stop slight impulses short and to allow only impulses of appropriate size through the system.

The transmitter migrates across the synapse to the dendrites. If enough transmitter is released to stimulate the dendrites, then the nerve impulse begins to flow along the membrane of the axon of the second neuron. Obviously, the transmitter substance cannot be allowed to accumulate in the synapse, or the second neuron might begin firing at random, not in response to a stimulus. *Deactivator chemicals* attack the transmitter almost as soon as it is produced, preventing it from causing any misfiring of the second neuron after it has transferred the original impulse.

Wouldn't the nervous system be simpler if neurons were simply connected one to another, independent of the chemical pathway of the synapse? Indeed, it would be, but it would also be much harder to control. If all nerve cells were intimately connected, then the tiniest impulse in any neuron would start the rest of them firing, and the whole system would be subject to one impulse after another. There has to be some kind of a brake or dampening device to stop slight impulses short and to allow only impulses of appropriate size through the system. Only a small proportion of the pulses in the first axon actually cross over the synapse, and not all of those that make it across actually trigger the second neuron. In some cases, the dendrites of one secondary neuron is in contact with the axons of many primary neurons. An impulse from any one of them—say, resulting from a single molecule of oxygen touching the surface of the skin—does not fire the secondary neuron. But if all the primary neurons fire at once—say, in response to a feather-light puff of breeze—then enough transmitter crosses the many synapses to send the impulse on its way along the axon of the second neuron. Mechanisms such as this provide the body with a sort of switch that selects among the stimuli traveling along a nerve and distinguishes the significant from the insignificant.

Certain drugs and poisons exert their effect by affecting the transmitter and deactivator processes in some way. The nicotine in cigarettes acts like a chemical mimic of a transmitter and causes neurons to fire, thus producing a stimulant effect. The class of powerful stimulants known as amphetamines causes excessive production of transmitters. Alcohol induces chemical changes in the neural membranes and slows the passage of impulses. Certain insecticides, like DDT and other chlorinated hydrocarbons, keep the deactivator substances from being produced; as a result, the level of transmitter chemicals builds up, and neurons keep firing madly, finally producing fatal convulsions and spasms. Some of the tranquilizers block the transmitter, thus amplifying the dampening effect of the deactivators and dulling the individual's response to stimuli—in other words, making him less nervous and slower to react. And the poison curare—long dear to the hearts of British mystery writers—mimics the deactivators and causes total paralysis. Some snake venoms work in similar fashion.

Concentrated nerve cells: the central nervous system

Figure 12–13 shows schematically the simple nervous system of the hydra, that potentially immortal freshwater relative of the jellyfish we first met in Chapter 9. Note that each area of the animal has roughly the same number of nerve cells. The nerve net of the hydra serves to unite the activities of the whole animal, but there is no centralization of its nervous functions. As one examines animals increasingly more advanced

Figure 12–13 The nerve net of the hydra, an extremely simple internal communication system.

than the hydra, the nervous system becomes more and more centralized into a nerve cord and a brain.

Man's brain and spinal cord, his *central nervous system*, acts like a special switchboard. The body contains two kinds of nerves, precisely the same in structure but different in function. The *sensory nerves* carry coded information about internal and external stimuli to the central nervous system, while the *motor nerves* direct and control responses to those stimuli. The central nervous system accepts information from the sensory nerves, interprets it, and directs the motor nerves to stimulate the glands and muscles to respond. And it mediates not only the responses we're conscious of but also a whole array of physical processes that we generally notice only when they don't function properly.

The spinal cord

Figure 12–14 shows how the spinal cord fits into the bony hollow of the spine and what a cross section of the cord looks like. Essentially it is a large and complex nerve comprising two basic pathways. One group of cell bodies and axons carries impulses from the body to the brain, while the other follows the opposite route. Axons and dendrites enter and leave the spinal column through channels in the vertebrae.

Some kinds of connections between sensory and motor nerves in the spinal cord or areas of the brain are relatively automatic. These reactions are called *reflexes*, and the path of nervous impulses coming into, passing through, and then leaving the spinal cord, for example, is called a *reflex arc*. Take the familiar and painful example of touching a hot stove. Sensory axons send a rapid flurry of impulses to the spinal

Figure 12–14 A cross section of the human spinal cord.

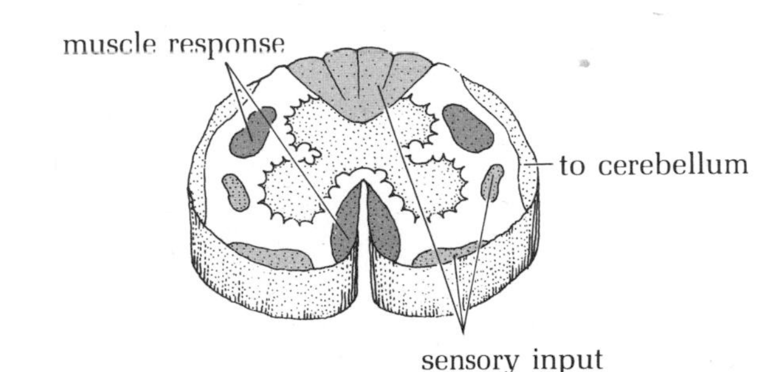

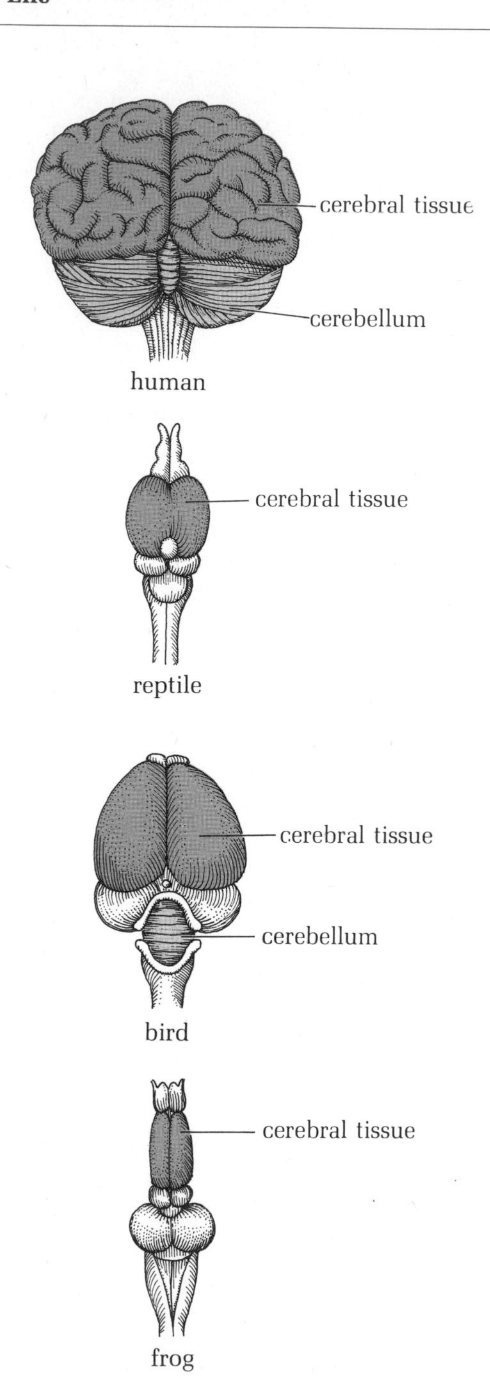

Figure 12–15 A comparison of four animal brains, showing the increase in proportional cerebral tissue as a major consequence of evolution.

cord that carry the alarm. A practically instant synapse with a motor nerve connected to the biceps sends an impulse to that muscle, causing it to contract and jerk the hand back before any harm is done. Only after the fact is the "hot" sensation felt. Some reflexes, like the many nerve signals that help to maintain balance, are more complex in terms of the number of synapses and muscles affected, but they all follow this basic pattern.

The brain

As one follows the progress of evolution, the brain becomes an increasingly important part of the anatomy. Figure 12–15 shows how brain size grows as life forms become increasingly complex. And, as we saw in Chapter 4, human evolution is really the evolution of the brain.

The first brains were really nothing more than a swelling or bulb at the end of the spinal cord, and that description essentially fits the human brain as well. Because of differentiation, the brain is divided into a number of structures, or regions, each of which acts as a control point for a certain part of the body or for a certain physical activity. Figure 12–16 details the major portions of the human brain.

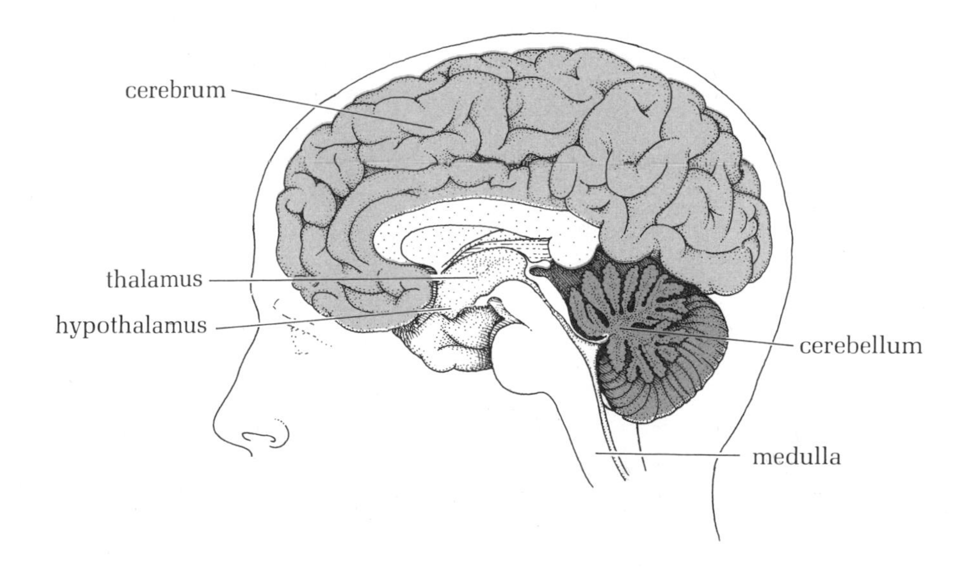

Figure 12–16 The human brain.

The *medulla*, located at the very base of the brain, is an extension of the spinal cord, and a remnant of the original primitive brain. It controls such basic physical rhythms as heartbeat and breathing, and contains the reflex centers for vomiting, sneezing, coughing, and swallowing. The *cerebellum*, set a little above and behind the medulla, is largely occupied with automatic controls of body position and motion. It monitors the tension of arm muscles, for example, and keeps track of where the arm is in relationship to the body and the environment. When you make a conscious wish to move your arm, the cerebellum acts as a kind of intermediary, translating the wish into a series of specific impulses and relaying them to the proper muscles.

The *thalamus* also acts as a sort of intermediary, but in the opposite direction. This small, bulbous region is the terminal for many of the sensory nerve fibers that enter the spine and travel up to the brain. In lower animals, sensory impulses stop here, but in man the thalamus relays them on to higher brain centers. The *hypothalamus* is the most important automatic control point, and it monitors particular needs and drives instead of specific portions of the body. Hunger, thirst, and sex are all regulated by the hypothalamus. It contains centers for the perception of cold, heat, touch, pain, and pressure, and it controls such basic but complex functions as sleep, appetite, hostility, pleasure, and so forth. Investigators probing the functions of the brain have been able to insert electrodes into the hypothalami of experimental animals and leave the rest of the brain undamaged and intact. Figure 12–17 shows a rat with a wired hypothalamus. Electric current sent through the electrodes has the same effect as a nerve impulse. By passing current through the electrodes, the rat can be made to attack aggressively, to flee in fear, to sleep, or to eat—the particular action depending on the position of the activated electrode.

The highest brain center, and the largest portion of the human brain, is the *cerebrum*. The very surface of the cerebrum, the *cerebral cortex*, is the site of conscious thought, of memory, and of intelligence, and the place where sensory impulses about sight, sound, taste, smell, and touch are sorted out and interpreted. The remainder of the cerebrum integrates

By passing current through the electrodes into the hypothalamus, the rat can be made to attack aggressively, to flee in fear, to sleep, or to eat.

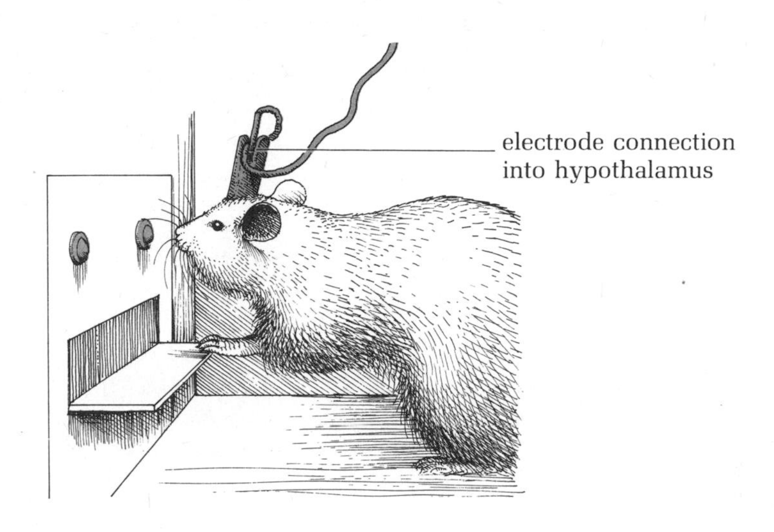

Figure 12–17 The hypothalamus apparently controls most major drives and needs. In the mid-1950s, Professor James Olds explored the correlation between electrical stimulation of the hypothalamus and the pleasure response in rats. As this rat presses the peddle in its cage, it triggers an electric current into its hypothalamus, thus creating a pleasurable response.

and controls the sensory and motor systems—it acts as a sort of final switchboard. Figure 12–18 shows the motor and sensory portions of the cerebral cortex associated with various parts of the body. By no means does the relative size of a particular body structure determine the amount of nerve tissue that controls it. Compare how much of the cortex deals with the hands and fingers with the amount that takes care of the whole of the body between the armpits and the ankles. One curious point about the cerebrum is that the nerve fibers entering and leaving it cross over each other deeper in the brain. As a result, the right side of the cortex controls the left side of the body, and vice versa. A blow to the right side of the head severe enough to damage the brain affects the left side of the body.

The two nervous systems

Although the central nervous system is one basic structure, there are two distinct sets of nerves within it. The sensory and motor nerves we've been discussing belong to the *voluntary nervous system*. This portion of the nervous system can be directed by conscious thought and usually is. Deciding to raise your arm or to take a step involves the volun-

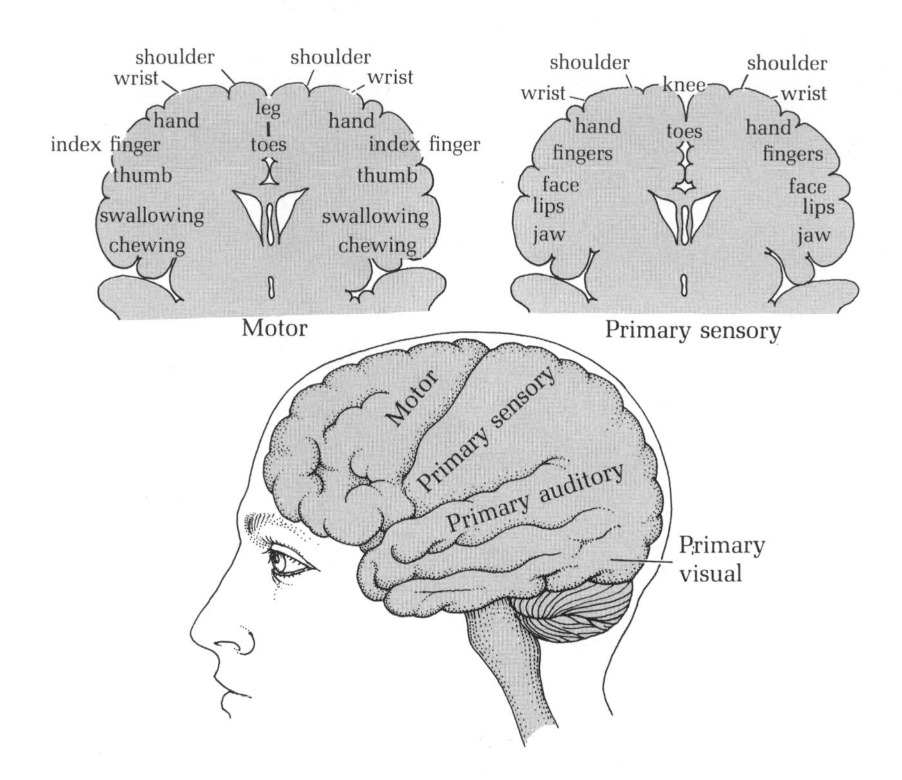

Figure 12–18 The motor and sensory portions of the human brain.

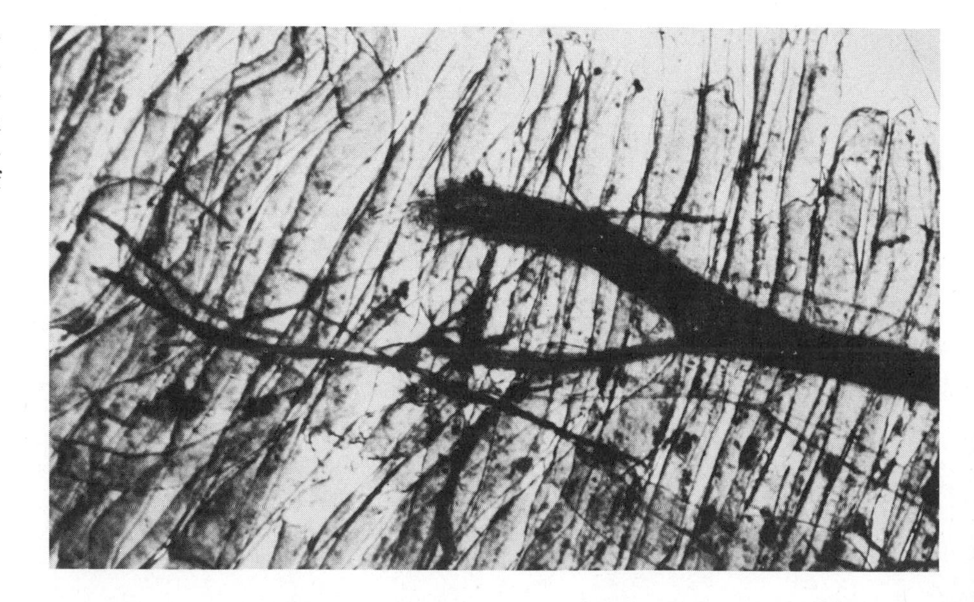

Figure 12–19 An effector nerve cell, connected to muscle tissue. A nerve impulse in this nerve cell stimulates the response of the muscle.

tary system. Figure 12–19 shows how a motor nerve is connected to a muscle. Each muscle cell in a large muscle is controlled by its own nerve fiber. The contact between the nerve fiber and the muscle cell is the *motor end plate.* The nerve impulse reaching the motor end plate triggers the release of a transmitter chemical that crosses the synapse between nerve and muscle and triggers contraction of the muscle cell. The more muscle cells that contract, the more contraction in the whole muscle.

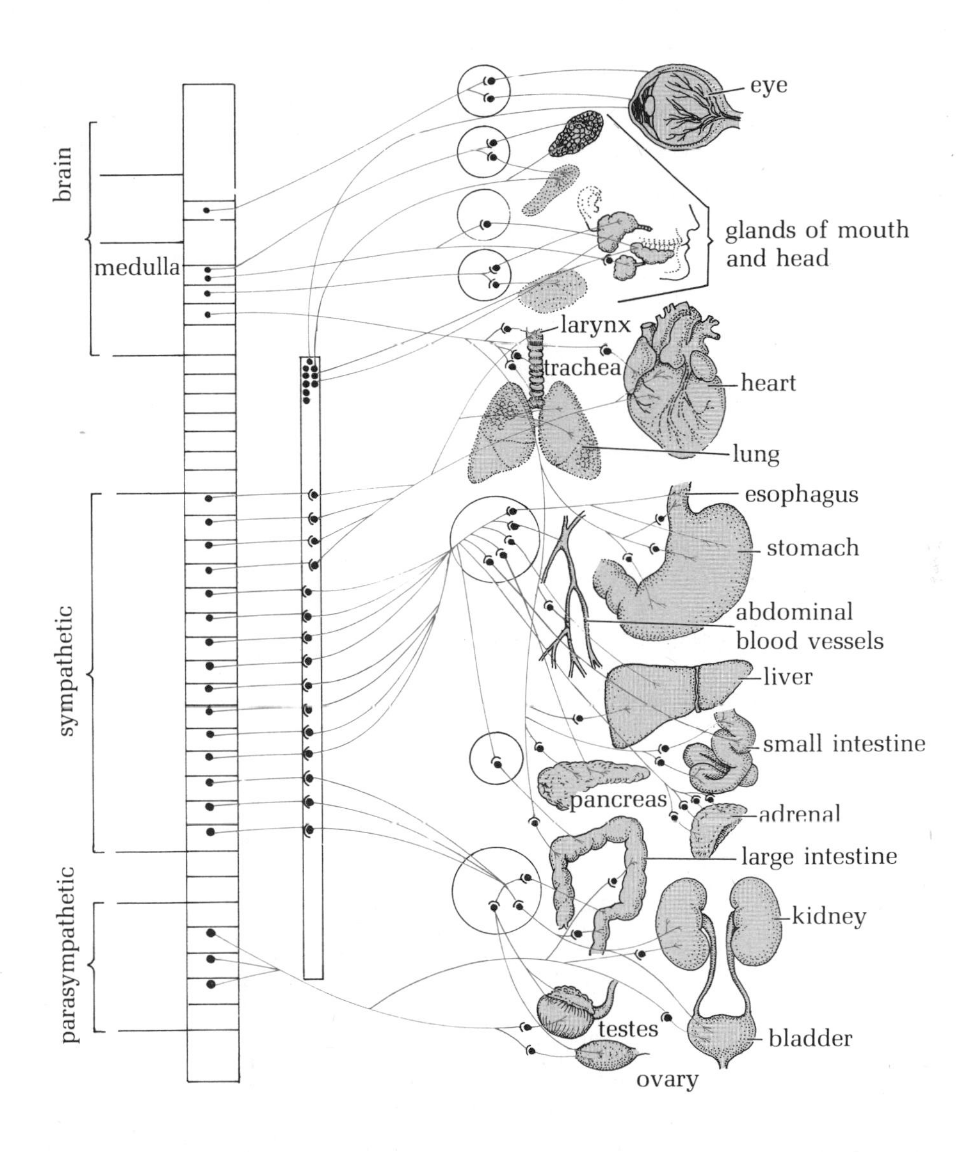

Figure 12–20 The coordination of the autonomic nervous system and the organs of the body.

Another portion of the nervous system, connected to the internal organs, normally functions apart from conscious decision, is called the *autonomic*, or *visceral, nervous system*. Figure 12–20 shows the autonomic nerves and the organs they affect. As you can see, almost every organ has two nerves, which, like the members of an antagonistic muscle pair, work in opposite directions. Impulses from one of the nerves stimulate the organ, while the other has an inhibiting or slowing effect. The two sets of nerves are analogous to the accelerator and brake of a car. If all the stimulating nerves fired all the time, the body would race along at top speed continually. By the same token, action by the inhibiting nerves alone would produce a state of constant standstill. The usual normal state of the body lies somewhere between these two extremes, and both sets of nerves serve to maintain this homeostatic balance.

In practice, the voluntary and autonomic nervous systems overlap quite a bit. Breathing is autonomically controlled, but you can change your rate of breathing consciously. There is experimental evidence that it may be possible to voluntarily reduce dangerously high blood pressure levels in heart patients. For centuries, Indian mystics have been able to reduce their blood pressure, heartbeat, and rate of breathing far below normal. The techniques for effecting these changes have become popular in this country under the name of transcendental meditation.

Putting hormones and nerves together

The endocrine and nervous systems differ generally in the speed of their responses, but they share a similar homeostatic task: ensuring that the parts of the organism function as a whole. How are these two systems themselves synchronized so that they work together?

By actual chemical and physical connections: The pituitary is situated very close to the hypothalamus, and its workings are under the control of that portion of the brain. The four tropic hormones are released in response to chemical messengers—essentially, short-distance hormones—sent by the hypothalamus to the pituitary. Two of the pituitary hormones, vasopressin and oxytocin, are actually manufactured in the hypothalamus and passed along nerve fibers to the pituitary to be released. The fight-or-flight response of the body is triggered by the nervous system. Nervous stimulation of the adrenal medulla promotes the release of adrenaline into the bloodstream. The same set of nervous signals increases the ACTH secreted by the pituitary, thus raising the level of cortical hormones available to the body.

These connections between the nervous and endocrine systems ensure that the most basic homeostatic task of all is accomplished: binding the millions of cells of the body into a coordinated and working whole.

Summary

Homeostasis refers to an organism's ability to maintain itself in the face of environmental changes.

Chemical and physical defenses keep foreign substances and microorganisms out of the body. Those that get inside are attacked by white blood cells, which manufacture antibodies. Each antibody is specific to a particular antigen, and the antibodies constitute the immunity response. Vaccination is a way of promoting immunity, either actively or passively. Passive immunity is conferred to a fetus before birth by means of the mother's antibodies and lasts for several months after birth. The immunity system can tell self from nonself. Immunity may go wrong, producing autoimmunity, allergy, or pregnancy complicated by the Rh factor. Immunity also complicates medical transplants.

Hormones act as chemical messengers, and they are the principal homeostatic mechanism of plants. Plants have several kinds of hormones: auxins, gibberellins, ethylene, cytokinins, and abscisic acid. The effect of each hormone depends on its amount and the target tissue. The principal hormones in man are secreted by the endocrine glands. The tropic hormones of the pituitary influence the secretions of the gonads, thyroid, and adrenals. The level of hormones is controlled by negative feedback mechanisms.

The nervous system is a high-speed communications network. The basic unit of the nervous system is the neuron, characterized by its dendrites and axon. A nerve impulse is a change in the electrical character of the axon. Special transmitter chemicals aid in the transmission of the impulse across the synapse separating neurons. In higher animals, nervous tissues are concentrated and specialized in the central nervous system, which connects the motor and sensory nerves. The spinal cord is a large nerve with numerous branches. The brain is a specialized enlargement of the spinal cord. Its various regions control and coordinate specific areas and activities of the body. Some of these regions of the brain are the medulla, cerebellum, thalamus, hypothalamus, and cerebrum. Both the voluntary and the autonomic nerves are under control of the central nervous system.

The endocrine and nervous systems are coordinated by the connections of the hypothalamus and the pituitary. Both systems work together to make the organism function as a whole.

Questions to consider

1. a. What is meant by homeostasis?
   b. Think about the various activities you perform and the environmental changes you meet in a typical day. What sort of homeostatic responses does your body commonly make?
2. a. The body's first line of defense against foreign substances and microorganisms is a variety of chemical and physical mechnisms. What are they?
   b. In what major way do these mechanisms differ from the immunity response?
3. a. How do white blood cells counteract invading microorganisms?
   b. How do antibodies perform their defensive task?
4. A developing fetus rarely comes into contact with microorganisms in the uterus, but blood tests of any newborn will show the presence of antibodies. Where have these antibodies come from?
5. Kidney transplants between two strangers chosen at random have a low chance of success, but a transplant between two closely related individuals, such as parent and child or brother and sister, often works. Between identical twins, a transplant nearly always succeeds. Why should this be the case?
6. a. What is a hormone?
   b. What sorts of homeostatic tasks do hormones perform?

7. a. What are the plant hormones, and what are the principal effects of each one?
   b. The effect of a hormone depends on the amount available and on the target tissue. How is this point shown in the plants?

8. a. What are the major glands of the human endocrine system?
   b. In the writings of a scientist given to a poetic turn of phrase, the pituitary gland is likened to the conductor of a chemical symphony. Why is this an accurate description? How does the pituitary do its conducting?

9. In man as in plants, the effect of a given hormone depends on the target tissue. The mechanism controlling the level of hormones in the body is built around this phenomenon. How does it work?

10. Plants maintain themselves almost entirely with hormonal regulation, but in man, as in many other animals, hormones alone are not enough. Why?

11. The most prominent feature of any neuron is the long protrusions branching out from it. What names are given to these branches, and what functions do they serve?

12. a. How does a nerve impulse pass through a neuron?
    b. What form does the code of the nervous system take?

13. What are the components and functions of the central nervous system?

14. A friend has gone on vacation and asked you to bird-sit his pet parrot, a particularly ill-tempered animal given to biting anyone who comes near. You open the cage to fill the parrot's water dish and see the bird preparing to attack by moving toward your hand. Instantly you jerk your hand from the cage and out of harm's way. What sort of nervous response is this?

15. a. What are the regions of the human brain?
    b. What role does each play?
    c. The human leg is a good deal larger than the hand. It seems logical to assume that the greater size of the leg earns it more space in the cerebral cortex than the hand gets. Is this the case?

16. How are the nervous and endocrine systems coordinated?

Glossary

abscisic acid a plant hormone that limits cell growth

abscission the natural loss of a portion of plant tissue such as a leaf or stem

active immunity the stimulation of an immunity response by the introduction of antigens

adrenal cortex the external layer of the adrenal gland, which produces a large number of hormones

adrenal gland one of two endocrine glands situated on the kidneys whose hormones serve to regulate the body's chemical functions

adrenalin an adrenal hormone which stimulates the body's response to quick energy needs

antibody a protein substance produced by the white blood cells which inactivates foreign substances introduced into the body

antigen a chemical substance that stimulates the body's production of a specific antibody

autoimmunity an abnormal condition in which the body produces antibodies which attack its own tissues

autonomic nervous system the portion of the nervous system that controls those body functions which are independent of conscious thought

axon the threadlike extension of the neuron which carries nerve impulses to other nerve cells

basal metabolism rate the total amount of energy from cellular respiration required to maintain minimal body functions

central nervous system the brain and spinal nerves which together form the major portion of animal nervous systems

cerebellum the portion of the brain above the medulla which controls movement and equilibrium

cerebral cortex the furrowed, outer layer of the cerebrum of the brain which is the site of thought and memory

cerebrum the largest part of the brain, housed within the skull, which is the site of thought, memory and sensory coordination *colostrum* the fluid secreted by nursing mammals which confers temporary passive immunity on the offspring

cortical hormones the group of hormones secreted by the outer layer of the adrenal glands

cytokinin a form of plant hormone which regulates new growth at buds

deactivator chemical a substance released at the synapse between nerve cells which halts the transmission of nerve impulses

dendrite the branched extensions of a nerve cell which conduct nerve impulses from the axon of another cell toward the cell body

ductless gland *see* endocrine gland

endocrine gland one of a complex system of glands which secrete their hormones directly into the bloodstream

gibberellin a form of plant hormone which regulates cell size and cell division

homeostasis the ability of living systems to maintain constant internal functions despite chemical and physical changes in the environment

hypothalamus the region of the brain which controls basic biological drives and relates them to specific senses

immunity response the consistent ability of an organism to produce specific antibodies against an antigen

ion pump one of the active sites along the axon of a nerve cell which regulates the internal-external balance of ions during impulse transmission

medulla (of the adrenal gland) the interior region of the adrenal gland which produces adrenalin

medulla (of the brain) the region of the brain which regulates basic rhythmic internal functions and some simple reactive mechanisms

motor end plate the juncture between a nerve cell and a muscle cell

motor nerve a nerve fiber that stimulates the contraction of a muscle

myelin the fatty substance that insulates the axon

negative feedback a form of control mechanism in which a substance that controls the production of a product is inhibited by that same product *nerve impulse* a burst of electrical energy which is transmitted by nerve fibers and which carries information

nervous system the organs and tissues that carry information and regulate internal functions in animals

neuron a nerve cell

passive immunity a ready-made immunity produced by the introduction of specific antibodies

reflex an almost automatic simple response to a stimulus which is controlled by the spinal nerves

reflex arc the path between the sensory nerve, the spinal column, and the motor nerve which results in a reflex

Rh factor a chemical antigen in the bloodstream which can have dangerous effects if there is Rh incompatibility between mother and fetus

sensory nerve a nerve fiber that detects environmental stimuli

synapse the gap between the dendrites of one neuron and the axon of another

thalamus the region at the lower side of the brain which receives some sensory information and relays it to the hypothalamus

thyroid gland the gland situated in the neck which secretes hormones that regulate the basal metabolism rate

transmitter substance a chemical substance released at the synapse which carries the nerve impulse between neurons

tropic hormone an endocrine hormone that regulates the functions of other endocrine glands

vaccination the induction of immunity by the introduction of the antigens of a disease

visceral nervous system *see* autonomic nervous system

voluntary nervous system the portion of the nervous system regulated by conscious thought

white blood cell the large blood cells which produce antibodies

13 Communicating: External

An important aspect of homeostasis is the ability to respond to changes in the external environment, an ability so intrinsic to all organisms that it was listed as one of the qualities of life in Chapter 1. Such responses are particularly dramatic and visible in organisms capable of movement. The simple, one-celled paramecium responds to a variety of environmental changes. If bacteria, which are one of the common foods of the paramecium, are distributed in a thin line in the organism's watery environment, the paramecium will follow the line of bacteria, consuming each one it meets. Paramecia are quite sensitive to temperature, preferring water at 80°F. If paramecia are put into water under this temperature and a small portion of the water is then warmed slightly, the paramecia will crowd into this warmer area. Should they venture out into the colder water, they quickly turn around and swim back to the warmth.

The remarkable thing about the responses of microorganisms like the paramecium is that they have no special organs for detecting or interpreting these changes. A paramecium has no nervous system, but when it follows the line of bacteria with the accuracy of a dog on a scent, it can somehow tell that there is more food at its leading end than at its trailing end. Likewise, without any special temperature detectors, it can tell that the water in one spot is warmer than the water nearby.

Man, like many of the multicelled animals, does have special sensory

In constructing its web, the spider uses its sense of taste to test the tautness of the filaments, its sense of balance to maintain equilibrium, and its sense of vision.

structures whose job it is to monitor the environment and inform the organism of any significant changes. Dependent as we are on our senses, we tend to take them for granted, but they are remarkable structures, not only from the point of view of anatomy but also in the way that they shape our behavior. In this chapter, we'll examine how the senses work and how they affect the way we and other organisms behave.

The nature and range of senses

The senses are essentially specialized receptors that convert particular changes in the environment into a pattern of impulses traveling along a sensory nerve and send them to the brain to be interpreted. Some of these receptors, like those for pain, are unspecialized, being simply bare nerve endings embedded in the skin. Some, such as the taste buds in the tongue, are specialized cells that pick up the stimulus and transmit it to a neuron. And still other receptor cells and the neurons associated with them are concentrated into elaborate sense organs like the eyes and ears.

It's important to remember that a given impulse traveling along a sensory nerve looks just like any other impulse. What distinguishes a taste impulse from a signal for sound is not the electrical or chemical nature of the impulse, but the nerve it is following and the part of the brain it leads to. A nerve cut off from its receptor cell may still have a working connection with the brain, and, so far as the brain is concerned, the receptor is still working. People who have lost an arm or a leg by surgery or accident often report that they feel what seems to be pain or itching in the missing limb, an example of "phantom pain." The cut nerves in the stump are irritated as the amputation wound heals. The irritation triggers the neuron, sending an impulse to the brain that is interpreted as pain or heat or whatever kind of signal the cut nerve carried when it was still intact.

Fortunately, you don't have to go through an amputation to experience the same effect. If you look to the left and then press gently on the right-hand corner of your right eye, a small oscillating spot will appear in the shadow cast by the bridge of your nose. Of course, the spot isn't actually there. The pressure is triggering receptor cells within the eye, and the brain interprets any impulse from these cells as visual, even when the signals are actually caused by pressure. A similar effect has troubled astronauts in space. They reported seeing white flashes that seemed to come from nowhere and that didn't appear on ground monitors. The cause of the flashes turned out to be high-energy, subatomic particles passing through the eye and triggering neurons.

Seeing

The very simplest eyes are nothing more than light-sensitive cells that tell the organism how much light is present in the surroundings.

What distinguishes a taste impulse from a signal for sound is not the electrical or chemical nature of the impulse, but the nerve it is following and the part of the brain it leads to.

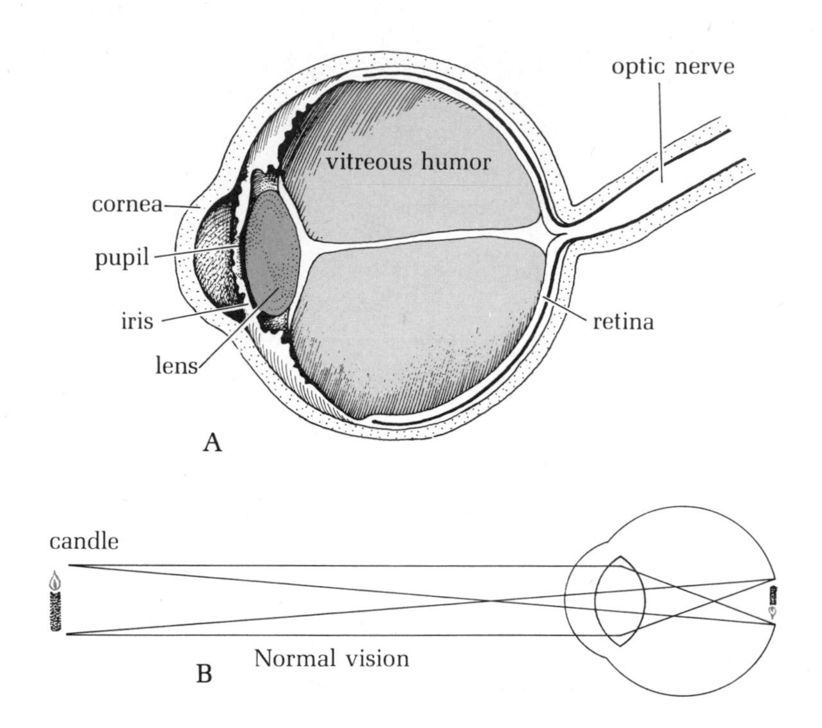

Figure 13–1 A. The human eye. B. The human eye creates a clear image of visual stimuli and constitutes one of our primary links with our environment.

More complicated eyes contain a spot of pigment that shades the eye from one direction; this kind of eye allows the organism to tell where the light is coming from. The sensitivity of a primitive eye is increased with the evolution of a clear lens that concentrates light on the receptor cells. The eyes found in man and other vertebrates are even more complex. The lens focuses light from different parts of the environment on different receptors to create not just a shadowy pattern of light and dark but an actual image of the surroundings (Figure 13–1).

The parts of the human eye

The round sphere of the eye is set into a socket of the skull. Muscles attached to the eye and to the skull move the eye within its socket and allow it to be pointed in various directions. The very front of the eye is a transparent bulge called the *cornea* that covers a fluid-filled chamber. At the back of the chamber is the *iris*, the colored portion of the eye, which surrounds the circular opening of the *pupil*. The iris can expand or contract to adjust the size of the pupil to the amount of available light. In a dim room at night, the iris opens fully, but in bright midday sunlight, the iris closes the pupil down to little more than a pinhole.

Behind the iris is a dense, transparent *lens* suspended in a cradle of muscles. The lens focuses the light coming through the pupil. The muscles change the shape of the lens so that the eye may focus on objects both far and near. What we experience as eye strain is often fatigue in the lens muscles. When the muscles are relaxed, the lens is focused for distance viewing, but the muscles must contract to focus the eyes for close-up work. Several hours of reading or typing tires eye muscles just as walking fatigues the legs.

Another fluid-filled chamber separates the lens from the back of the eye, or *retina*, where the cells sensitive to light are located. A network of nerve endings connecting the receptor cells gather into the large cable-like bundle of the *optic nerve*, which carries the visual impulses from the retina to the brain.

Rods and cones

The eye contains 120 million receptors divided into two kinds, each sensitive to different stimuli. They are given the names *rods* and *cones* because of their shapes (Figure 13–2). The cones are sensitive to color, but only in bright light. Most of the cones are concentrated near the center of the retina. As one moves away from the center, rods become increasingly common, finally replacing the cones altogether. The rods are sensitive even to very dim light, so much so that they respond to the smallest measurable unit of light. However, they cannot tell one color from another. The differing sensitivities of these two kinds of cells explains why human vision varies so much from night to day. At night, with the moon and the stars the only sources of light, you cannot tell one color from another or read fine print, and you are more likely to notice motion from the corner of the eye than from the center.

The chemistry of color detection is still a mystery. Black-and-white vision is somewhat better understood, but the whole picture is by no means complete. One thing known for sure is that vitamin A, or similar compounds called *carotenes*, are necessary for vision in vertebrates. Vitamin A cannot be synthesized in the body and must be included in

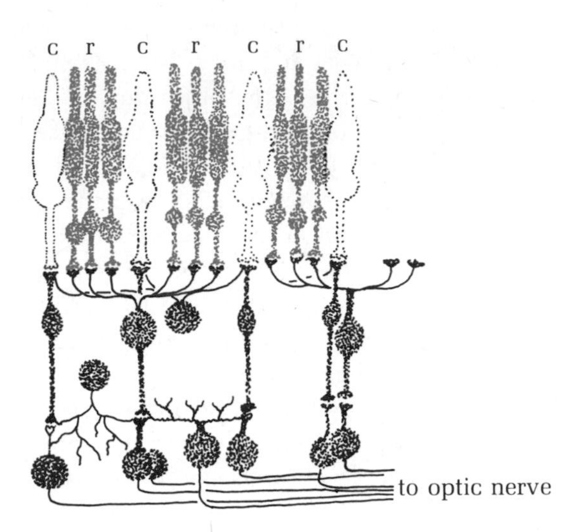

Figure 13–2. A portion of the cross section of the retina, showing the alignment of the rods and cones. The rods (r) are sensitive to all intensities of light, including dim light, but do not detect color. The cones (c) detect and differentiate between colors. Both rods and cones are connected to minute nerve fibers that lead to the optic nerve and then to the brain.

the diet in the form of fruit, leafy greens like lettuce and spinach, or the yellow vegetables such as carrots, winter squash, and sweet potatoes. Many animals, including man, whose diets lack carotene develop night blindness, an inability to see in dim light.

Seeing different colors

Man is not the only creature with color vision, but ability to see color is not widespread. Animals active only at night—a very great number of species—see only black and white. In addition, animals that do perceive color don't necessarily see the same colors that we humans do. Bees, for example, are known to be sensitive to color, and most of their foraging for pollen and nectar is guided by a sense of color and shape. However, bees can see only a limited portion of the spectrum, ranging from ultraviolet (which the human eye cannot perceive) to orange. The average human, however, can see some 60 colors and blends of color, running from violet to red. The bee's ability to see ultraviolet light gives him a picture of the world very different from the one we get. Figure 13–3 shows two photographs of the same flower, one taken with standard film and the other with film sensitive to ultraviolet light.

Figure 13–3 Seeing the world differently. The same flower looks very different to the bee, whose eyes are sensitive to ultraviolet light (right), and to a human, whose eyes do not detect ultraviolet light (left).

Compound vision

Animal eyes differ not only in sensitivity to color but in structure. Certain arthropods—among them grasshoppers, dragonflies, fireflies, lobsters, and crayfish—have *compound eyes*, aggregations of many smaller, simple eyes (Figure 13–4). Each eye has its own cornea and its own receptor cells. These eyes cannot rotate or move, and they see only the light that falls directly on them. As a result, each eye sees only a small part of the world around it. What the insect or shellfish perceives is a composite of all these tiny, individual bits of vision. The result is probably something like a newsphoto, a great number of fine dots so close together and so tiny that they appear to be only one image. Obviously, the greater the number of dots, the sharper and more detailed the picture. Insects

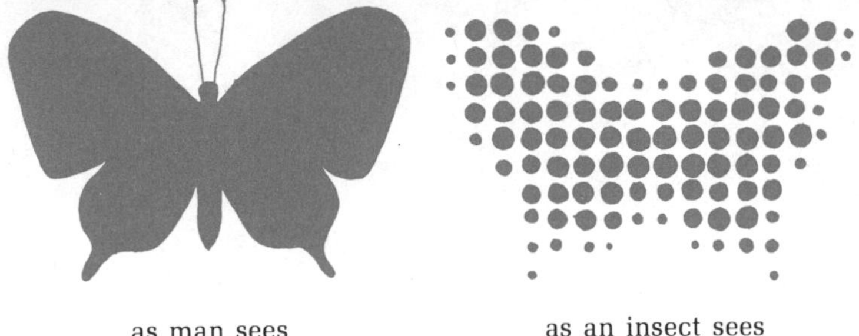

Figure 13–4 A. The compound eyes of the housefly. Each facet is a separate lens system that perceives a small fraction of the total visual space around the insect. Thousands of additional images are coordinated in the insect's brain to form a single picture of the outside world. B. As seen through the compound eyes of a bee, the butterfly appears as many separate dotlike images.

with acute vision, like the dragonfly, have as many as 28,000 simple eyes in each of the pair of compound eyes, while the eye count in an insect with poor vision runs under 1,000.

Man's eye can detect more detail than even the dragonfly's, but the compound eye is a superior instrument by far for detecting motion and for seeing at high speed. A film, for example, consists of single pictures projected in sequence at a rate so rapid that the human eye cannot tell one picture from the next. As a result, the image on the screen seems to move. A dragonfly, though, would not see motion at all, but a series of individual frames. In fact, the film would have to be shown five times faster than normal before the illusion of motion would appear to the dragonfly. So far as seeing is concerned, flying at high speed is the same

The dragonfly, who perceives motion at a speed five times faster than that of man, would have little patience with a human projectionist.

thing as watching a film, except that the observer and not the observed is moving. The compound eye of the dragonfly and other fast-flying insects provides a way of seeing accurately even at high speeds; thus a dragonfly zipping over its habitat can make out each object clearly.

Fitting eyes to lifestyle

Although all mammals have roughly similar eyes, their vision is affected markedly by how those eyes are positioned in the head. If the eyes are set into the front of the head, then the field of vision of one eye overlaps with the other. As a result, the visual image has three dimensions, and the animal can estimate distances accurately up to some limit. However, this front-of-the-head arrangement reduces the effective field of view (Figure 13–5). If the eyes are set so that one is on each side of the head, then there is little or no overlap. The visual image is flat, with little perception of distance, but the animal has an extremely wide field of vision encompassing almost the whole area around him. Given this trade-off between distance judgment and wide field of view, which kind of mammal is likely to be equipped with which eye arrangement?

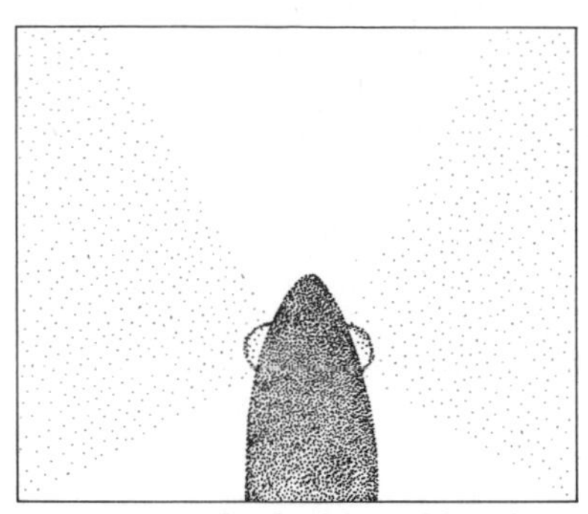

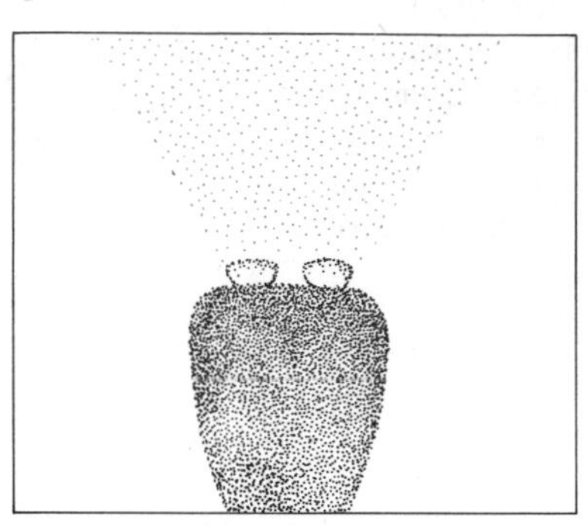

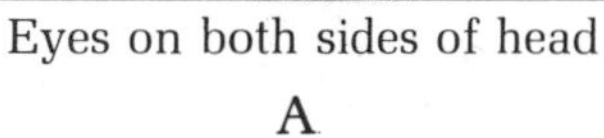

Eyes on both sides of head
A

Front-of-head arrangement
B

Figure 13–5 Field of vision. Eyes mounted on each side of the head (A) provide a large field of view, but do not offer three-dimensional vision. Eyes mounted at the front of the head (B) provide a smaller degree of peripheral vision, but offer the advantage of three-dimensional vision and accurate perception of distance.

Figure 13–6 compares the faces of a wolf and a cat with those of a deer and a bighorn sheep. Predators, who have to rely on stalking prey, are greatly aided by vision that allows them to estimate distances ac-

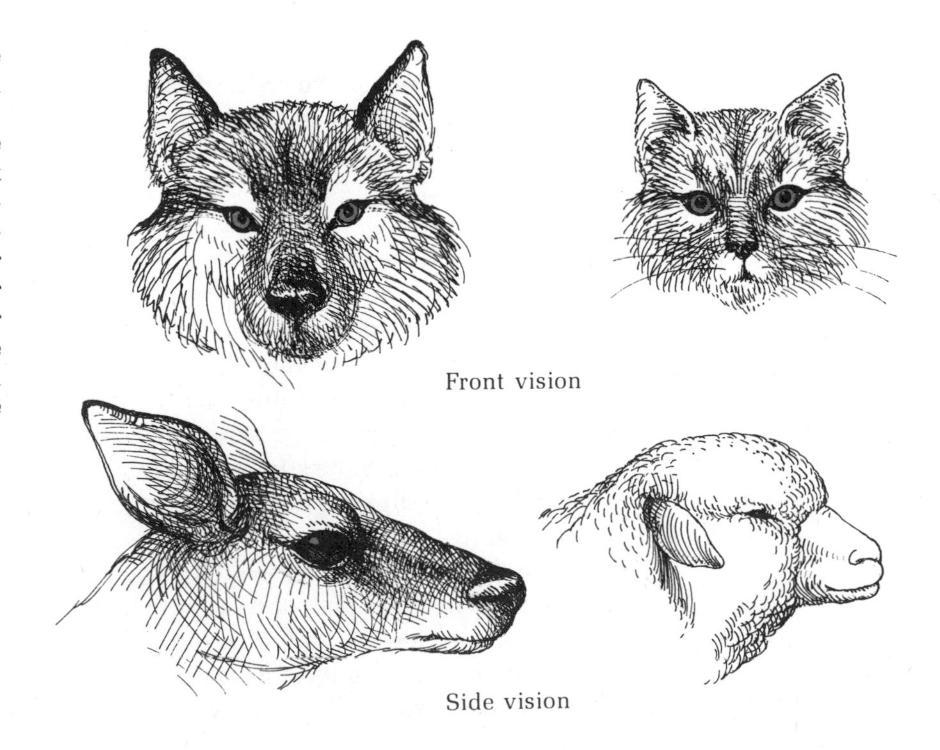

Figure 13–6 Some common mammals and their fields of vision. The wolf and cat have eyes placed at the front of their heads, thus providing three-dimensional vision. The deer and sheep, on the other hand, have a far wider field of view, because their eyes are mounted on each side of the head.

curately. A lion, for example, can maintain top speed for only about 150 yards. Its success at catching an antelope often depends on whether the prey is 50 or 25 yards away at the start of the chase. To a grazing animal hunted by predators, three-dimensional vision is of less value. But a wide field of view allows the animal to keep a close watch on all avenues of attack, even those to the back and the sides, and to flee at the first sign of a hungry predator skulking in the grass.

Seeing selectively

Given the great sensitivity of our many rods and cones, it's a wonder that our brains aren't overwhelmed with a constant rush of visual stimuli. Simply inspect the area close by you, and notice the huge variety of things there are to see. Yet we see only a small proportion of them. We select among the great welter of stimuli and concentrate only on certain aspects of them. Take as an example the reading of this page. By no means do we notice each and every letter. Instead, we see groups of letters, or words. For that matter, we don't even see every word; *the* and *a* seem to slip unnoticed into the background, as do marks of punctuation. A similar process occurs in all our seeing, and in all the seeing of other animals.

Part of this effect is due to the way neurons work. Some of the synapses in the eye inhibit rather than stimulate. A message jumping the synapse from one neuron to another may actually prevent the next neuron from firing. The overall pattern of synapses, some inhibiting and some stimulating, obliterates some images and accentuates others, concentrating the relevant details in the mind.

In addition, we learn to pay more attention to some things than to others. The details we pay the most attention to depend partly on inborn knowledge, or instinct, and partly on experience. Specialized looking produced by instinct, learning, or both can produce some comparatively amazing feats. It seems that migrating birds and homing pigeons find their way to their destination by sight. They fly in relation to the position of the sun or the stars in much the same way that a navigator keeps a plane or a ship on course. But man's navigation is a recently learned and complex skill, involving knowing the time of year and day, determining how far north or south of the equator one is, and making accurate sightings of the sun or stars with instruments. Somehow the birds do a very

Figure 13–7 The English robin defends his territory primarily by the detection of the red breasts of intruders.

passable job without instruments, relying only on instinct. And that is rather remarkable, when you recall that all hereditary information is transmitted on DNA molecules. What precise pattern of bases tells a canvasback duck the route from British Columbia to Mexico and provides it with the right visual clues to guide its way? That is a big question, and an unanswered one.

Another curious fact about looking is how little the animal has to look for. In other words, a visual clue that may seem insignificant can cause the animal to change its behavior. Take, for example, the red-breasted English robins shown in Figure 13–7. Each spring, as the mating season approaches, the male robins stake out precise territories and drive off any males that violate the boundaries. How do the robins tell who is a male of their species and who is not? If a bundle of red feathers is put into a robin's territory, he will attack it just as if it were another bird. But a mounted specimen of an immature robin, which does not have the red breast, will be ignored. Obviously, all the robin defending his territory looks for is the red breast of an intruder.

Hearing

We are all familiar with what happens when a pebble is tossed into a still pond. Circle after circle of waves flow from the pebble's splash toward the shore. Sound results from a very similar physical phenomenon. A vibrating surface, like the tines of a tuning fork, transfers its vibrations to the air molecules touching it. These molecules transfer the vibration to their neighboring molecules, and so on, so that the sound moves out in a wave. If the vibrating tuning fork is touched to other objects, the vibration will travel through their molecules as well as through those in the air. In fact, the sound may even move more quickly. Sound traveling through the air hits about 720 mph, the so-called speed of sound. In water, the waves move four times faster, and, in steel, 15 times faster.

The human ear collects sound vibrations, translates them into nerve impulses, and sends them to the brain to be interpreted.

The workings of the human ear

The skin-covered cartilage of the *outer ear* serves to catch sound and channel it into the *ear canal* (Figure 13–8). There the sound waves strike a thin membrane—the *eardrum*, or *tympanic membrane*—and set it vibrating. The slight movements of the eardrum are amplified some 20

Figure 13–8 The human ear.

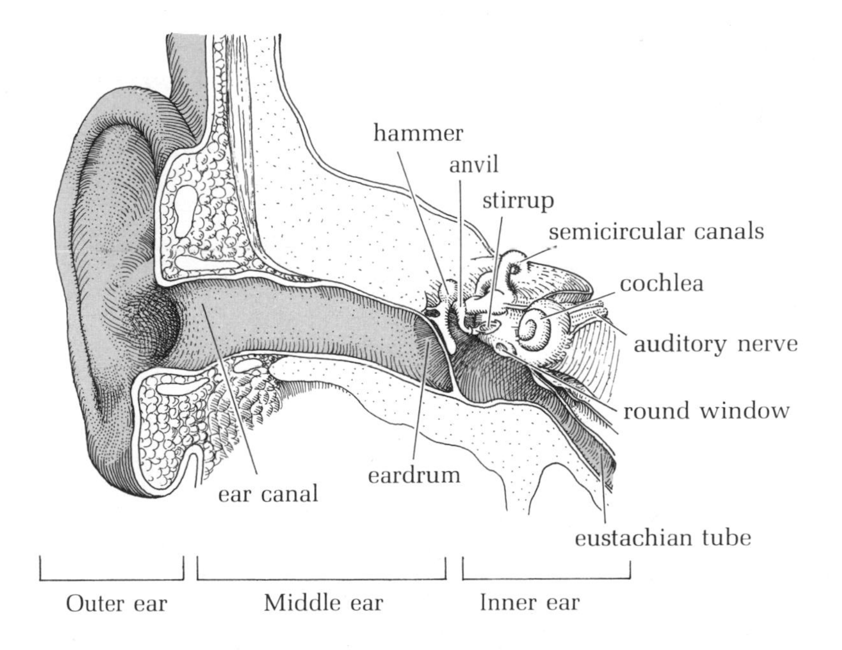

times by three small bones in the *middle ear*. Their names come from their shapes: the *anvil*, the *hammer*, and the *stirrup*. Set into motion by the vibrations of the tympanic membrane, the stirrup taps against a smaller eardrum leading into the fluid-filled *inner ear*.

The middle ear is filled with air, and it has to be kept at the same pressure as the air in the atmosphere. If there were no way for the air pressure inside to equalize with the pressure outside, then any great change in the atmospheric pressure could do severe damage to the tympanic membrane. Air at higher altitudes has less pressure than air at sea level. When you gain altitude rapidly in an elevator or an airplane, that sensation of swelling in the ears comes from the higher-pressure air in the middle ear pressing out against the eardrum. A narrow channel called the *eustachian tube* connects the middle ear to the back of the throat. Swallowing adjusts the pressure by drawing air from the middle ear through the eustachian tube.

Sound waves are translated into nerve impulses in the fluid-filled inner ear, or *cochlea*. Normally coiled tightly, something like a snail's shell, the inner ear is unwound in Figure 13–9 so that its workings are easier to see. The tapping of the stirrup against the small membrane sets the fluid of the inner ear in motion, thus changing sound energy in air to sound energy in liquid. The pressure wave flows through the cochlea's outer canal, and then is dissipated by the movement of the *round window* at the end. Without this escape hatch, our ears would never stop ringing. The movement of fluid in the outer canal sets the fluid in the central canal moving. The actual receptor cells line a membrane on the bottom of the central canal and look something like a patch of bristles. These cells are

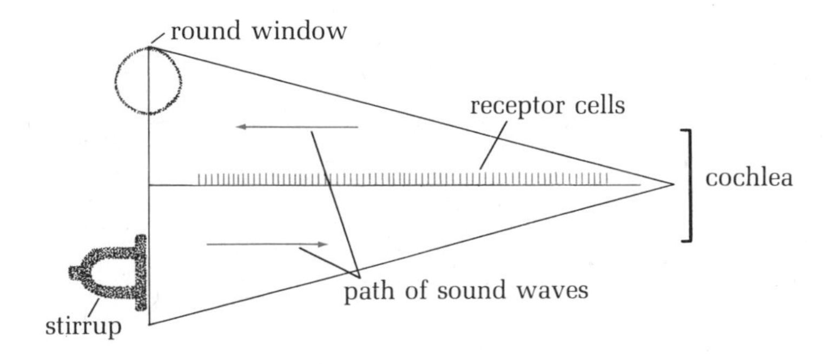

Figure 13–9 The inner ear, where sound waves are converted into nerve impulses. As the stirrup vibrates, it sets the fluid in the lower portion of the inner ear in motion. This energy is released at the other end of the canal with the movement of the round window. The receptor cells, distributed along the central membrane, detect the passing waves and convert them into sound waves, then pass the message to the auditory nerve, which leads to the brain.

not sensitive to all sounds but only to specific ones. The particular pressure wave sets off a chemical reaction in specific receptor cells that, in turn, stimulate neurons leading ultimately to the brain. The intensity of any particular sound is indicated by how often the receptor cell reacts: the more frequent its reaction, the louder the sound.

How much can we hear?

The receptor cells respond to sounds of specific *pitch*, or *frequency*. Frequency is simply a measure of the number of vibrations striking the ear each second. By convention, pitch is stated in cycles per second, or cps. The lowest note of a piano is at some 28 cps, while the highest has a much greater frequency, 4,186 cps. The normal human speaking voice produces sounds that range from about 80 to 300 cps. In general, humans can hear sounds ranging from 20 to 20,000 cps. Children are sensitive to even higher frequencies, while people past middle age miss high-pitched sounds because, with the passing years, the ear membranes lose their elasticity. At any age, humans are most sensitive to sounds running from 500 to 4,000 cps.

How sensitive is that? How does man's hearing compare with that of other animals? In terms of low-frequency sounds, the human ear is about as sensitive as it can be and remain useful. Sounds produced by the workings of the body and traveling through the tissues are at a low frequency. The ear is just sensitive enough to pick up the crunch of eating a carrot and the words of our own speaking, sounds that travel through the bones of the skull. But if we could hear sounds below 16 cps, we would very likely be surrounded by such a racket put up by our own bodies that we could hear few sounds coming in from outside. In terms of higher frequencies, though, we miss quite a bit. Dogs hear sounds up to 30,000 cps, and bats can actually sense tones of 100,000 cps.

If we could hear sounds below 16 cps, we would very likely be surrounded by such a racket put up by our own bodies that we could hear few sounds coming in from outside.

But frequency is not the only measure of sound. There is also its *intensity*, or loudness. Loudness is measured in units called *decibels*, abbreviated to db. By convention, the softest sound that the average young adult can detect is assigned the value of 0 db, and all other sounds are rated in comparison to that beginning point. The loudest sound that the brain interprets as sound and not pain lies between 120 and 130 db. Figure 13–10 shows a decibel scale and rates some common sounds in terms of it. It's important to note that the decibel scale is what is known as a logarithmic scale. That is, a move from one number to the next is not just an addition of one unit, but a multiplication by 10. Thus 6 db is 10 times as loud as 5 db and 100 times louder than 4 db. The decibel scale covers quite a range; the upper end of 130 db is 10^{130} times as loud as 0 db.

The bat's sonar system

Hearing is important to practically all vertebrates, but probably no animal's hearing is put to more remarkable use than the bat's. Most

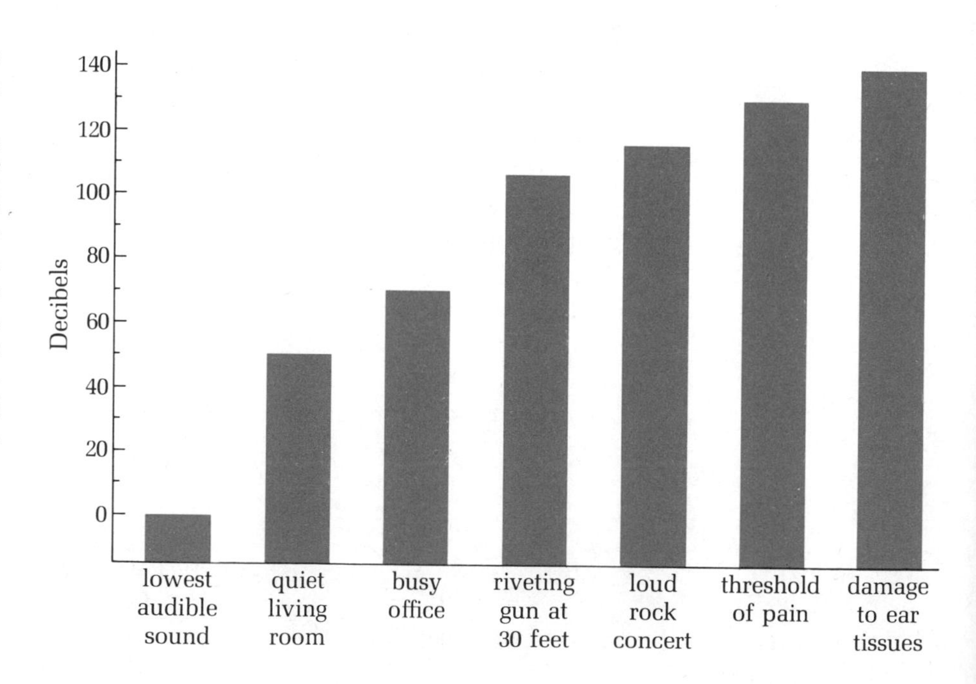

Figure 13–10 The decibel scale relates a physical quantity—sound intensity—to the human perception of that quantity—sound loudness. It is a logarithmic scale; that is, each increment of one decibel represents a ten-fold increase in loudness. The graph indicates the decibel ranges of some common sounds.

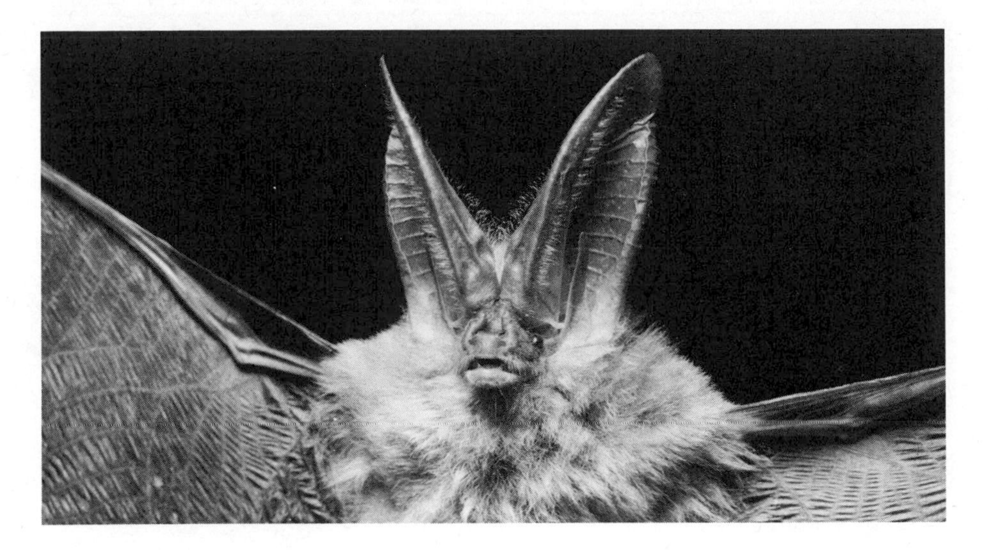

Figure 13–11 The huge ears of the bat. Living in a dark cave, the bat depends on his sense of hearing as much as humans depend on the sense of sight.

species of bat are active at night, hunting for the flying insects that make up the bulk of their diet. Obviously, sight is of little value. How can a flying hunter find a flying prey in the dark of night?

As it flies, the bat emits high-frequency chirps at the rate of about 30 per second, analogous to sonar transmission. Figure 13–11 shows the bat's ears, which act like huge traps to catch and channel even the tiniest sound. If the bat's chirps hit an object, an echo bounces back and is picked up by the bat's ears. Guided by the echoes, the bat can accurately locate the prey and snatch it out of the air. Experimental work has shown that bats use their hearing alone, unaided by sight. Blind bats can negotiate an obstacle course just as well as bats with normal vision.

The simplicity of hearing cues

We saw in the case of the English robin that it often takes only a seemingly small visual clue to elicit a certain behavior. The same holds true of hearing signals. Consider a turkey hen that has just hatched out a brood of chicks. Such a hen is very protective and aggressive, attacking any living thing that comes near her nest. In fact, the only thing she doesn't attack is her own offspring. Curiously, the chicks chirp constantly, and one re-

Figure 13–12 Noise from high performance cars, planes, heavy construction machines, and other motorized equipment can cause damage to the delicate internal structure of the human ear.

searcher wondered whether the hen recognized the chicks by their noises. To find out, he deafened a hen. Unable to hear her offspring, she attacked them. The researcher then replaced the usual brood of chicks of another hen with a litter of kittens outfitted with small tape recorders emitting the sounds of chirping turkey chicks. This hen did not attack them. Obviously, the hen was instinctively programmed to attack anything, and only the sound – not the sight – of the chicks stopped her aggression.

Simple sounds can also serve to aid in reproduction as well as in aggression. Insects can hear, although their sense organs for sound are quite different from the human ear. The male mosquito uses his bushy antennae to pick up sounds. He listens particularly for the sound produced by the wing vibrations of the female mosquito – which to us sounds like an annoying whine rather than a love song. Sound alone draws the male to the female and stimulates him to mate. A net full of male yellow

A net full of male yellow fever mosquitoes will be put into an absolute sexual frenzy by an artificial pitch of 250 to 500 cps.

fever mosquitoes will be put into an absolute sexual frenzy by an artificial pitch of 250 to 500 cps, and they will attempt to mate with anything around them, including the netting and other males.

You are what you hear

One of the curious things about human hearing is that what one hears is very much related to how one feels. It has long been known that people who are subjected to loud, whining noises of a single frequency lose the ability to hear that pitch. Apparently the loudness of the sound destroys the receptor cells. This kind of ear damage is a common affliction in workers in noisy occupations, like jet mechanics and boilermakers, and it also befalls rock musicians fond of amplifiers turned up to full volume. However, exposure to loud sounds for a long period of time has even more pernicious effects, and not just on the ears. Exposure to a daily, 8-hour dose of noise between 80 and 130 db produces definite hearing loss over the whole range of sound. And a steady diet of sounds over 75 db actually puts the body into a state of alarm. Medium and small arteries constrict, raising blood pressure. Respiration and pulse rate increase. The individual may become irritable and jumpy. Noise at these levels may contribute to heart disease, high blood pressure, deafness, enlarged adrenal glands, and ulcers. Look again at the decibel scale of Figure 13–10. Many people living near busy streets and airports or working in certain industries are subjected every day to damaging noise levels (Figure 13–12).

Why should humans possess such a debilitating sensitivity to noise? Probably the answer lies in the course of human evolution. When city people are introduced to wilderness, one of the first things they notice is the silence of the backcountry. In fact, a loud noise is usually a sign of danger: the clap of thunder, the roar of an avalanche or waterfall, the growl of a bear. Primitive man's reaction to any loud or violent noise is the alarm reaction, the fight-or-flight response brought on by an increase in adrenaline. The same reaction occurs in modern, urban humans. They are still equipped with the sensitivity to noise of one who inhabits wilderness, but live instead in the noisiest environment of all.

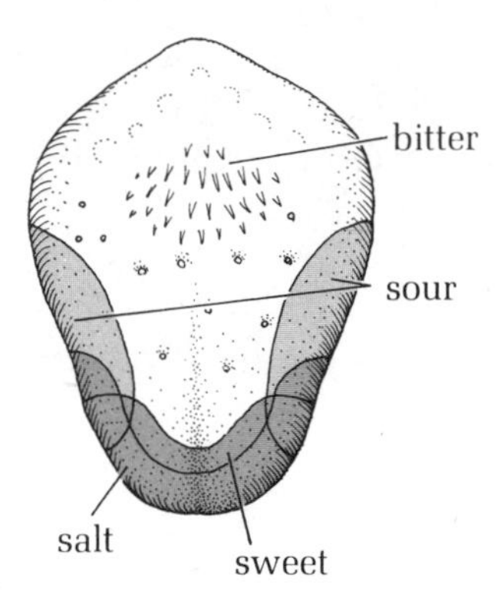

Figure 13–13 The distribution of the taste buds on the tongue.

Tasting and smelling

Tasting and smelling are so closely related that separating them is oftentimes akin to splitting hairs. In man, each sense is located in a different place, the nose and the tongue. The receptors for taste are specialized for liquids, while those for smell detect gases. But the signals sent to the brain are interpreted together, not separately. You have no doubt noticed that when the stuffiness of a cold has put your sense of smell out of commission, food tastes very flat. The reason is that the brain is getting only taste signals. Much of what we think of as flavor is actually smell. The nasal passages are connected to the throat near the back of the mouth, and the odors of food in the mouth travel to the smell receptors in the nose.

The bitter and the sweet

We humans can taste only four flavors: sweet, sour, bitter, and salty. The receptors of taste are located in little packets, or *taste buds*, scattered over the upper surface of the tongue. Apparently there is some division of labor among the taste buds. As Figure 13–13 shows, the buds near the front of the tongue are most sensitive to sweetness, while those to the back and sides react primarily to bitterness and saltiness, respectively. By no means are we equally sensitive to each flavor. It takes a goodly dose of sugar in a cup of coffee to register "sweet" in the brain, but only half that much salt will trigger a response. We are 75 times more sensitive to sour than to sweet, and 10,000 times more adept at recognizing bitterness.

The sense of taste serves as a sort of watchdog determining what should or should not be swallowed. Curiously, though, different organisms react differently to the same tastes. Man, like most vertebrates, dislikes bitter-tasting foods, but many birds feed on bitter seeds. Nearly all animals from insects up respond positively to sweets. The cat, however, shows no preference for sweets and has few receptors for that taste.

Man has a good sense of taste in comparison to some animals, but a poor one with regard to others. Flies, for example, have their taste receptors on their feet. Odd as this arrangement may seem to us, it allows the fly to quickly sample every food source it happens across. One species of fly is so sensitive to sweetness that it can detect a single tablespoon of sugar mixed into 125,000 gallons of water. In humans, the upper limit of sensitivity is about one heaping teaspoon of sugar in a quart of water. In other words, the fly is 10 million times more sensitive than man to sugar. But man is not a dullard in this respect; he is about three times more sensitive to sugar than is the honeybee.

Taste cells on the outside

Although both flavors and odors are mixed together in the water fish inhabit, they have different receptors and different brain centers for each sense. The taste receptors are located on the outside of the fish's body, so that it can literally taste the water as it swims along. Among the bottom feeders, taste plays as major a role in determining behavior as sight does in man. The most developed part of the brain of the carp, a bottom scavenger, deals with taste. And the long, trailing "whiskers" that give the catfish its name are studded with taste receptors that guide the fish to its food.

Testing the air

In many ways, smelling is tasting at a distance. In humans, and other mammals, the sense of smell is much more sensitive than that of taste,

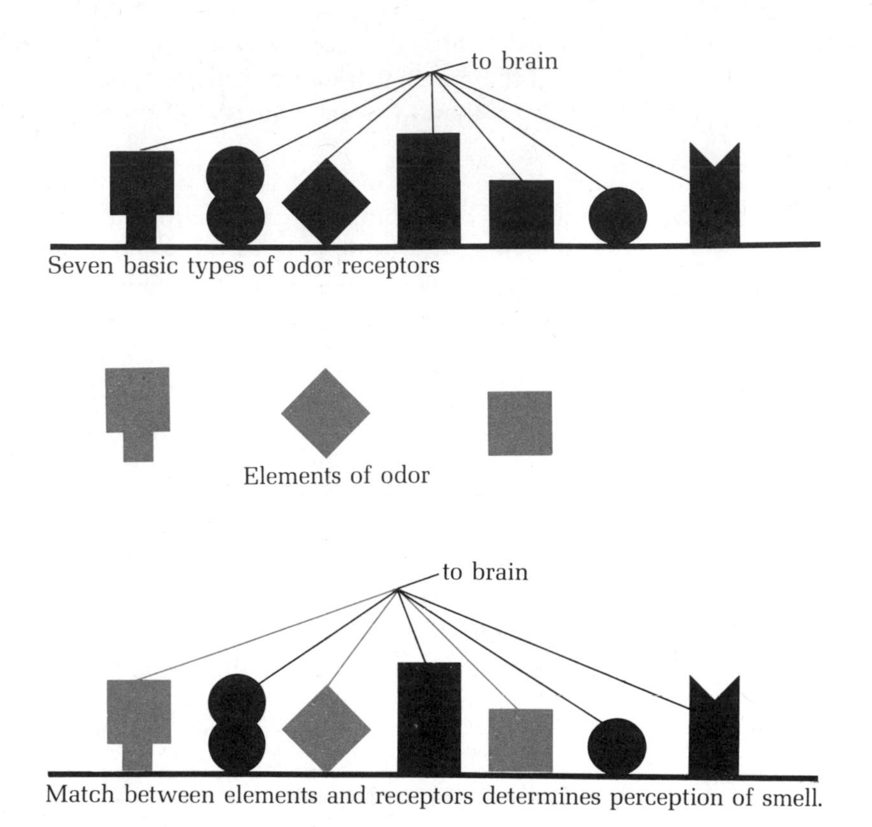

Figure 13–14 A theoretical explanation of the sense of smell suggests that all smells can be classified according to the presence or absence of seven basic odors. The nasal receptor membranes are also divided into seven basic types that match the seven basic odors.

and we can detect very small amounts of scent. Man's sense of smell is comparatively unsophisticated, but we can still distinguish some 10,000 different odors and shades of odors. And we are quite sensitive to some of them. It takes only 18 parts per million (ppm) of oil of peppermint in the air to trigger a reaction, and musk, an ingredient of many perfumes, can be picked up at a concentration of a mere 0.3 ppm.

The odor receptor cells are clustered in two thumbnail-sized membranes in the nasal passages behind the nostrils. Oddly enough, there are relatively few odor receptors, a fact that leads one to wonder how so few cells can distinguish so many odors. Does each cell specialize in only one odor, the way taste buds seem to do, or are they equally sensitive to all 10,000 odors? According to one theory, these 10,000 odors are actually combinations and blends of seven basic odors; receptor cells also come in seven types, each corresponding to one basic odor. The receptors rec-

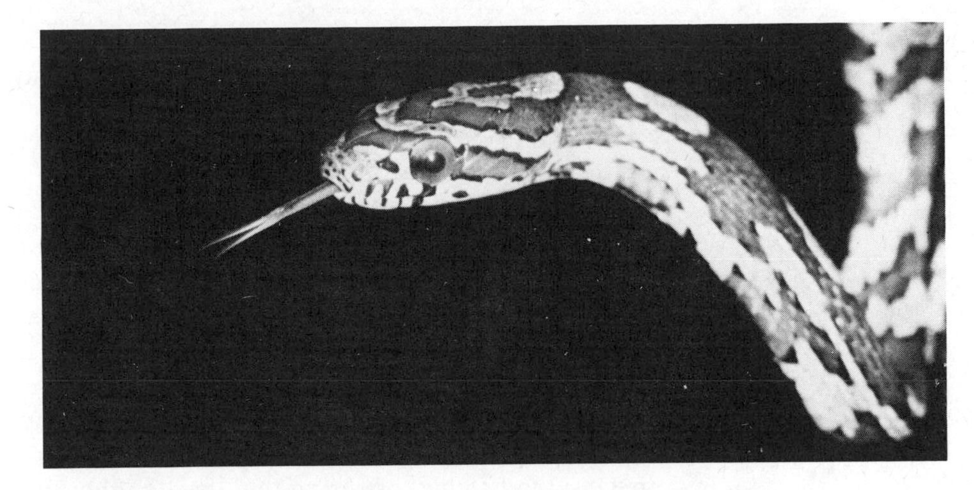

Figure 13–15 The corn snake smells his environment with his tongue, one of the many variations of the sense of smell.

ognize smells by the shapes of the molecules, molecules and receptors fitting together like a lock and key. The brain then blends the varying signals of the receptors into one smell (Figure 13–14).

Our sense of smell actually helps to prepare us for eating. When the brain receives signals indicating good-smelling food, it relays the message along the autonomic nervous system to the stomach, which responds by secreting digestive fluids. But this reaction, or any other reaction to a particular smell, doesn't last forever. After we are exposed to a smell for more than a few minutes, the nerve cells stop sending messages to the brain, and we no longer notice the smell. This frees the smell receptor to notice any new odors (Figure 13–15).

Different ways of smelling

As discriminating as man's sense of smell is, its accomplishments pale when compared to the sensitive noses of other mammals. In fact, most mammals rely on their noses much more than their eyes. Mammals evolved originally as nocturnal animals, and, in the night, smell is a surer guide than vision.

Mammals are by no means the only vertebrates with a good sense of smell. As was the case with taste cells, smell receptors are often scattered over the entire exterior surface of a fish's body. Smell receptors are also concentrated in small pits in the front of the fish's head. Water swirls about in these pits, and the fish can detect smells in the water. In some fish, these pits extend into the mouth as nostrils that allow the fish to smell small objects in its mouth.

The characteristic flicking of a snake's tongue is not a way of tasting the world but of smelling it. The roof of the mouth contains cavities lined with smell receptors. The snake samples the air with its tongue, then inserts the tongue and the sample into the smelling cavity to be analyzed.

The smell receptors of insects are usually located on the antennae. Insects can generally pick up the same range of smells as man, and some of them are quite sensitive to odors and are able to discriminate among them with great selectivity. Hunters love to brag about the fine noses of their dogs, but no pointer or setter has come anywhere near the feats of certain parasitic wasps. These wasps lay their eggs in the larvae of other insects; the hatched offspring then feed on their unwilling host. Some of these parasitic wasps lay their eggs in the larvae of another wasp, the wood wasp, which they locate by means of smell. The amazing thing is that the wood wasp larva generally lies dormant deep inside a beam or the trunk of a tree, protected by several inches of wood. The parasitic wasp locates the larva, drills through the inches of wood covering it, and lays its eggs within the host's body—all guided by the sense of smell!

Sending messages by smell

Among animals sensitive to smell, odors may trigger behavior. Substances that produce such odors are called *pheromones*, and they are in effect chemical signals. The pheromones secreted by animals serve a great many purposes: to warn intruders, to identify members of a colony or family group, to find prey, to mark trails, to ward off enemies, to attract mates. Male dogs urinate on the boundaries of their territories. The scent of their urine tells all other male dogs that they enter at their peril. Brown rats are cooperative and helpful toward members of their own pack, but murderous toward strangers. They recognize their own tribe solely on the basis of smell. A fire ant worker who finds a rich source of food marks his trail back to the nest with a special scent. Workers who follow the scout back to the food source add their scents to the trail and keep it fresh and strong. Once the food source is exhausted and the workers stop traveling back and forth, the chemical trail disappears rapidly, thus saving the ants from being misled by out-of-date information.

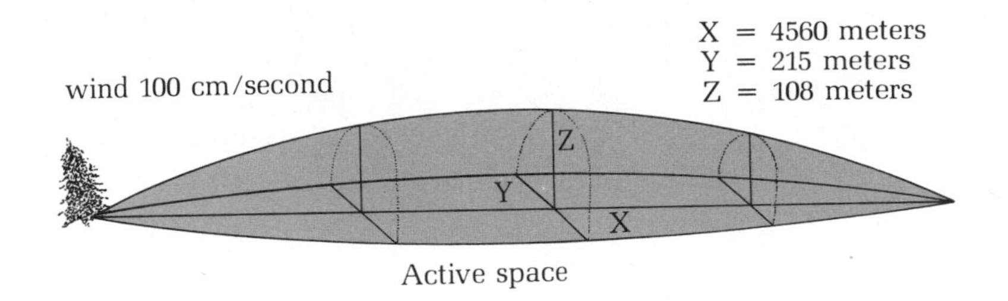

Figure 13–16 The pheromones of the gypsy moth. Within the area shown in gold, a male gypsy moth will detect the pheromones of a female in the tree at the left.

In some species, the animal is so acutely sensitive to a pheromone that one molecule of the substance is enough to trigger a response. Females of most moth species emit pheromones that attract the males for mating. The male gypsy moth, for example, can detect a female at well over a mile (Figure 13–16). Obviously, if chemicals potent enough to elicit such dramatic responses remained in the environment too long, they would cause great confusion. As it turns out, many pheromones, like the moths' sex attractant, last only a relatively short time before dissipating.

Some chemical signals are produced by the nonliving part of the en vironment. Such a signal triggers one of the most impressive sights in the natural world, the migration of salmon upriver to the spawning beds. The fish move in great schools, the surface of the water thrashed white by their excited rolling, leaping and throwing themselves over any obstruction in their upstream path. Salmon belong to a class of fish known as anadromous, which migrate from salt water to fresh water to spawn. The fish hatch in fresh water and grow to fingerling size before swimming downstream to the ocean. They spend from 2 to 8 years at sea, then return to the pool where they were first hatched to spawn and die. Tagging and marking individual fish showed that, as suspected, they do indeed return to their birthplace. How could these fish navigate faithfully as much as several hundred miles, choose among the many seemingly identical tributaries of a main river, and then actually find the same small pool where they themselves were spawned? Hatchery workers had long

How could the salmon navigate several hundred miles, choose among the seemingly identical tributaries of a main river, and then actually find the same small pool where they themselves were spawned?

known that salmon have a very acute sense of smell. If someone dips his hands into a river several hundred yards upstream from a school of salmon, the fish almost immediately show alarm, even though the amount of scent in the water is infinitesimally small. On the basis of this, investigators wondered whether the salmon smelled their way upstream, following a scent remembered from their early days as fingerlings. Indeed, this did prove to be the case. Experimental fish whose nostrils were plugged and could not smell were unable to find the correct spawning pool.

Touching

We think of touch as but one sense, but the skin actually contains different sensory receptors for pain, pressure, heat, touch, and cold. By no means are these receptors distributed evenly over the whole of the body. The hands and face, for example, have many more receptors than the back does, and, as a result, the hands are about 20 times as sensitive to all varieties of touch as the back.

You can demonstrate this point with a simple experiment requiring two sharp pencils and a friend. Have your friend take off his shirt and sit with his back toward you. Hold the pencils so that the points are only a fraction of an inch apart and touch them to your friend's back. Ask him whether the sensation came from one point or two. Most likely, he will say, "One." Repeat the touching, separating the pencil points a little more each time, and ask the question again. In the majority of cases, the pencil points will have to be more than 2½ inches apart before the person can tell that there are two instead of one. To put the back's insensitivity into perspective, have your friend close his eyes, and repeat this experiment on the tip of his finger.

Another version of this experiment provides a way of mapping the receptors in the skin. Mark off 1 square inch of skin on the inside of the forearm, then test this area for receptors. Use a thin, sharp needle for pain, a hot needle for heat, and a pointed sliver of ice for cold. Every time you locate a receptor, mark it with ink, using a different color for each sensation. Now count and tally the various colored dots. You'll find that there are about 10 receptors for cold to every one for heat and that pain receptors outnumber the others.

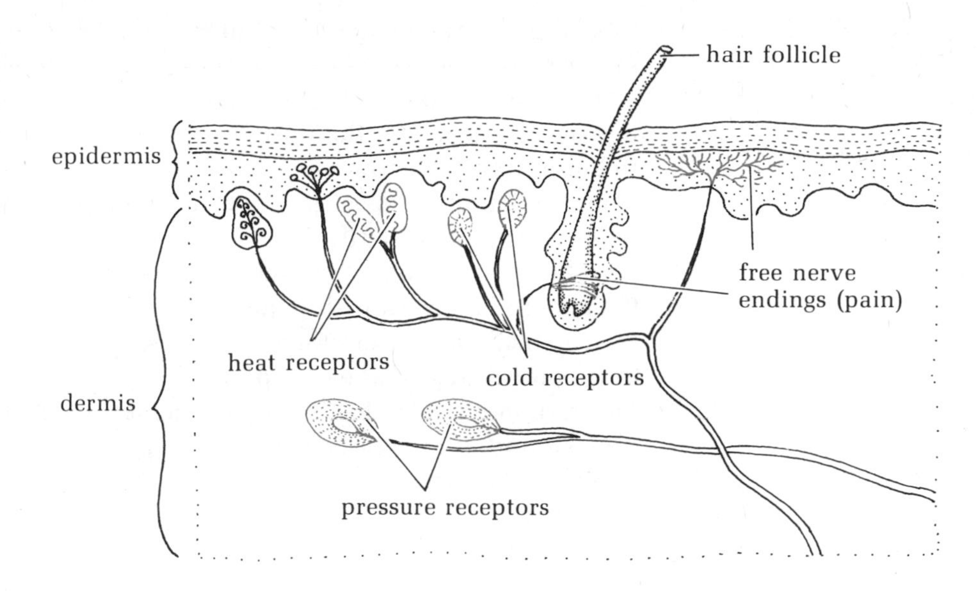

Figure 13–17 The specialized receptors of the skin. Four different types of receptor cells detect pain, pressure, heat, and cold.

In man, touch serves as a way of analyzing all the various stimuli that come into contact with the skin. The sensitivity of our fingers and hands adds greatly to our ability to use them for very delicate tasks. Pain, heat, and cold receptors are all connected into reflex arcs, allowing us to react very quickly to physical danger (Figure 13–17). But touch can serve other purposes as well.

Touching without contact

The rattlesnakes belong to a group of snakes known as pit vipers, which have evolved a special sense organ useful for finding prey. The pits in the front of the snake's face are lined with heat receptors sensitive to changes in heat so small that a human would not even notice. The snake uses its smell along with its eyesight to find prey, estimate the distance to it, and make an accurate killing strike. A rattlesnake will even rely on its sense of heat alone if need be. If the snake's eyes are covered, it can be induced to strike at a warm light bulb.

A group of parasitic insects, the ticks, also use heat to find victims. The ticks live in long grass or brush and wait in ambush for a mammal to pass by. After the tick hops onto its victim, it buries its head into the victim's skin and begins to suck blood from the wound. The tick may stay attached for weeks, and can cause serious infection or disease. The remarkable thing about the ticks' sensitivity to heat is their ability to distinguish a very narrow range of temperature very quickly. Most of the common American species of tick usually infest deer, which have a body temperature of about 101°F. Ticks are most likely to hop onto an animal with this body temperature than onto one that is either warmer or cooler. If a man and a dog walk through the same tick-infested brush, the dog will pick up many more ticks than the man will. Why should ticks prefer dogs to people? Actually, they have no such preference; it's a matter of what they respond to. Man has a body temperature of 98.6°, but a dog's usually runs around 101°, nearly the same as a deer's. The tick is much more likely to sense the dog and hop onto it than it is to perceive the presence of the man.

The dancing of the bee

We humans are accustomed to using touch as a method of communication. For example, we use a wide variety of caresses to indicate affection or sexual interest. The bee performs the marvelous feat of learning flight directions by touch. Students of insect behavior wondered just how a worker bee who had located a rich find of nectar-laden flowers was able to tell the other workers where to find it. It was known that bees didn't use pheromones as the ants do. What kinds of cues could they employ? Research showed that when the worker arrived at the hive, she performed a waggling, rhythmic dance on the wall of the hive. As Figure 13–18 shows, the pattern of the dance provides navigational directions to the newfound source of food. The other workers in the hive crowd around the dancer and follow her movements by touching her with their antennae. The bees follow the dance not by sight but by touch.

The coming of the locust

Easily the most fascinating—and destructive—case of the role of touch appears in the transformation of certain kinds of grasshoppers into locusts. One of the first records of a locust invasion appears in the biblical book of Exodus as one of the plagues Moses visited on the pharaoh of

Figure 13–18 The meaningful dance of the honeybee. By a specific pattern of rhythmic, circular movements on a vertical surface within the hive, a returning bee is able to communicate the location—relative to the hive—of a source of food.

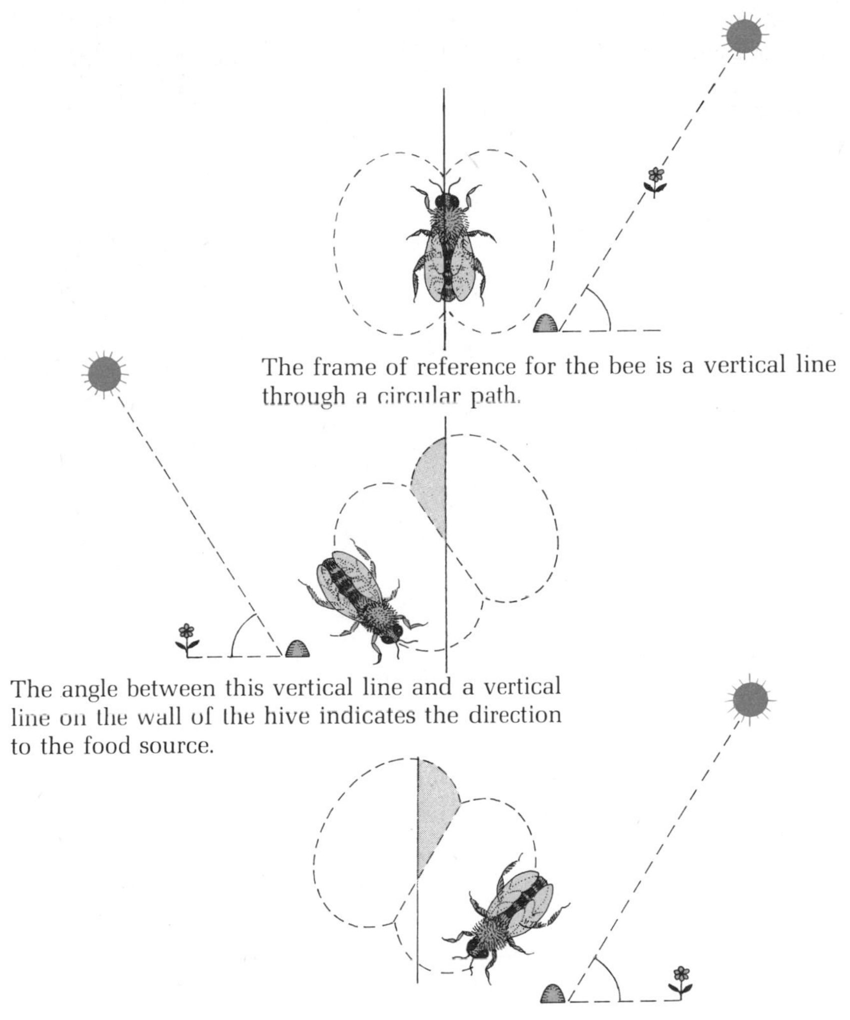

Egypt for his refusal to let the Hebrews go. According to that account, the locusts covered the land of Egypt and ate everything that was green. The biblical description almost understates the kind of destruction locusts can do to man's crops. The insects swarm in such great numbers that their weight breaks limbs off trees, and they may cover the ground to a depth of several inches. They eat everything that appears even remotely edible: leaves, straw hats, hoe and shovel handles salty with sweat, leather goods,

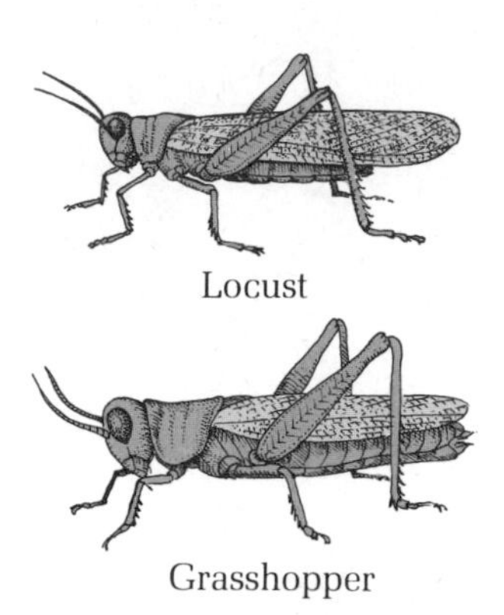

Figure 13–19 The physical resemblance between the locust and the grasshopper is an important clue to the origin of locust plagues.

plus any and all green plants. In the 1800s, locusts caused great economic damage in the American Midwest and very nearly wiped out the fledgling Mormon community at Salt Lake City. For some reason, the locust became extinct in America around 1900, but the swarms have continued to appear again and again in many parts of Africa, Asia, and South America.

For a long while, the sudden appearance of locusts was a mystery. Where did the insects come from? Even in those areas most subject to plagues, solitary locusts were never found, and the plagues usually came several years apart. Where were the insects in the meantime? How could such a huge population of insects appear, seem to vanish altogether, and then arise again?

The amazing answer, discovered some 60 years ago, is that the grasshopper of today may be the locust of tomorrow. Grasshoppers and locusts are related (Figure 13–19), but they appear to be different insects. Like the locust, grasshoppers feed on crop plants, but they are solitary creatures who don't migrate. But some, not all, species of grasshoppers turn into locusts if their senses feed them the right information. And most of the right information comes from the sense of touch.

If several years with conditions favorable to the survival of grasshoppers follow in a row, then the population of grasshoppers grows steadily. However, the grasshoppers aren't spread evenly across the countryside. The insects like to sun themselves on exposed patches of ground, and on these open bits of earth more and more grasshoppers congregate. They touch each other constantly, and the size of the congregation and the amount of touching increases steadily. The touching

Figure 13–20 The tickling jar experiment. The development of a locust from a grasshopper can be recreated in the laboratory with this simple apparatus.

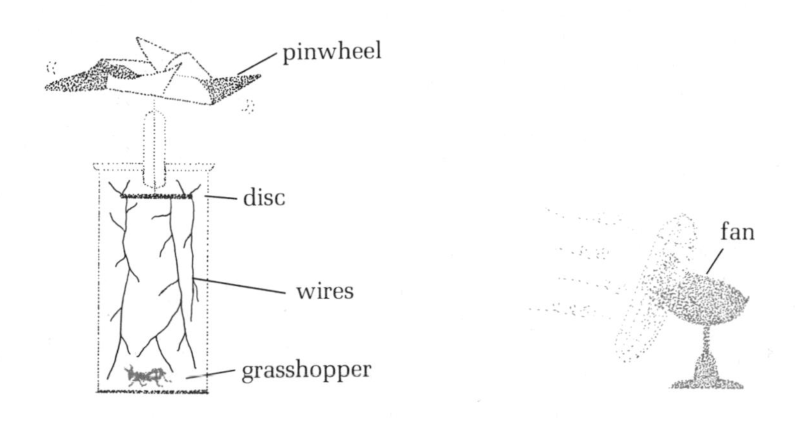

triggers a change in how the grasshoppers behave and even in what they look like. No longer do the grasshoppers prefer a solitary existence; instead, they actually seek out new contacts with other grasshoppers. Their skin color darkens. And the eggs laid by these dark grasshoppers hatch out a new offspring: The males are larger, and both sexes have longer wings and shorter hindlegs. This generation is even more restless and prone to gathering in groups, and their tendency to congregate grows so strong that a locust swarm results. The migration begins with the great cloud of marauding locusts. As the migration travels farther, the swarm thins out, spreading the species to new, previously unpopulated areas. Uncrowded, the survivors of the migration begin again to behave like grasshoppers.

Saying that such a massive change in both behavior and appearance is brought on by the sense of touch seems like a bit of scientific leg-pulling. Is there any experimental evidence? Indeed there is, the product of a rather ingenious demonstration. Normal grasshoppers of species that may become locusts were put into the "tickling jar" shown in Figure 13–20. As long as the fan blew, fine wires constantly touched the grasshoppers in much the same way that other grasshoppers would have. When these grasshoppers were released from the jars, they sought out contact with other insects, even ones that weren't grasshoppers, congregated, and tended to march.

Are there other senses?

With vision, hearing, taste, touch, and smell, we have completed the traditional list of senses. But, in considering any list, one should always ask whether it covers everything. What other kinds of stimuli do we humans or other animals respond to? For example, if you spin around very fast for about 30 seconds and then stop, you'll feel dizzy and find yourself unable to walk a straight line. Why does the brain think the body is moving even when it is still? If you are blindfolded and thrown end over end into a pool of water, you'll probably find your way to the surface in relatively short order. Why is it unlikely that you would mistake bottom for top and start swimming down? How do we tell up from down? Given that the earth's gravity is the force that pulls everything down, do we have a sense of gravity? And, if we can indeed sense something as diffuse and unchanging as gravity, what about such physical forces as atmospheric pressure and magnetism?

Sensing gravity and motion

If the brain didn't know all the time which way was down, we'd have a very hard time maintaining stability and balance. But our sense of gravity generally works so well that rarely are up and down confused. It works so well that in the course of evolution it has not been discarded. Man's special organ for sensing gravity is practically the same as the one found in jellyfish. Each inner ear contains a number of small crystals of the dense and hard compound calcium carbonate. These crystals are fastened to a bed of hair cells very similar to the hearing receptors in the cochlea. As the head moves with the body, gravity pulls the crystals down from varying angles, bending different hair cells at each angle. The bending of the hair cells generates nerve impulses, and the rate of firing depends on the amount of bending. These nerve impulses constantly remind

Man's special organ for sensing gravity is practically the same as the one found in the jellyfish.

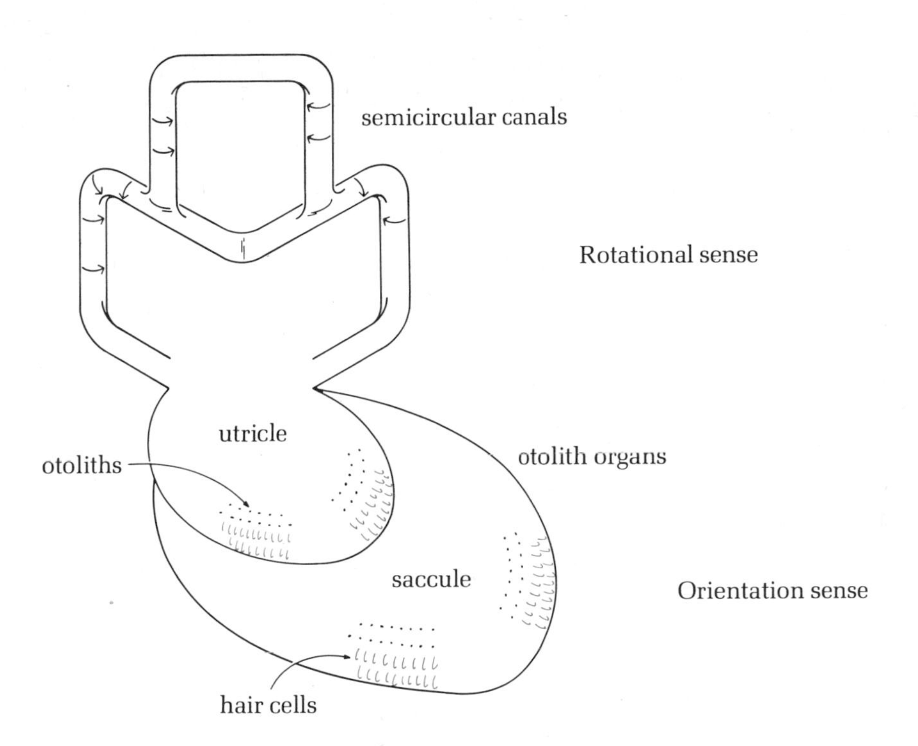

Figure 13–21 A schematic view of the balance sensory organs of the inner ear. The three semicircular canals, oriented at right angles to each other, are filled with thick fluid which moves as the head rotates or turns. However, the canals cannot distinguish between up and down. In the two outpocketings beneath the canals—the saccule and the utricle—minute crystals of calcium carbonate called otoliths align according to the vertical position of the head. Hair cells, the receptors, detect the position of the otoliths and transmit this information to the brain.

the brain which way is down, an important reference point in maintaining balance (Figure 13–21).

The organ for detecting rotation or movement is likewise located in the inner ear, in the structures known as the *semicircular canals*. Much like the central canal of the cochlea, the semicircular canals are filled with fluid and are partially lined with hairlike receptor cells. The three canals run in three directions, one for each of the dimensions—height, width, and depth—of our three-dimensional world. Movement of the head sets the fluid in the various canals in motion, thus triggering the receptor cells. By interpreting the rate and origin of the impulses, the brain can tell which way the head is moving and how fast. However, since the fluid has its own momentum, it stops moving a little after the head does, and this is what causes dizziness. For example, if you spin in a revolving chair or on your heels for a few quick turns, the fluid in the horizontal canals is set in motion. When you stop spinning abruptly, it takes a while for friction to slow and finally stop the fluid. All the while the receptor cells send impulses signaling rotation to the brain. Your eyes tell you that you are standing still, but, according to your brain, you are still spinning around.

The outer reaches of perception

We rarely number gravity and motion among our senses because there are no obvious and visible organs to perceive these stimuli. Instead, these organs are concealed within the ear. Are there any other phenomena perceived in any equally subtle way?

As a matter of fact, there is a rather long list of poorly understood curiosities. For example, a number of species of eel, fish, and ray use electrical shock to ward off their enemies; they can manufacture electricity chemically, much as an automobile battery does, and can store it for release at a later time. These species can sense changes in the electrical field about them, and tell whether these changes mean danger or food or are of no consequence. Also, there is evidence that at least a few plants and animals respond to changes in the earth's magnetic field. Certain grain seeds germinate better if their ends point north and south than they do if pointed east and west. Flatworms respond to light no matter which way they are traveling, but how much they change course depends on whether they're headed north or west at the time. And psychologists have found that the fluctuations in the pressure and electrical charge of the air that accompany changes in the weather can affect the moods and mental outlook of humans. Even though these various effects have been observed any number of times, no one knows how these environmental signals are perceived.

If this is the case, then isn't it reasonable to assume that a whole realm of information may come to us by pathways other than the traditional senses—that is, by *extrasensory perception*, or ESP? Generally the term ESP refers to feats more spectacular than simply responding to magnetism or air pressure, and includes such things as thought transference or mind reading; perceiving past, present, or future events at a distance; and speaking or writing through a medium. Because ESP often seems to go hand in hand with some kinds of fakery and fraud, it has long been a scientific unmentionable. But a few scientists, particularly at Duke University and the Stanford Research Institute, have begun studying the phenomenon carefully. The researchers have shown that some people can send mental messages over a distance to other people simply by thinking.

Besides these experiments, there are well-documented studies of people who seem to have had extraordinarily sensitive ESP. Consider the case of Edgar Cayce, who lived in the first half of this century. Of average education and skill, Cayce made his living as a photographer. He was a religious man, who enjoyed golf and fishing. At first glance, Cayce would seem a relatively typical American. But when he hypnotized himself into a state resembling sleep, Cayce assumed the personality, vocabulary, and knowledge of a gifted physician. Supplied only with the location of a sick person hundreds or even thousands of miles away, Cayce would proceed to diagnose the disease and prescribe treatment in great detail. After the trance was over, Cayce said he could not remember anything that had happened. Between 1903 and 1944, he went into thousands of these medical trances. Thorough checks by skeptical doctors showed most of Cayce's diagnoses to be accurate and his treatments to be successful.

What was the source of Cayce's medical knowledge? How did he analyze the health problems of people about whom he knew nothing other than where they lived? And, how could he do it over such distances? No one knows the answer to any of these questions, nor has anyone even been able to come up with a good explanation. The only thing known for sure is that extrasensory phenomena like Cayce's have occurred.

The question of internal clocks

A similar problem arises with what is known as *endogenous rhythm*. This term is applied to any activity that is repeated at regular intervals. Earlier we've seen the example of the menstrual cycle, which usually runs on a 28-day interval. Most animals and plants have definite periods of rest and activity that change little from one day to another. No doubt you tend to fall asleep and to become awake at about the same time each day. Many migrating birds begin or end their yearly journey with exceptional accuracy; the swallows of San Juan Capistrano are perhaps the most famous case. In nature, these rhythms are correlated with changes in the temperature, the amount of light, or the season. Shore crabs, for example, emerge from a rest period in their burrows at high tide and scurry about in search of food. But the surprising thing is that if the crab is put in a laboratory where conditions are held constant and there is no tide, it will scurry about at the time of the high tide back on its home beach. The rhythm seems to arise within the crab: that is, to be endogenous. Similar experiments performed on plants have turned up the same results, as shown in Figure 13–22. Experimentation has shown that endogenous rhythms exist

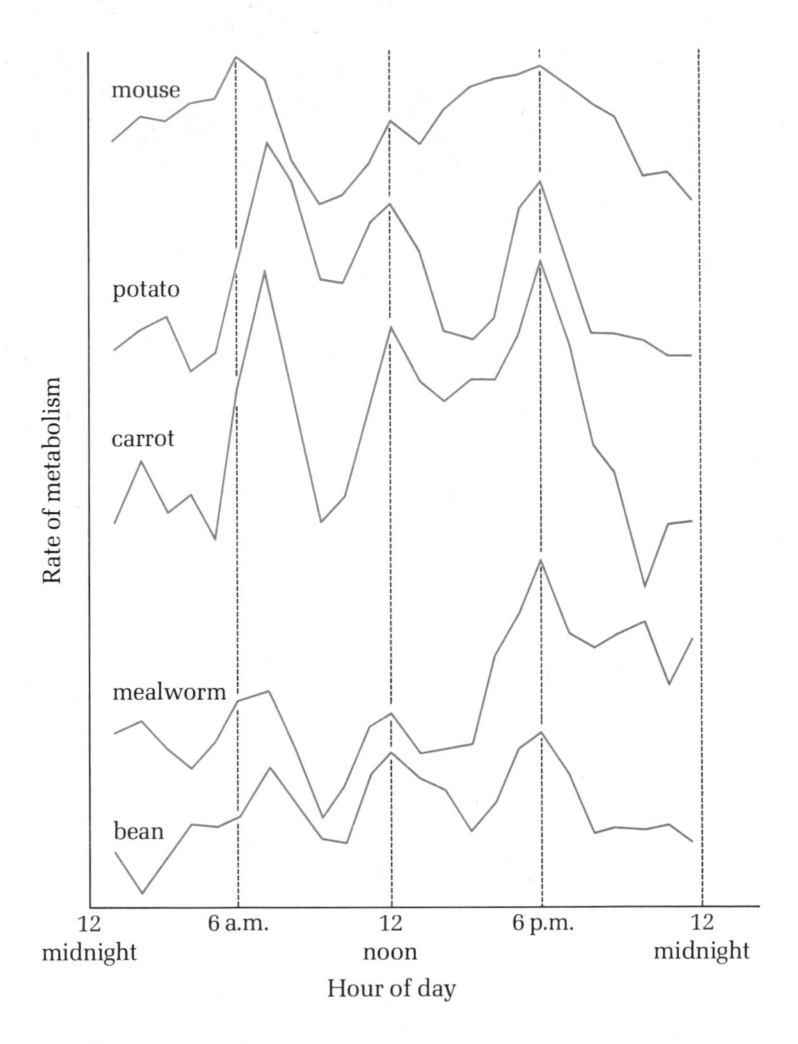

Figure 13–22 Biological rhythms. In support of those scientists who believe that many biological rhythms are endogenous, it has been demonstrated that the metabolic activity of many common plant and animal tissues seems to peak at the same three times of the day, despite the maintenance of uniform conditions around the clock.

in thousands of species representing every major group of plants and animals. In fact, endogenous rhythms seem to be as characteristic of life as reproduction or growth.

But some researchers wonder whether these rhythms are truly endogenous or if instead they may be a response to changes in the environment. Tests on countless organisms under carefully controlled conditions have shown that the rhythms persist outside the natural environment. These experiments kept the laboratory environment constant in terms of light, temperature, and humidity, but not in terms of gravity, magnetic and electrical fields, air pressure, or background radiation. We have already seen that some organisms are definitely sensitive to atmospheric pressure and to electrical and magnetic fields. And it is now known that these factors can vary with a remarkably accurate and rhythmic cycle, covering days, months, seasons, and years.

The question of whether endogenous rhythms are truly endogenous will remain until someone can figure out a way to hold all the possible stimuli constant and see how various organisms react. Creating such a controlled environment is a very difficult problem, one that no researcher has yet solved. But the question is an important one: How responsive are we, and other living things, to all those physical changes that we can't see, hear, taste, smell, or touch?

Summary

The sense organs are specialized structures that translate changes in the environment into nerve impulses.

The eye responds to light. Light entering the human eye is focused on the retina, which contains the receptors and the nerve endings. Not all animals with color vision see the same colors. Many arthropods have compound eyes, collections of a great number of simple eyes, which aid in high-speed vision. In mammals, the position of the eyes in the skull is related to whether the animal is predator or prey. What is selected from the great number of visual stimuli depends on cues set by inheritance or learning.

The human ear responds to sound waves. Liquid in the cochlea set into motion by sound waves stimulates special receptor cells and the neurons associated with them. Man can hear a relatively wide range of sounds, but the hearing of some other animals is even more acute. Like vision, hearing provides important cues for behavior. Because of man's sensitivity to sound, exposure to continuous loud noise can affect health.

Taste and smell are often difficult to distinguish because they work together much of the time. In man, the organs of taste are the taste buds located on the tongue. The smell receptors are found in the membranes lining the nose. Smell is generally more discriminating and sensitive than taste, and it serves as an important guide to behavior in many species. Pheromones are special chemicals secreted by animals that trigger particular behaviors. The salmon's sense and memory of smell guide him from the ocean to the pool where he was born.

What we call touch actually comprises five different kinds of receptors in the skin: touch, pain, pressure, heat, and cold. Some organisms, like rattlesnakes and ticks, use sensitivity to heat to help locate prey. When a bee dances out the way to a newfound source of food, her hivemates follow her movements by touch. Touch triggers the change of grasshoppers into locusts.

The human inner ear contains a special organ for sensing gravity and three semicircular canals that detect rotation or movement of the head. It has also been found that various organisms react to other stimuli, such as magnetism, without any known specialized receptors. Extrasensory perception has been shown to exist, but there is no adequate biological explanation of it. How much these still uncharted senses may affect life is unknown, but there is the possibility that they are the cause of endogenous rhythms.

Questions to consider

1. a. Is there any difference between a nerve impulse in the optic nerve and one in a sensory nerve leading from the foot?
   b. Can you cite examples that lend support to your answer?
   c. Where are the various sensory impulses sorted and interpreted?
2. a. What path does light follow from the outside environment to the optic nerve?
   b. What names are given to the actual light receptors?
   c. Do all these receptors work in the same way, or are they sensitive to different stimuli?
3. a. What sort of eye is characteristic of many arthropods?
   b. What advantage does this eye confer?
4. Dogs have their eyes set into the front of the face, while the eyes of a deer are on the side of the head. How does the positioning of the eyes in each species fit the animal's ecological niche?
5. How do the structures of the human ear translate a sound into a set of nerve impulses?
6. What are some examples of behavior elicited by sounds?

7. The Environmental Protection Agency, an arm of the federal government concerned with implementing legislation to lower pollution, numbers noise pollution among the undesirables it wants to eliminate. Is noise pollution a real danger, or is it just so much political hocus-pocus?
8. Disease or damage to the face occasionally results in a loss of the sense of smell. People so afflicted often find eating a chore to be endured rather than a pleasure to be enjoyed. Why is this the case?
9. a. What is the name given to odors used as messages between individual animals?
   b. If the trail left by an ant scout is not renewed frequently, the odor quickly dissipates. Wouldn't the ants be better off if the odor lasted a long while? Why?
10. A mouse put into the cage of a blind and hungry rattlesnake will very quickly be found, killed, and eaten. How has the snake located its prey without being able to see it?
11. What sense serves as the trigger for the transformation of a grasshopper into a locust?
12. How does the human body tell up from down?
13. a. What evidence is there for senses other than the five traditional ones?
    b. What phenomena fall within the range of extrasensory perception?
14. a. What is an endogenous rhythm?
    b. People who travel across the continent go through a 3-hour time change. A New Yorker arriving in Los Angeles may think it is time for dinner, but to the locals it is the middle of the afternoon. Oftentimes the traveler finds himself irritable and unable to sleep for a day or two, a phenomenon called jet lag. Can jet lag be explained in terms of endogenous rhythm?

Glossary

anvil one of the small bones of the middle ear, so named because the hammer beats against it *carotene* one of a group of orange plant pigments which contributes to visual perception in animals

cochlea the coiled tube in the inner ear which is lined with receptor cells

compound eye the visual receptor in most arthropods, characterized by multiple, separate corneas and receptors

cone one of the two types of receptor cells in the retina, specifically sensitive to color

cornea the transparent tissue which forms the outer surface of the eye

decibel a unit of measure of sound intensity, arranged on a logarithmic scale

ear canal the long passageway that connects the outer ear with the eardrum

eardrum *see* tympanic membrane

endogenous rhythm a regular, internal biological rhythm not apparently related to environmental cues

eustachian tube the passageway that connects the middle ear and the throat

extrasensory perception the perception of stimuli not detectable by the five senses

frequency the measure of the number of cycles of a sound wave each second, perceived as pitch

hammer one of the small bones of the middle ear, which transmits sound by beating against the anvil

inner ear the innermost portion of the ear, in which receptor cells transmit the sensation of sound to the brain

intensity the amount of energy carried by a sound wave, perceived as loudness

iris the colored portion of the eye, which regulates the amount of light entering the pupil

lens the transparent, flexible structure in the eye which focuses incoming light onto the retina

middle ear the center portion of the ear, in which sound waves from the outer ear are transmitted to the cochlea *optic nerve* the bundle of nerves which transmits the sensation of light from the retina to the brain

outer ear the portion of the ear which traps external sound waves and funnels them to the eardrum

pheromone a chemical substance secreted by one member of a species which triggers a behavioral response in other members of the same species

pitch the component of a sound determined by the frequency of the sound wave

pupil the opening of the eye through which light is directed toward the retina

retina the inner lining of the eyeball in which the receptor cells are located

rod one of the two types of receptor cells in the retina, specifically sensitive to light intensity

round window the membrane-covered opening in the cochlea which vibrates to equalize the internal pressure in the outer canal

semicircular canals the fluid-filled cavities in the inner ear whose receptor cells determine position and balance of the organism

stirrup one of the three small bones of the middle ear, which vibrates against the outer wall of the cochlea

taste bud a taste receptor found on the tongue, specific to one of four types of tastes

tympanic membrane the flexible membrane that transmits sound waves from the outer ear into the bones of the middle ear

Part Three

The Community of Life

Up to this point, we have seen how living systems organize matter and energy to maintain themselves and their species. But biology has also given us a fascinating perspective on the concept of community, the patterns of interrelationships which characterize groups and societies. In this final part, we will see how living things relate to their environments on a large scale. In Chapter 14, *Living Together*, we will look at the specialized structures of different societies in the animal world and at the dynamics of population growth and stability. Chapter 15, *Living in Man's World*, applies some of these same principles to man's unique position in the biosphere and his crucial impact on his environment. And finally, in Chapter 16, *The Future*, we will see what biologists and other scientists can confidently tell us about our future, and what broad and important questions remain to be answered.

14 Living Together

When we first looked at ecology in Chapter 3, we noted that a single ecosystem, even a relatively small one, is home to a great many organisms in terms of both numbers of individuals and numbers of species. Consider, for example, a typical hardwood forest of the kind found throughout the East, Midwest, and South. It takes about 2 acres of ground to support a nesting pair of songbirds like robins or thrushes. That area is covered with a wide variety of plants, ranging from the tall maples and oaks down to the tiny and barely noticeable mosses. And the birds are by no means the only animals present. Those 2 acres probably support about 70 mammals, predominantly small organisms like mice and shrews and a few large ones like deer; 25,000 snails and slugs; 20,000 centipedes, millipedes, and sow bugs; 70,000 assorted spiders, mites, and daddy longlegs; and over half a million large insects like ants, praying mantises, beetles, and moths. And this is only scratching the surface. Were one to count the organisms in the soil, the numbers would quickly hit astronomical proportions. A cubic yard of soil in a grassy area swarms with 165,000 mites. The tiny worms known as nematodes number in the tens of thousands, and earthworms are likewise common. The total population of bacteria and fungi, if it could ever be counted, would run into the trillions and perhaps beyond.

These organisms and individuals don't live in isolation from each other. Individuals of the same species and of different species come into contact with each other constantly. How do they interact? How does one species affect another? Since the number of species in a community and the number of individuals in the species tend to remain relatively constant, there must be some kind of definite patterns to these interactions. What are they?

In this chapter, we'll look at the answers to these questions. First we'll examine the relationships between species that share the same turf. Then we'll turn our attention to how members of the same species get along, and pay particular attention to animal forms that live in societies. Finally we'll look at the forces within the environment and within organisms that keep populations constant.

What happens between species

Many species coexisting within the same ecosystem are hardly even conscious of the existence of many of the others (Figure 14–1). A grazing deer pays no attention to the earthworms only an inch or two under its nose. These two animal organisms are related, but only in an indirect way, as members of the same ecological community. The earthworms' many tunnels help break up the soil and keep it rich and suitable for plant

Figure 14–1 Animals sharing the same space and functioning as part of the same ecosystem do not necessarily interact with each other. The giraffes, elands, and rhinos on an English game preserve can live their entire life cycles without really taking notice of each other.

Predation is one of the simplest and most common forms of interaction between species: (A) the alligator and its prey,the gar; (B) the sparrowhawk, named for its favorite catch; (C) the osprey with its next meal.

life. If the earthworms suddenly disappeared, the quality of the soil would suffer, grass would grow less abundantly, and food would be scarce for the deer. In this sense, every organism in the community—plant or animal and large or small—is related to every other living thing. But what of more direct interactions between various kinds of organisms?

Who eats whom

Perhaps the most common relationship among animal species is that of *predator* and *prey*—the predator kills and eats the prey for food. With the exception of the few carnivorous species like the Venus's flytrap, plants manufacture all their food by photosynthesis and have neither the need nor the capability to eat other organisms. But among nonphotosynthetic organisms, from the tiny one-celled microorganisms to the mammals, many species get their food by killing and eating other animals, and those species most likely to be preyed on expend considerable energy avoiding predators.

Although humans are often predators, we tend to think of predators as bloodthirsty and vicious and to see their prey as exploited and pitiful. In this country, for example, it has been standard procedure until recently to pay bounties for killing coyotes, wolves, and mountain lions, while deer and rabbits are protected by laws regulating the hunting season, the number of animals killed, and so forth. Actually this view is shortsighted. Although the animal that provides a meal for another obviously doesn't benefit from its own death, its species as a whole may actually profit. But how can a group benefit if its members are killed off? Isn't that a contradiction in terms?

Not really. Because of evolution, the prey and predator species native to a given ecosystem are pretty evenly matched. If this were not the case, then the prey species would be eliminated, and the predator would have to find a new food supply or die out as well. Prey animals are not totally defenseless. For example, a moose can stand at bay and face down a pack of wolves, and a deer can generally outrun a mountain lion. In each case, the predators go without a meal. What animals then can the predators kill? Usually the animals that fall to the carnivores are the ones with weakened defenses: the aged, the sick, the crippled, or the immature. Only occasionally do predators kill healthy adults, and usually that's with a bit of luck. As a result, predation acts to cull the prey, eliminating hurt, sick, and old individuals. Later in this chapter we'll look at the effects of predation on populations.

Living at someone else's expense

While a predator actually kills the prey to extract energy from it, a *parasite* lives off its *host*, appropriating for itself part of the food resources of the host. Parasites come in a variety of sizes. For example, the bacteria and viruses responsible for many animal diseases are parasites. Fleas and lice are tiny insects, but they usually appear in large, irritating groups. The intestinal tract and related internal organs are often home for a variety of worms, ranging from the relatively small ascarids, or roundworms, to the tapeworms, some of which may reach a length of several yards. Parasites come even bigger. The so-called lamprey eel, which is not an eel at all but a primitive, jawless fish, is nearly as large as the fish it parasitizes. The lamprey cuts a hole in the side of the fish with its rasping tongue and sucks out the fish's body fluids (Figure 14–2).

Obviously it is to the benefit of the parasite not to kill the host, because if the host dies the free ride is over. In man, for example, the only invading microorganism that regularly causes death if left untreated is the rabies virus. So far as is known, only one person has ever survived rabies. Although many other microorganisms may cause death, a significant number of people with these parasites survive. And this is true of almost all

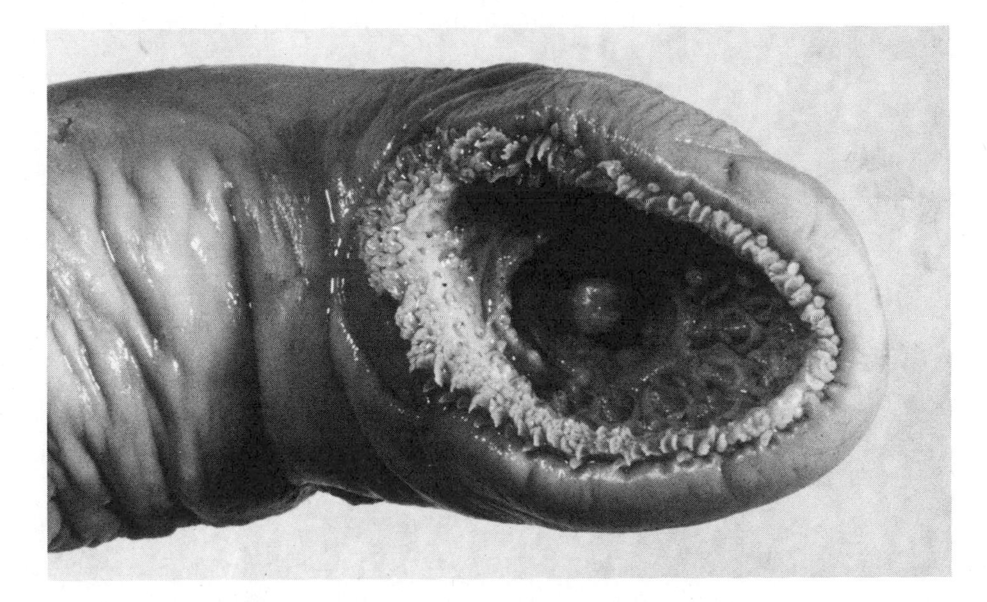

Figure 14–2 The lamprey parasitizes fish by biting a hole in the fleshy portion of the fish and sucking its host's blood.

parasites. Most wild animals are infested with parasites of one sort or another. However, these infestations are usually smaller than what would harm or kill the host.

But how do parasites become dangerous? Why do diseases, for example, suddenly become rampant and nearly wipe out a whole population? Usually what happens is that the natural relationship of the host or parasite is upset. The life cycle of many parasitic worms, for example, usually prevents massive and fatal infestation of a single individual. These worms usually lay their eggs in the host's feces. Grass or other food contaminated with the eggs can carry the parasite into another host organism. Wild animals often wander, scattering their feces and the parasite's eggs in them far and wide, thus greatly reducing the chances of multiple infestations of the same animal (Figure 14–3). But this life cycle, which protects wild animals, may prove dangerous to domesticated species. Since domesticated animals are concentrated in fixed locations, it's possible for the animals to become infected again and again by the same batch of parasites. Thus wild dogs and wolves are often infected with a controllable number of roundworms, but these same parasites often prove deadly to domesticated dogs, particularly puppies, because the animals are infected again and again.

A similar effect occurs when the environment somehow changes, upsetting the balance between host and parasite. The Great Lakes were free of the lamprey until the St. Lawrence Seaway was completed, con-

Figure 14–3 The wandering of wild animals, for instance, these wild dogs, permits the distribution of parasitic organisms over a wide geographical area.

necting the lakes to the Atlantic and providing the lampreys with an open door to a new environment. The lampreys invaded the lakes in great numbers and did considerable damage to the fish populations. A similar situation is occurring in many parts of Africa as a result of damming rivers like the Nile and the Zambezi. For many centuries, people along portions of these rivers have been subject to a serious disease called schistosomiasis, which can be fatal. Schistosomiasis is caused by the schistosome, a worm that parasitizes the liver. The schistosome spends part of its life cycle in the body of a snail that lives only in sluggish waters, and thus can be no more common than the snail. The disease was practically unknown along those parts of the river where the current ran fast, because the snails were restricted to a few backwaters and stagnant pools. But the dams have turned the swift rivers into sluggish reservoirs, and the snails have suddenly become common. So has schistosomiasis.

New parasites, ones that the host has never before experienced in its history, can have catastrophic effects. When Europeans first came to the Hawaiian islands, they carried the parasites that cause measles and syphilis, and these microorganisms were unknown to the Hawaiians. As a result, the Hawaiians had no natural immunity. Measles, which was considered a mild childhood disease in Europe, proved highly fatal to the Hawaiians, and syphilis seemed doubly potent. When Captain Cook first came to Hawaii, the islands' population was 300,000. A hundred years later, the native population numbered under 50,000, most of the loss due to the new parasites.

Sharing space

But organisms don't always exploit each other. In some cases, two different species may live together in such a way that both benefit. This relationship is called *mutualism*. In Chapter 3, we noted that certain kinds

When Captain Cook first came to Hawaii, the islands' population was 300,000. A hundred years later, the native population numbered under 50,000.

Figure 14–4 The cattle egret and its host. The cattle egret follows cattle in a herd and feeds on the insects stirred up in the cattle's path. The cattle in turn enjoy protection from disease-causing insect pests.

of plants, most notably the legumes, contain nodes that are the homes of nitrogen-fixing bacteria. Another example of mutualism among plants is the lichen, which is actually a complex of fungi and algae, two different kinds of organisms acting like one. It is a very successful and beneficial relationship, for lichens can inhabit hostile environments where neither algae or fungi alone can survive.

Animals also engage in mutualistic relationships. The tick bird spends much of its time feeding on parasitic fleas, lice, and ticks that trouble the rhino. Thus the bird gains a ready food source, and the rhino gains relief from the annoying insects. Also, the tick bird's elevated perch protects it from attacks by ground-level predators, while the rhino, a notoriously nearsighted animal, is warned of any approaching danger by the bird (Figure 14–4). The hermit crab protects itself by carrying on its back the empty shell of a dead sea snail. Often the crab will coax a sea anemone to attach itself to the top of the shell. The anemone's stinging tentacles provide the crab with added protection, and the moving anemone gets a better chance of finding food than a sedentary anemone. And man's long-standing relationship with dogs is mutualistic, probably based on a primitive hunting partnership where man benefited from the dog's sharper senses of smell and hearing, and the dog gained the advantages of man's brain. The two killed more game and ate better working together than either did on his own.

Commensalism is the name given to a relationship where one organism benefits, and the other is neither helped nor harmed. Such a relationship, for example, exists between orchids and trees in tropical rain forests. The tree foliage grows together in a tight canopy that lets little sunlight through to the floor of the forest. The orchids escape competition for sunlight by growing in the upper branches of the trees. The trees, though, gain no advantage. Commensalism is particularly common among marine organisms. Barnacles hitchhike to new food sources by attaching themselves to crabs and even to the vast sides of whales. A burrowing worm has earned the nickname "innkeeper" because the burrows it digs afford shelter for various fish, scale worms, crabs, and clams.

Relationships between plants and animals

We saw in Chapter 5 that all the energy in living systems is trapped in chemical bonds by photosynthesis and then passed from trophic level to trophic level in the form of these chemical bonds. Therefore it's hardly surprising to find that plants serve principally as a source of food to animals and that animals are, depending on the particular case and on one's point of view, either predators or parasites on plants. By evolution, plants

An epiphytic orchid growing on a palm tree shares the sunlight and nutrients of its host.

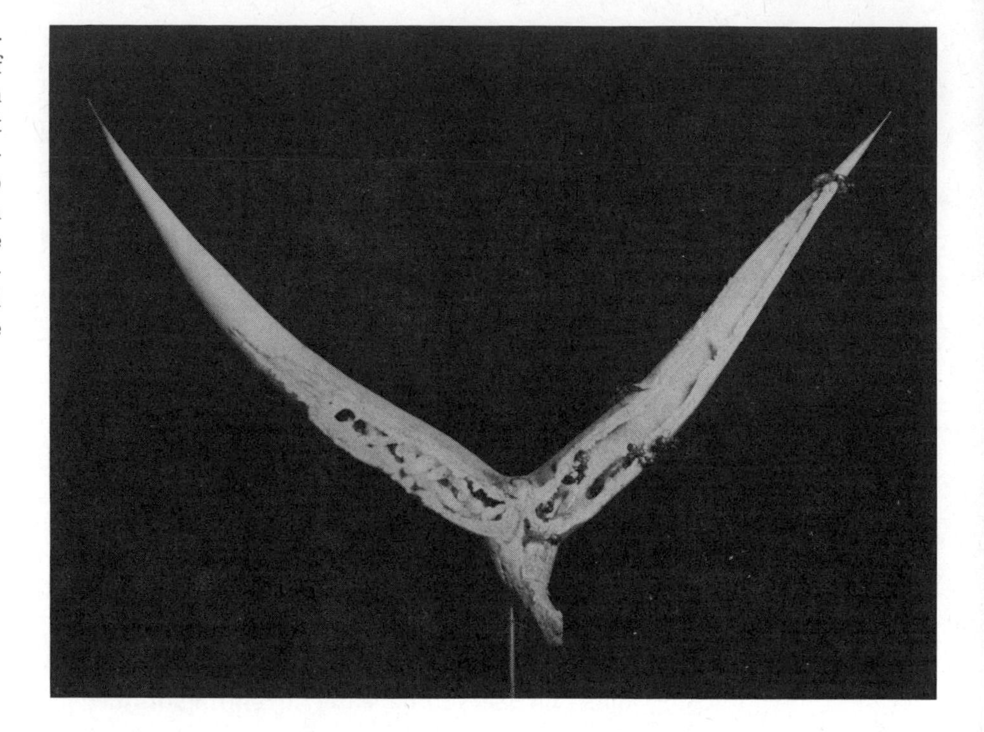

Figure 14–5 The interior of some leaves of the bull-horn acacia tree and its resident ants. This complex mutualistic relationship provides the ants with food and a home, while the ants serve as a complete maintenance and protective force for the tree.

have developed a great number of adaptations which serve as defenses against herbivores, ranging from the cactus' sharp spines to the deadly nightshade's poison. In a few cases, though, plants and animals have developed mutualistic relationships. For example, in Chapter 9, we noted the interplay between plant and pollinator, where the animal gains a food source, and the plant's reproduction is ensured success.

Easily the most developed and fascinating mutualism between a plant and an animal is that of the bull-horn acacia, a tree species, and an ant that lives on the acacia (Figure 14–5). The acacia contains a number of special hollow thorns that the ants inhabit. The ants get all of their food from the acacia, primarily from the nectar-producing structures at the base of each leaf. By no means, though, are the ants simply exploiting the acacia for food and shelter. The acacia is native to the tropical rain forests of Mexico, where its survival depends on escaping herbivorous mammals and insects and on getting an adequate supply of sunlight. The ants patrol the surface of the acacia constantly, like so many sentries, attacking and stinging any animal large or small that tries to feed on the tree. As the growing acacia comes into contact with the branches of other plants, the ants strip the bark off these branches and remove them, providing the acacia with a tunnel of sunlight through the jungle canopy. The relationship between ant and acacia is so complete that one simply cannot survive without the other.

Figure 14–6 Many animal species form societies, permanent groupings that provide for common protection, division of responsibilities, or care of offspring. Lions group in prides, fish in schools, and birds in flocks.

A general concept

All of these relationships between two or more organisms of different species involve a close living relationship of some kind, which may or may not involve benefit to one or both parties. This phenomenon is collectively called *symbiosis*, and it is found throughout the natural world.

All of a kind: animal societies

Cooperative associations exist not only between such different species as hermit crab and anemone or ant and acacia but also between members of the same species. A great many kinds of animals, particularly insects and vertebrates, come together in groups, or *societies*. These groups are of such importance that even our common nonscientific language contains a great many words for the groupings of particular kinds of animals: packs of wolves, swarms of bees, flocks of ducks, schools of fish, herds of deer, coteries of prairie dogs, coveys of quail, gaggles of geese (Figure 14–6).

How about beds of oysters? No, oysters are not social animals, even though they live in very extensive colonies. A society is not simply a collection of individuals gathered together in one place; it also entails communication and cooperation. In warm waters where sharks are plentiful, fresh blood will bring a great many sharks searching for food. However, this is simply a number of individuals responding to the same stimulus, like so many moths fluttering about a naked light bulb, and not a society. The sharks never cooperate to form a hunting pack as do wolves, which truly are social animals.

The pyramiding units of societies

The simplest social unit is the *family*. In some cases, the family unit is temporary. This is the case with most species of bear. The male stays with the female during the mating season and for a month or two thereafter, until she suddenly turns on him and drives him away. After the cubs are born in the spring, the female and her young make up the social unit for the 2 years needed for the cubs to mature. In a few species, though, the family unit and particularly the *pair bond* between male and female are quite permanent. Canada geese are such a species. A goose and a gander will remain with each other year after year, usually until one dies. A pride of lions is actually a polygamous family, comprising the adult male, his several mates, and their many offspring.

Families can be extended into large groups of more-or-less related animals. These *extended families*, as they are called, represent the social organization of chimpanzees and baboons. Some of these extended families are quite large; a troop of baboons may number into the hundreds. Another primate species, man, has also often been organized into extended families. That social unit, called the *tribe*, is based on family ties, some near and some quite far removed.

Many social animals form groups far larger than individual families or even than extended families. This is the social pattern of many larger grazing animals, an aggregation of many mated and unmated adults along with immature animals of all ages. Figure 14–7 shows a typical scene on the East African savanna with its extensive herds of wildebeests, gazelles, impalas, and zebras. These large groupings are also common among fish and birds, where they are called schools and flocks, respectively. The curious thing about flocks, schools, and herds is the way that all the individuals can act as one. Anyone who has watched enough Westerns has seen examples of the devastation wrought by stampeding horses,

Figure 14–7 The East African savanna is noted for the beauty of its wildlife as well as that of its scenery. The many graceful herbivores found in this region—the wildebeests, impalas, and zebras—all travel in herds.

Figure 14–8 Anyone who has ever watched migrating birds has noted the intricate and regular patterns they form in flight, like these laughing gulls.

cattle, or buffalo. What provokes a stampede is fear spreading through the herd; one animal starts running and all the rest run in response even though they may not know what they're running from. Birds and fish are even more adept at such synchronized behavior, so that thousands of flying or swimming animals all change direction at once with the cadence-perfect precision of a well-trained drill team (Figure 14–8).

What do they stand to gain?

A mutualistic relationship like that between the ant and the acacia is maintained because both parties benefit greatly from it. Social organization must also profit the species in some way, or else the forces of evolution would never have selected for it in so many organisms, from bees to man. What is the survival value of a society?

Generally, social groups confer the following advantages: an aid to mating; protection against predators, particularly for the young; and a division of labor.

Society and mating

One of the biggest problems to successful sexual reproduction is getting a ready male together with a ready female at the right time for fertilization to occur. Herding of one type or another ensures a good supply of both sexes. Some species, such as the elephant seal, form social units only during the mating season. In other species, the social organization may change markedly as the mating season approaches. Elephant herds, for example, are composed of cows and calves, with the bulls off on their own either singly or in small bands. Only when the females are coming into season do the bulls enter the herd and become a part of it. Many marine mammals form breeding herds at particular geographical locations at certain times of the year. For example, during the summer, small groups of gray whales feed in the Arctic Ocean, but in the winter they migrate south and gather together at a few sites on the coast of Baja California where calving and breeding take place.

The protection of society

Many animals, particularly herbivores, spend considerable time and energy trying to avoid becoming the next meal of a hungry predator. Being in a herd makes the task easier. Many pairs of eyes and ears working together are more likely to sense a predator's approach than is one animal watching its own flanks all by itself. The first animal to sense danger can send up a warning and tell the others to escape.

Oftentimes the characteristic size of the group of a certain species depends on that species' habitat and the best tactics for avoiding the dangers the habitat poses. Large, hoofed animals like deer, which live in the heavy cover of timber, usually do best if they simply try to escape the predator's notice. Thus deer tend to run in small family groups rarely numbering more than a dozen, a strong contrast to the huge herds typical of plains animals. But the plains animal faces a very different problem. On the unbroken and flat expanse of a grassland, concealment is out of the question; there's simply no place to hide. Instead, safety exists in numbers. The trick to surviving is to locate the predator before he gets too close and then keep him at a safe distance. Sometimes two separate spe-

For the plains animals, the trick to surviving is to locate the predator before he gets too close and then to keep him at a safe distance.

cies will herd together for mutual protection. Baboons have keen eyesight, and wildebeests have sensitive hearing. When these two animals feed together, a lion or a leopard or a pack of hunting dogs cannot approach without being detected. Aware of the predator, the baboons and wildebeests can stay a safe distance away and have a good head start on the predator if it decides to attack.

Some species have group defenses other than simply running off in a stampede. A covey of quail hiding in brush or high grass will not break into flight at the first sign of an approaching bobcat or a hunter. Instead, they sit absolutely still until the predator is nearly on top of them, and then all the birds burst simultaneously from cover, the whirring of their beating wings sounding like an exploding hand grenade. The noise is so loud that it usually flusters and panics the predator. Musk oxen make an impressive defensive array when wolves approach. The males stand shoulder to shoulder in a tight ring with the females and young safe inside. The wolves cannot breach the circle without facing those awesome horns and hooves. They usually choose not to.

The defensive attitudes and advantages of a herd's behavior depend on its environment, and if the environment changes, that defense may be of little use. In part, this is what happened to the American buffalo. A grazing buffalo herd was usually arranged in a ragged circle, with the bulls on the perimeter standing guard and the cows and calves to the center. If wolves or Indians approached, the guard bulls either gathered the rest of the herd into a circle or started a stampede. Although the herd regularly lost individuals to both wolves and Indians, the buffalo as a species survived quite well. But then came the white man with his guns, and he stumbled across a curious weakness in the buffalo's defenses. A buffalo herd will stampede madly if a dying animal groans and cries out in agony. Oddly enough, though, the buffalo will not stampede if an animal drops and dies suddenly and silently. Neither wolves nor Indian arrows can kill quickly enough to keep a dying buffalo from squealing, but a gun is another matter. A bullet through the lungs and heart can kill almost instantly. The white hunter could stay back at a good range and pick off buffalo one by one, starting on the outside and working in. As long as the hunter killed each one quickly, the herd would remain where it was until the last animal was dead. Only if the hunter muffed a shot and wounded an animal without killing it did the remainder of the herd stampede. In one afternoon a single man could kill several hundred buffalo. In less than 15 years, a few thousand such men killed nearly all the American herds.

Divvying up the tasks

One of the key aspects of many animal societies is the division of labor, whereby different individuals perform different jobs for the community. In some cases, the social roles are relatively temporary or short-lived. When a flock of crows is feeding on the ground, one or two of the birds stand guard in the top of a nearby tree. Periodically the sentry is replaced by one of the other birds, so that he too may have a chance to feed. The several females in a pride of lions may work as a team to find and kill game. One of the lionesses will flush the prey in the direction of another waiting in ambush. And, while the lionesses are off hunting, the lion cares for the cubs. Many social rodents, such as prairie dogs, live in elaborate runways and burrows obviously constructed by teams of cooperating animals (Figure 14–9). And, of course, man's industrial civilization is based on such a specialization of labor. Some people grow food, some people work in factories, some people care for children in homes and in schools, some people attend to health needs, and so forth.

Figure 14–9 The prairie dog is an example of a social rodent, living in shallow underground tunnels, built as a community effort.

The role of love

In some species, social interactions do something more than simply pay such immediate benefits as ease in finding a mate. With these animals, society may be as necessary as proper nutrition for normal development of the individual. It has long been known, for example, that puppies raised in isolation from other puppies and humans tend to be untrainable and either very shy or very fierce. But no one really studied these effects until Harry Harlow, a researcher at the University of Wisconsin, set about to raise a strain of germ-free rhesus monkeys to be used in research. Newborn monkeys were separated from their mothers shortly after birth and raised alone in cages. The infant could see other monkeys, but it couldn't touch them. Harlow achieved his goal of keeping the monkeys free from disease, but he found that the adults grown from these infants displayed very abnormal behavior. None of the monkeys mated; often the male and female fought violently when put together. By contrast, wild male monkeys only rarely attack a female, and usually they demonstrate great affection toward all the females in the group. Wild monkeys also spend considerable time grooming and cleaning each other. Harlow's monkeys, however, ignored one another, and the only interchanges between them were displays of fear and aggression. Alone in their cages, Harlow's monkeys acted in ways that would be judged neurotic in humans—staring into space, walking about like possessed robots, holding their heads in their hands, and rocking back and forth for hours on end. Some of them even bit and chewed on themselves, inflicting serious injury (Figure 14–10).

Without realizing it, Harlow had somehow upset the normal developmental pattern of these monkeys by isolating them. He then devised a series of experiments where he cut infant monkeys off from some social interactions, but allowed them others. He compared the behavioral effects of the various experiments to find out which social interactions were important and what aspects of behavior and personality each influenced. He found that while the relationship between mother and child

Alone in their cages, Harlow's monkeys acted in ways that would be judged neurotic in humans.

Figure 14–10 Harry Harlow's experiment in maternal deprivation in Rhesus monkeys is one of the most famous accomplishments of behavioral psychology. The monkey on the left seeks the contact of its soft terry-cloth mother surrogate, even though it will not provide him with food. Monkeys deprived of maternal care, such as the one on the right, become unstable, reflecting behavior similar to that of a neurotic human.

is important, particularly in the first 3 months of life, the relationship between *peers*—that is, monkeys of approximately the same age—is even more crucial to normal development. Infants removed from their mothers but allowed to play with each other develop normally, but infants left with their mothers and denied access to their peers until 6 months of age are sexually and socially abnormal.

Harlow's experiments have shown that social arrangements among animals, particularly primates, are not simply relationships of convenience based on benefit to the individual. With the monkeys, as with man, the relationships between individuals are the cement holding the society together. The other immediate benefits are very nearly secondary.

The specialized societies of the insects

However intricate human society and the division of labor characterizing it, by no means is it as elaborate and specialized as those societies found among the social insects—the bees, ants, wasps, and termites. Used as we are to equating social structure and intelligence and to thinking of insects as something less than brilliant, it may be a little hard to believe that insect societies are more specialized than ours. What is the meaning of such a statement?

To understand it we must remember the biological definition of "specialization," which means that something is specially fitted to one particular task and not to any other. Thus a bird's wing, fitted for flight and flight alone, is a highly specialized limb, while the human hand, which can be used equally well for a fine and delicate task like assembling a watch or as a weapon for attacking a would-be mugger, is relatively generalized. The same analogy applies to the comparison of insect societies and human societies. Human society has appeared in many forms, ranging from small hunting bands to very large and complex nation-states. Although human roles may be difficult and hard to learn—becoming an atomic physicist, for example—the differences between social roles is largely one of training, or education, and opportunity. While special ruling or priestly classes have dominated certain human societies, the distinctions separating these groups from the masses were not physiological. Biologically, there is no difference between a prince and a pauper. Just the opposite is true of an insect society. Each individual is specially outfitted for its job, oftentimes to the point where it is physically different from the other insects of that society. Also, the individuals are by no means unique. If one insect can no longer perform its role—if it falls to disease or mishap, for example—it can be replaced by another insect that is its carbon copy. Among the social insects, the individual exists solely for the benefit of the group as a whole.

Class and caste among the insects

The social division of labor has reached its most extreme development among the termites. All termite colonies—from the wood-eating

Figure 14–11 The results of the worker termites' labors—a nest built on a tree limb.

termites of the continental United States to the African species that construct 30-foot-high mounds—have at least three specialized forms, or *castes:* reproducers, workers, and soldiers. Each caste has a different body shape, a different behavior, and a different function for the colony (Figure 14–11).

Only two termites, one male and one female, make up the reproductive caste. Because of their singular position, they are called the king and the queen. The king is nothing more than a sperm producer for the prolific queen. Sequestered in a special chamber and attended to by worker termites, the queen produces as many as 30,000 eggs a day. In addition to caring for the queen and her eggs, the workers, who are the second caste of termites, build and repair the nest and forage outside for food. Unlike the king and queen, who may live for 15 years, the workers are short-lived. The third caste, the soldiers, performs the task their name implies: guarding the nest. The bodies of the soldier termites are adapted to make them especially suited for fighting, but the adaptation varies from one species to another. In some species, the soldiers have enormous jaws capable of delivering a powerful bite. Others develop glandular "squirt guns" that shoot sticky, entrapping threads or irritating chemicals. And some soldier termites have plug-shaped heads that they use to block the tunnels leading into the colony. While both workers and soldiers may be of either sex, they are sterile.

How caste is determined

Obviously, a termite colony, which may comprise 3 million individuals, can survive only if the numbers of each caste remain constant. A disproportionate number of soldiers would mean that the colony would be well defended, but it would soon fall into disrepair. Likewise, too many workers might result in a colony too large for the soldiers to defend effectively. But such imbalances rarely occur in nature. All the nymphs hatched from the eggs laid by the queen look the same up through their first molting phase. Then they begin to differentiate into soldiers or workers or, more rarely, reproducers. The amazing thing is that the nymphs become what the colony needs. If workers are removed from a colony, then the nymphs become workers. If the removed termites are soldiers, then soldiers come from the nymphs' final molt. And when the king or queen dies, one nymph becomes a reproductive form.

Remember that all the termites in a colony come from the same two reproducers, that they are, in human terms, brothers and sisters. First of all, how can such different individuals—queen, worker, soldier—develop from the same genetic information? Second, how do the nymphs "know" what the colony needs in terms of replacements? The same phenomenon answers both questions. Recall that the chemical environment of cells shapes what happens to them in development, as shown in tissue induction (see Chapter 10), and what role the sex hormones play in maintaining sexual characteristics. Each caste of termites produces a special chemical that is passed to other termites by touching the antennae and by exchanging food. These chemicals make up an important part of the environment surrounding the developing nymphs. If a significant number of the members of one caste are removed from the colony, then the chemical they produce becomes rarer. Such a change in the chemical environment causes a corresponding change in how the nymphs develop. These chemicals are called *social hormones.* Like all hormones, these messengers are produced by one unit to affect another unit, but in this case the units are individuals rather than organs.

Experiments in laboratories have shown how these hormones are passed through a colony and what effect they have. A typical termite colony containing one queen and one king was divided in half by a wire mesh that prevented the two half-colonies from passing food back and forth but allowed them to touch each other's antennae. The colony was divided in such a way that the king and queen were together, thus leaving one of the half-colonies without reproducers. In the half-colony without a king and a queen, fertile males and females developed from nymphs, but the other termites killed them. Normally, reproducers would not have developed at all; apparently, whatever stopped the development of reproducers was passed around by means of food exchange, which the mesh prevented. The absence of food exchange caused reproducers to be produced. But apparently the workers in the half-colony with the new reproducers could sense the presence of the former king and queen by contact with the antennae of their counterparts on the other side of the mesh, and they killed the new reproducers. Further experimentation showed this to be the case. When the two half-colonies were separated so that they could neither exchange food nor touch antennae, then a new king and queen developed and reproduced normally. Cut off from the social hormones of the other half-colony, the one that had been left without a royal couple developed into a normal and complete colony all its own, thus restoring the typical chemical balance.

A living thing all its own

Throughout this book we've seen how a number of simple things can be organized into a single complexity. Simple molecules can be aggregated into complex molecules, which in turn make up cells, which are the building blocks of whole organisms. In each case, the complexity that results is more than just a sum of its parts. Very much the same thing is true of termite society. A termite alone is not really a termite. It exists only as a part of a complex social organization, as if the colony were an organism and the individual only a single cell.

Populations: how many individuals are there?

In understanding how an ecological community works, we have to know not only the way that the various species interact with their own kind and with other organisms, but also how many living things there are. And that means counting populations.

A *population* is simply a group of individuals of the same species that live close enough together that they may breed with each other. A population can be described in a variety of ways. The simplest is quantity, the number of individuals in the group. Numbers in themselves, though, mean relatively little without keeping the kind of organism in mind. For example, a population of 100 elephants is one thing, 100 fleas quite another. To help overcome this disparity and allow more accurate comparisons between populations of different organisms, groups of animals or plants are often measured in terms of the total weight of living material they comprise—their *biomass*. Biomass allows for a better picture of the relative sizes of equal-numbered populations of fleas and elephants (Figure 14–12). Biomass measurements, though, can also be misleading, particularly when considering the food value or food requirements of a population. For example, 1,000 pounds of living oysters are mostly shells, while 1,000 pounds of chickens represent a goodly quantity of meat.

It can also be important to relate a population to the space it occupies. A population of half a dozen mice in an acre of corn field is living under very different conditions than another half dozen confined to a cage that measures a foot square. Populations can be related to the space they occupy by means of *density:* the number of individuals in some unit of

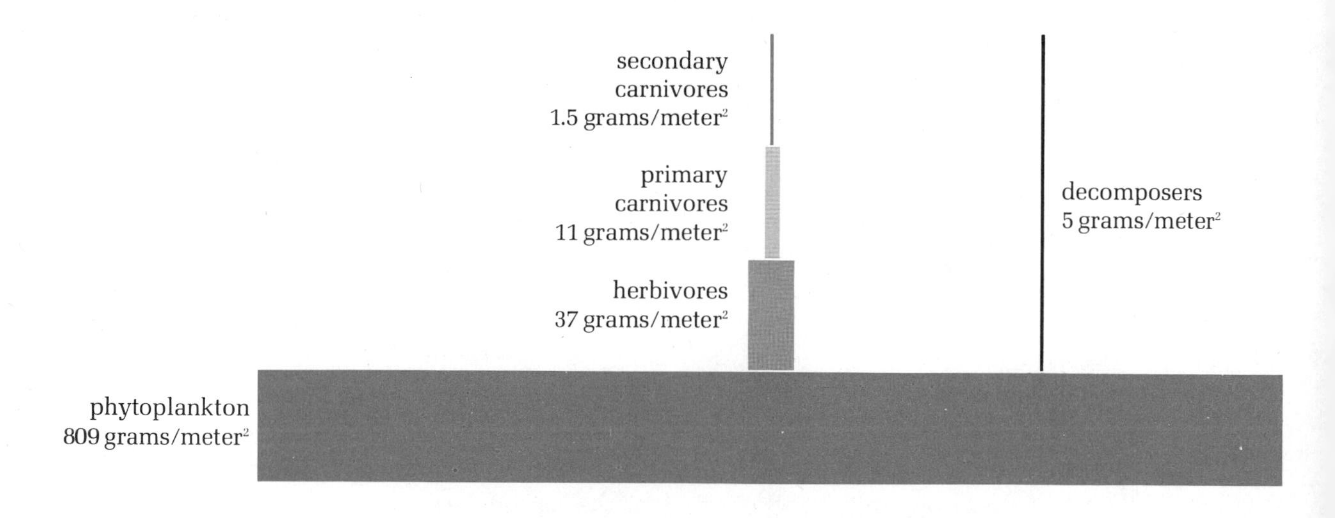

Figure 14–12 One of the many useful ways of looking at an ecosystem is the biomass of its occupants at each trophic level. Each level will comprise a greater total weight of living things than the level it supports directly above it.

measurement, such as an acre or a square mile. The density of a species can vary greatly from one environment to another, perhaps most remarkably in the case of man. New York City has a density of 24,700 per square mile, while the state of Wyoming averages 3 per square mile.

In addition, populations exist in time, and the size of a population may vary greatly from one period to the next. Some populations grow, some decline and perhaps even vanish, and others vary greatly in a cyclic pattern. A cyclic pattern holds true of any species that has a well-defined breeding season. Take, for example, the ups and downs of a population of king salmon in a good-sized spawning pool at the headwaters of a northern California river. Usually these fish come upstream in October or November, and, in such a pool, 250 pairs, or 500 fish, may breed. After breeding, all the adult salmon die, so that the population drops from 500 to 0. But each female has produced about 5,000 fertilized eggs. Even if only a third of these eggs escape predation by trout, the hatching of eggs jumps the population to about 315,000. Preyed on constantly by other fish, the number of tiny fry steadily declines. By the time these fish have made it to the ocean, matured, and return upstream to spawn and die themselves, the population will very likely number 500 again.

Such cyclic variation in a population shows why it is also necessary to determine the kinds of individuals that make up a population. Two populations of the same species may occupy similar areas, be distributed at about the same density, yet be very different from each other. If a population comprises mainly young individuals, it is likely to grow in the future, as this large juvenile population reaches sexual maturity. Such a population is said to be expanding. A stable population, one likely to remain the same size for the foreseeable future, contains members of all age groups, with immature individuals slightly outnumbering older ones. If the greatest proportion of individuals in a population are over the reproductive age, then the population is said to be diminishing (Fig. 14–13).

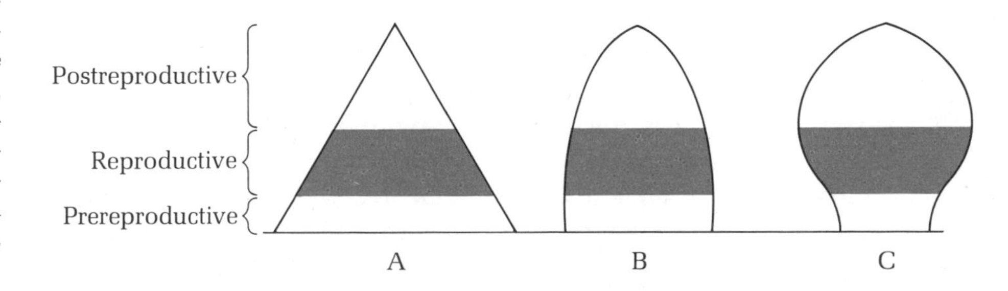

Figure 14–13 The distribution of the members of a population between the prereproductive phase, the reproductive phase, and the postreproductive phase will determine whether the population is increasing in size (A), stabilizing (B), or decreasing in size.

Populations: stability, growth, and decline

In Chapter 4 we noted that one of the biological facts that Darwin wanted to explain with his theory of evolution was the stability of natural populations. The example that Darwin posed to illustrate the problem involved a population of eight pairs of birds. Suppose half of these birds mate, each pair producing one offspring. Half of the increased population —10 pairs—breeds, again producing one offspring per breeding pair. Continuing at this rate for 7 years, the original population should swell to 2,048. But, as Darwin knew, this growth was only hypothetical. All other things being equal, a natural population will remain more or less the same size over time.

Birds are by no means the only organisms to have such a high reproductive potential. Assuming that all flies survive one year and that all females reproduce each generation, a single pair of houseflies can produce over 6 million descendents—6,182,442,727,320 to be exact—in 7 generations! A tapeworm sheds up to 120,000 eggs a day. A female dog can produce two litters of 6 to 10 puppies each year. And the testes of one young human male in good health produce enough sperm cells in a month to impregnate almost every woman of reproductive age on the face of the globe. All species of organisms have the potential to take over the world.

Occasionally this potential is realized in nature when an organism is introduced into a new and favorable environment. Presented with all the necessities of life, the population soars for a while, then levels off at some more or less stable level.

Australia has been the unfortunate setting for several such population bursts. In 1859, an English gentleman living in Australia imported a dozen rabbits from Europe to add a touch of old-country charm to his estate. The rabbits multiplied rapidly and soon seemed to be taking over the grounds. In 1865, the gentleman killed 20,000 rabbits and figured that he had eliminated only two-thirds of the population. Of course, the rabbits journeyed beyond the confines of his property, and their numbers grew by leaps and bounds. Less than 30 years after the first rabbits were introduced, 20 million were killed in but one Australian province. Still the population increased phenomenonally, cropping so much vegetation that the sheep industry was in trouble. In 1950, a fatal disease called myxomatosis was purposely introduced into the rabbit population to see whether sickness could succeed where guns and poisons had failed. For a while the number of rabbits declined steadily because of the disease, but then the rabbits made a comeback, until today they are nearly as common as they ever were. What happened? For one thing, occasional rabbits survived, and passed their immunity on to their offspring. More importantly, though, natural selection favored the less potent strains of the virus. The highly virulent forms quickly ran out of hosts because the rabbits died off so fast. Less potent strains left some rabbits alive, thus preserving a supply of hosts for future generations of the virus. Today, most rabbits contract myxomatosis, but only 5 percent die (Figure 14–14).

In this country, too, such population explosions have occurred. The starling is one of the most common birds in the eastern United States, but its vast population came from a mere 100 original birds released in New York City in the early 1890s. San Francisco Bay is home base for well over a million striped bass, a valuable food and game fish. Unknown on the West Coast before 1879, the modern stripers are all descended from 435 fish brought from their native Atlantic and deposited in the bay.

If such population explosions can occur, why aren't we simply inundated by organisms of all shapes, sizes, and types? All organisms have such a reproductive potential, but populations typically remain stable. The rabbit, for example, was stable in Europe, yet it overpopulated Australia. Its reproductive rate—the number of offspring produced by a given number of adults—was the same in both places. In the European population, though, the death rate must have matched the reproductive rate, so that the number of rabbits reaching sexual maturity approximately equaled the number of adults dying. As with the termites, reproduction served to replace missing individuals, not to swell the total population. In Australia, though, the rate of reproduction far exceeded the death rate, and the result was a rapid increase in the population. Why should this be? What factors affect the rate of reproduction and the rate of death?

Figure 14–14 The European rabbit, imported to Australia in the nineteenth century. After an extraordinary population explosion, a deliberate epidemic of myxomatosis was used to limit the number of rabbits in the country. The selective pressures in favor of immunity to the disease, however, produced a population in the mid-twentieth century that is almost entirely immune to the disease.

A population and its environment

Biologists who study populations divide the factors into two groups. The first group, *density-independent factors*, is not related to the density of the population in question. These factors affect a population, usually by changing the number of animals killed, no matter what the population size. Small populations are affected just as much as large ones. Density-independent factors are a sort of grim reaper that chops back on population size without taking the absolute or relative size of the population into account. The second group is known as the *density-dependent factors*. Their effect varies with size, so that they least affect the smallest populations.

Chief among the density-independent factors is climate. Severe storms in the Great Lakes often dislodge clams and toss them ashore, where they dry out and die. The absolute number of clams killed depends on the number of clams on the bottom, but the proportion or percentage of dead clams will be about the same no matter what the absolute size of the population. Small clam beds will be hit just as hard as large ones. Climactic catastrophes can take huge tolls of animal populations. A deeper than average snowfall can mire deer, leaving them helpless and subject to starvation or predation. An unusually cold spring and summer may keep lakes below the 65°F black bass need to spawn. Because no offspring are produced, dead fish are not replaced, and the population drops. The rising waters of a new reservoir kill all sorts of animals, both the commonplace and the rare, and drown every submerged land plant. Density-independent factors decrease the stability of a population and may cause alternating rises and crashes.

The density-dependent factors tend to emphasize stability. When a population is at its upper limit in size, those factors cut growth down, but they also promote growth in undersized populations. All in all, the density-dependent factors are more important to population stability, and we'll look at several in turn.

Density-independent factors are a sort of grim reaper that chops back on population size without taking the absolute or relative size of the population into account.

The food supply

A given environment usually produces a relatively constant amount of food. As the population of the animals using that food supply increases, the amount of food available to each animal decreases. For example, imagine 1,000 acres of good range land given over to the exclusive use of two cattle, one cow and one bull. Each of these animals has an abundant food supply of 500 acres of grassland, more than enough to support them well. But with each addition to the herd, the food supply available to each animal decreases. With 100 cattle, the food supply is down to 10 acres apiece, hardly adequate. If the population doubles again, many of the cattle will starve. The cattle have exceeded the *carrying capacity* of the environment, the population it can sustain without damage to the population or the environment.

Starvation can keep a population in line with its food supply, but natural populations usually fall short of the theoretical carrying capacity of the environment. Why?

The predator

One reason is predation. To a predator, the animal it preys on is its food supply. But the relationship between predator and prey is more complex than that of cattle and grass. The population of predators depends on the population of prey, which, in turn, depends on the supply of plant food. In a stable environment, all three are in balance: the predators crop the excess herbivores, the herbivores don't outgrow the food supply, and enough plants escape being eaten to produce for the next season.

In some cases, however, the populations of prey and predator may go through great cyclic variations, such that a species common one year may be rare the next. The Hudson's Bay Company, which bought furs from trappers in Canada's Arctic north, recorded the numbers and kinds of skins it took in, and these records serve as a kind of rough census of the populations of fur-bearing animals. Figure 14–15 shows the variations in the populations of snowshoe hares and of lynx, large predators similar to the bobcat, which depend heavily on the hares for food. As the number of snowshoe hares increased, the lynx, too, became more populous after a short time lag. It is doubtful that all these lynx resulted from new births; many of them may have migrated from areas where hares were scarce.

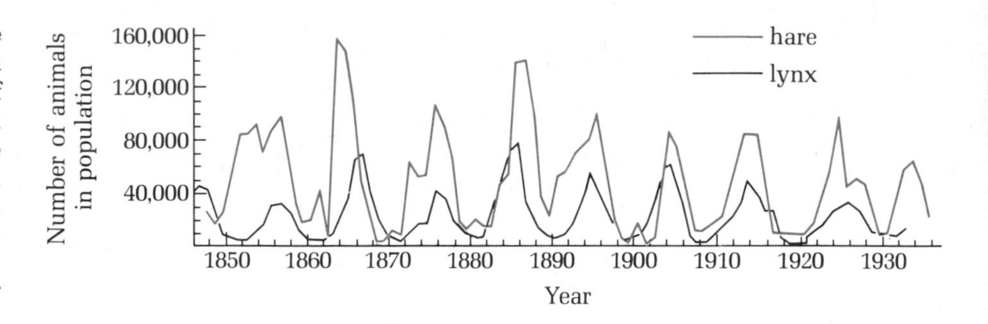

Figure 14–15 The cyclic relation between the populations of snowshoe hare and lynx according to the records of the Hudson Bay Company. This simple predator-prey relationship dictates a predictable pattern of increase and decrease in the populations of both species.

Partly because of increased predation, the hare population declined, cutting down the lynx population as well. In this case, too, many of the lynx may have migrated.

Cyclic changes such as this one are atypical, since most populations of both predator and prey tend to stay about the same size from one year to the next. If either population changes suddenly, the other is affected immediately. Consider, for example, what happened on the Kaibab Plateau in Arizona. In 1906, this area was made part of the Grand Canyon National Game Preserve. At the time, the plateau was home to a population of 4,000 mule deer, several hundred mountain lions, a few wolves, about 7,000 coyotes, and some horses and cattle. As a result of the plateau's new legal status, the domestic animals were moved to other ranges, all deer hunting was banned, and federal hunters began to systematically eliminate the predators. The deer were presented with an increased food supply, because they no longer had to compete with the horses and cattle. But even more importantly, the predators that fed on the deer were removed. Thus the deer population blossomed. By 1924, there were 100,000 deer, and food was getting very scarce. With the coming of two hard winters in a row, starvation struck with a vengeance. By 1930, only 30,000 deer remained, and most of them were in poor health. Another 10 years passed, and the herd was reduced to 15,000 animals, where it has remained until the present day (Figure 14–16).

As paradoxical as it might seem, removing predators from an ecosystem does not always increase the prey population. Norwegian game management officials once felt that they could increase the grouse population—and attract more hunters to buy licenses—if they cut back the number of hawks, the major predator on the grouse. Many hawks were shotgunned, with the curious result that the grouse grew scarcer. Some

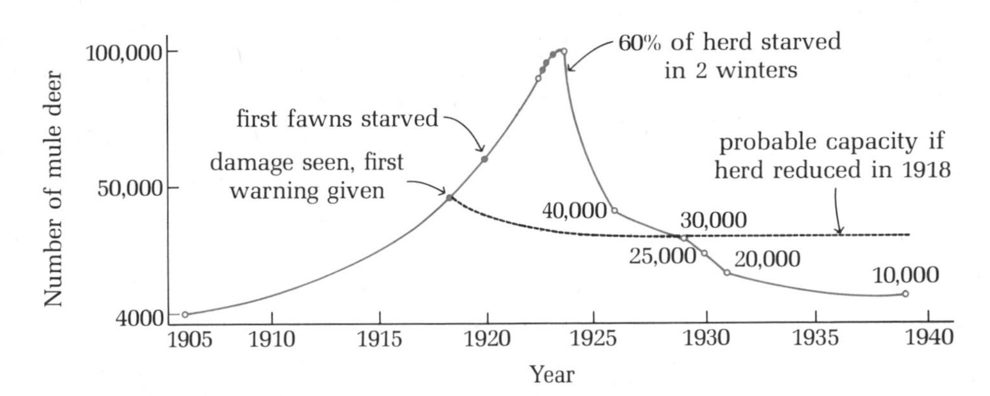

Figure 14–16 The importance of predators in controlling the size of the prey population was dramatically demonstrated when the mule deer of the Kaibab Plateau were protected from their major natural enemies by federal government programs. The expanded population put a severe strain on the plateau's food resources, leading to a sharp die-off rate within ten years.

Removed from Kaibab Plateau:

				Totals
Cougars: ←	600, 1907–17	74, 1918–23	142, 1924–39	816
Wolves: ←	11, 1907–23		(last wolf 1926)	30
Coyote: ←	3000, 1907–23		4388, 1923–39	7388

careful research showed that the grouse were dying from a disease that had been in the population all along. When hawks were plentiful, the disease spread slowly, because the diseased grouse couldn't run or fly well and the hawks caught them more easily than healthy birds. With the hawks gone, the sick grouse lived longer and spread the disease farther. To restore the grouse populations, the game officials had to import hawks, which began cropping the sick birds.

Rarely does a predator hunt a species into extinction. The reason is not that predators are enlightened game managers, but that they are opportunists. Barracuda living in a coral reef, for example, will tend to eat one species of prey fish. If the barracuda are very successful and kill so many of the prey that the population is in danger, they have so much trouble finding any of these fish that they shift their efforts to another species. The pressure of predation is removed, and the first prey species can rebuild its population. The one exception to this general rule is man. Armed with modern weapons that give him the advantage, man has elim-

Rarely does a predator hunt a species into extinction . . . the one exception to this rule is man.

inated or nearly eliminated one species after another, ranging from the American buffalo to the carrier pigeon, and put many more in danger. But this tendency to hunt a species into extinction is not a phenomenon only of modern man and his advanced weapons. There is evidence that Australian aborigines exterminated the giant kangaroo with weapons no more sophisticated than the spear and the boomerang. And the first human inhabitants of North America killed off several species, among them a kind of giant sloth and the American camel.

Regulating the population from the inside

Animals subject to heavy predation, such as the rabbit and the deer, seem to produce great and unchanging numbers of offspring. The population remains stable primarily because of predation; if the predators are removed—as happened to the Kaibab deer—or if there are none to begin with—as happened to the Australian rabbits—the population quickly outgrows the environment's ability to sustain it. Some animals, though, are able to keep their numbers constant through some kind of internal, physiological mechanism.

Predation on the Canada goose often varies considerably from one year to the next. The goose's main enemy is man, and all hunting pressure on the goose is concentrated in the fall, when the flocks make their way from the summer grounds in Canada to the winter habitat in the southern United States and northern Mexico. The number of geese killed each year along the Pacific flyway, one of the main routes from north to south, depends on the timing of the winter rains in Oregon and California. If the rains are late, there is no standing water except in large reservoirs and marshes, and the migrating geese congregate in great bunches that hunters can easily locate. If the rains are early, though, many farm fields are flooded, and the geese spread far and wide, making them hard to find and kill. The goose population, however, doesn't rise or fall cyclically like the snowshoe hare's. Instead, it remains remarkably constant. How can this be? In those years when hunters score heavily, there should be fewer breeding pairs, and therefore fewer offspring. When the hunters do poorly, there should be more pairs and more offspring.

So it would seem. But the geese adjust the number of eggs they lay in the spring to the number of nesting pairs. If the hunters in the fall killed relatively few geese and there are many breeding pairs, the size of the clutch is small, but it may double when the flock was heavily decimated and nesting pairs are few. Barring a catastrophe or an outbreak of serious disease, the geese retain a stable population. But how do the geese "know" how many eggs to lay? Do they remember the number of geese killed, subtract that number from the total, and then lay enough eggs to make up the difference? Obviously the geese use no such accounting method, but the precise mechanism remains a mystery.

The common house mouse also regulates its population through some internal method. Laboratory populations of mice confined to pens of a specific size grow until they reach a certain density. Then, two strange changes occur. First, many of the new offspring die, sometimes even all of them. Then breeding stops; the adults show no sexual interest, and none of the females come into season. This reaction is not due to a decrease in the food supply. Even if the food supply is increased far beyond the population's needs, the number of mice still stops growing at some definite point. An internal change has taken place, for if individual mice are taken from the pen and put into a new environment, they become fertile immediately. What's happening with these mice? Apparently, crowding causes some physiological change that shuts down the hormone cycle inducing ovulation.

A similar process is part of the cause for the cyclic changes in the number of snowshoe hares. Apparently the fact of overcrowding puts the animals under stress, and this stress alters the hormonal balance of the body so much that even a small additional stress, like a loud noise, can send the animal into fatal convulsions. In the hares' case, though, this mechanism hardly aids population stability, contributing instead to the rises and falls of population shown in Figure 14–15.

Populations and societies

Physiological controls are not the only method found in the animal world for ensuring that a population doesn't outgrow its food supply and that, should catastrophe threaten, at least some members of the species will survive. Many animal societies serve just such a function.

Dividing the landscape

Many different kinds of animals, ranging from crickets to antelope, display a kind of behavior known as *territoriality*. The characteristic of this behavior is that one animal, usually a male, sets up a territory with definite boundaries and then defends it against invasion by any other male of the species. The male will allow a female to enter, though, and the territory becomes the mating ground. The precise size of the territory and the activities that go on in it depend on the species. A pair of golden eagles, for example, maintain a territory of about 40 square miles, which is not only a breeding area but the eagles' hunting ground. In the case of the stickleback and many other fish, the territory comprises the nest and a small area around it. There the fish mate and attend to the young, but they hunt for food elsewhere, on neutral ground. Previously, in Chapter 9, we looked at the arenas of the Uganda kob. These territories serve solely as places for mating. Away from the arena, all the animals feed together, and the young are cared for by the whole herd (Figure 14–17).

What purpose does territoriality serve? For one thing, it helps keep a population's density constant, by limiting the breeding population to a certain number of territories. Among the kob, for example, only those males that take and hold a territory mate. The males who fail to win a territory don't get a chance to breed, and, curiously, show no interest in sex. Because the number of breeding males remains constant, the kob cannot overpopulate. But if the kob suffer severe losses, say because of increased hunting, there remains a surplus of sexually mature males who can take over the breeding tasks of the killed animals. Among those species where the territory also serves as a hunting ground, territoriality ensures that, if there is a food shortage, only the surplus animals, those without territories, will starve. In some cases, the size of the territory varies with the availability of food, growing in lean years and shrinking when times are good. For example, the arctic jaeger, a large predatory bird that looks much like a seagull, adjusts the size of its breeding territories to the supply of lemmings, its main food source.

Territoriality keeps a population's density constant, by limiting the breeding population to a certain number of territories.

Figure 14–17 The Uganda kob. This species has evolved a territoriality pattern that limits the population by assigning mating rights to those males that can hold and protect an area of land.

Rank ordering in the group

Among birds and mammals, many societies more complex than a simple family unit are organized by means of *dominance*, where one animal leads the group. African wild dog packs are usually led by a dominant male, but when he dies, his role may be assumed by his mate. Wolves are usually led by a dominant male. Except for the breeding season, elephants are organized into sexually segregated groups, each led by a dominant individual of that sex.

The dominance hierarchy among chickens, called the *pecking order*, has been studied more closely than any other. One hen, for example, hen A, dominates all the other hens. She can peck any hen in the flock, and the pecked hen will not peck back. Hen B can peck any hen other than hen A, while hen C can pester any one but hens A and B, and so on through the flock to hapless hen Z. The top hens usually have first choice of food and shelter, and they look healthier than the hens at the bottom, who get nothing but the leftovers. The cocks have their own pecking order, independent of the hens.

Like territoriality, dominance also provides for a floating reserve in case the population meets some kind of trouble, and it maintains a high-quality gene pool. Dominant cocks breed the most, while for the lower-order roosters, mating is infrequent. If the population is overcrowded, the lowest ranked males don't mate. The same thing holds true of rats. As the population grows and pushes against the carrying capacity of the environment, the males stop breeding, starting with the bottom order and working up until the population is in balance. And, if food is in short supply, the lowest ranked animals of both sexes are the first to die of starvation.

Keeping fighting to a minimum

Both territoriality and dominance also serve to limit fighting and violence between members of the same species. Territorial species usually fight in a symbolic way. The kob fence with their horns, pushing each other back and forth. Although the males could hurt each other severely, they don't. Hummingbird males engage in aerial aerobatics, buzzing each other like possessed jet planes, but only rarely do they actually make contact with their sharp claws. However, in societies organized by dominance, the fights establishing the order of dominance are often bloody. But once the order is set, though, life proceeds harmoniously, with little fighting. When two animals meet, the lower animal indicates its submission to the higher by means of gestures. Figure 14–18 shows a dominant wolf "ambushing" a submissive member of the pack, a ritual that reminds the lower animal of its status.

There are a few major exceptions to the general rule that animals don't usually kill their own kind. Colonies of certain species of ant will raid other colonies. Brown rats will attack and kill rats from an alien band. And man kills on both a small scale, which we call murder, and a grand scale, which we call war.

Figure 14–18 The dominant wolf in a pack is able to impose his will on those wolves of lower status in the group.

Summary

Any ecosystem comprises numerous species of both plants and animals. These organisms interact in definite and regular ways. Predators kill prey and eat them. Parasites live in or on a host, and, although they do not generally kill the host, they live at its expense. Mutualism describes a situation where two or more kinds of organisms live side by side and each benefits from the association. In commensalism, only one benefits, while the other is neither helped nor harmed. The usual relationship between plants and animals is that of food and consumer, but, in a few cases, plants and animals have evolved mutualist associations.

Animals of the same species are often associated in societies, groupings that involve cooperation and communication among individuals. Societies may comprise any of a number of units: the simple family of a mating pair and its offspring; the extended family of many related individuals; and the herd, a very large group of related and unrelated animals. Societies provide many benefits for the individual and the species. Animals can find mates easily, they are protected against predators, and complex tasks can be performed by many individuals instead of one alone. Among some species, particularly primates, social interaction, love, and affection are necessary for the normal development of the individual.

The most highly specialized societies are those of the insects. Termite colonies are organized into castes, each physically and behaviorally different from the others. The king and queen, the two reproducers, fertilize and lay eggs, respectively. The workers attend to the eggs, care for the nest, and gather food. The soldiers protect the colony. Each group produces social hormones whose presence or absence determines how the nymphs hatched from the queen's eggs develop.

A population is a group of organisms of the same species within a certain area, close enough together to breed. A population can be measured in terms of numbers of individuals or of its biomass. Density relates a population to the area it occupies.

Populations tend to remain stable even though they have the reproductive potential to grow greatly. Two kinds of factors control this growth. Density-independent controls affect all populations no matter what their size. An example is severe change in climate. The effect of density-dependent controls increases or decreases with the size of the population. Among these factors are the food supply, predation, and physiological control of population size by hormonal change.

Certain social arrangements also serve to keep populations constant. Territoriality limits the breeding population, and in some species spreads the individuals out over the available habitat. Dominance hierarchies provide much the same safeguards. Both social structures reduce violence within the society as well.

Questions to consider

1. How does the relationship of predator and prey differ from that of parasite and host?
2. If you observed two organisms of different species living in close association, what criteria would you use to determine whether the relationship is commensal or mutualistic?
3. Both wolves and oysters are found in groups. Why is the wolf said to be social and the oyster not?
4. Why do animals group together into societies? That is, how do societies add to a species' ability to survive?
5. Many orphanages that care for abandoned infants hire women who do nothing but cuddle the children for an hour or so each day. Administrators of these institutions maintain that such care is necessary for the normal development and mental health of the children. Is there any scientific evidence that lends credence to the administrators' position?
6. a. No termite has ever performed brain surgery, written a novel, or split the atom, but it is proper to say that termite society is more specialized than human society. Why?
   b. What sorts of tasks are performed by each termite caste?
   c. What number of members of each caste tends to remain stable. What mechanism controls the development of the nymphs?
7. a. As a rule, populations remain stable over time, but there have been cases of extreme population explosion. How can the disastrous growth of Australia's rabbit population be explained?

b. On the basis of this example and others like it, would you say that introducing new species into an ecosystem is generally a good idea?

8. Which of the following population controls are density-dependent and which are density-independent: a severe blizzard, the food supply, long-term drought, predation?

9. Like the wolf, the lion, and the barracuda, man is a predator, but he behaves differently toward his prey than the others. What is this difference, and why is it important?

10. a. What is meant by territoriality?
    b. How does territoriality differ from dominance?
    c. What social functions do these behaviors serve?

Glossary

biomass the total weight of the living organisms within a specific area

carrying capacity the maximum potential population that an area can support without damage to the population or the environment

caste one of the physiological and occupational groups within an insect society

commensalism a symbiotic relationship in which one organism is benefited and the other organism is not directly affected

density the concentration of living organisms within a given area

density-dependent factor a population-controlling factor which is directly related to the number of individuals

density-independent factor a population-controlling factor which is not related to the number of individuals

dominance the influence or leadership of an animal society by a single individual

extended family a large group of animals related by heredity which functions as a social unit

family the male-female reproductive unit of a society and the offspring of that unit *host* an organism that provides some or all of the life necessities of an organism of another species

mutualism a symbiotic relationship in which both parties benefit

pair bond the reproductive bond of a male and female which can be either permanent or temporary

parasite an organism that obtains some or all of its life necessities at the expense of an organism of another species

pecking order the dominance hierarchy among some species of birds

peer one of a group of animals all of the same species and approximately the same age

population a group of organisms of the same species related by habitat and breeding

predator an organism that meets its energy needs by killing and eating members of another species

prey an organism that is killed by an organism of another species for food

social hormone a substance secreted by an organism or group which regulates the growth and function of the society

society a group of organisms of the same species which are organized into a unified living unit characterized by cooperation and coordination

stable population a population whose composition and size remain fairly constant

symbiosis the phenomenon in which members of differing species establish a close living relationship over a period of time

territoriality the defense of a geographical area by an individual or population

tribe a human form of the extended family, which functions as a social unit

15 Living in Man's World

From our meager beginnings as an African hunter and scavenger, we humans have pushed farther and farther until we occupy almost every corner of the world, even the inhospitable ones. Although the discernible physical evolution of man ceased some 35,000 years ago, the centuries since then have been witness to the stupendous growth and evolution of human culture. Our places of dwelling have grown from little circles of huts in forest clearings to the hundreds of square miles of concrete and glass that is New York and Los Angeles. While other animals must depend on the natural ecosystem for food, humans have created their own ecosystems of domesticated animals and plants. Our societies have grown larger and larger, reaching the high point of China's 800 million people. And human science has cracked the atom, unraveled the double helix of DNA, and made it possible to travel to the moon.

But many success stories have their dark sides, and ours is no exception. Our cultural evolution, particularly the technology spun off from science, has created an array of serious and seemingly insoluble problems. These problems are often referred to together as the ecological crisis. Some of these dilemmas are so inescapable that many observers wonder whether what appear to be man's crowning glories actually amount to elaborate funeral preparations for the human species. Our success and the means we have used to achieve that success have brought us face to face with biological realities that we once either ignored or dismissed as irrelevant. Now they won't go away so easily.

In this chapter, we'll explore our present world and look at the problems we face. We'll be concerned with the size of the human population, with our sources of food and energy, and with our effect on the ecosystem as a whole.

How many of us are there?

In Chapter 3, we saw how the interactions of organisms in an ecosystem tend to keep the populations of all species constant and in balance. Man, though, has been an exception to this general rule, particularly in the last 100 years. Figure 15–1 graphs the rise in the human population from 10,000 years ago—about the time agriculture began—until the present. Back then, in the first days of human settlement and culture, the earth was home to probably 5 million people. Today our numbers have reached 3.8 billion. However, this growth from 8000 B.C. to 1973 A.D. has not been even. In 600 A.D., the time of the fall of the Roman Empire and the first

Figure 15–1 The growth of world population. Estimates place the total human head count at 6 billion by the year 2000.

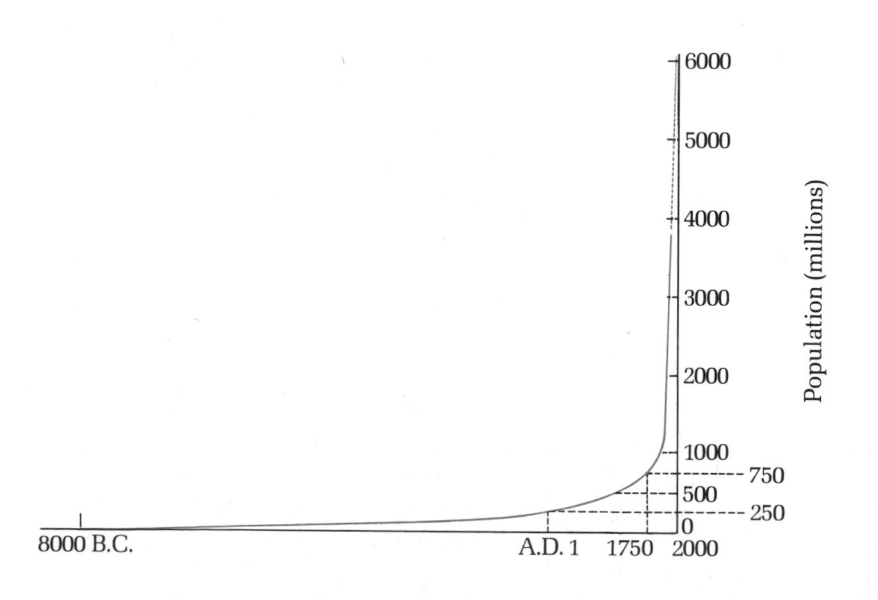

barbarian invasions of Europe, the population of the earth is estimated to have been about 250 million. It took about 1,000 years for this population to double to 500 million in the year 1650. But, by 1850, only 200 years later, the population had doubled again to 1 billion. Add 80 years, and the population had doubled again, to 2 billion. The present rate of doubling appears to be about 35 years, which means that the present world population will double shortly after the year 2000. Instead of achieving the stability and balance emphasized in Chapter 14, the human population has multiplied with the speed of Darwin's hypothetical birds. How has man outgrown the limitations affecting other species?

The roots of population growth

The prime cause of today's population upsurge actually occurred 10,000 years ago, with the simultaneous evolution of agriculture in the Middle East, Southeast Asia, and Central and South America. Rather than relying on nature's whim, humans began to clear the land and to plant grain and vegetables developed selectively from wild stock. Hunting can be a hit-or-miss proposition, but raising cattle, sheep, goats, and dogs provided humans with a steady source of meat. Agriculture quickly spread from one group of humans to another, until most of the human race became farmers. The change in man's way of life—in what was his ecological niche—was profound.

When humans depended on hunting animals and gathering wild seeds and fruits, the carrying capacity of the land was an important factor limiting the size of the human population. It probably took from 2 to 10 square miles to support a small, family-sized hunting band, depending on the productivity of the land. Also, the population had to be spread out. Too many people gathered into too small an area would quickly kill off all the game and starve themselves out. With agriculture, the amount of available food was increased. The population as a whole rose in size, so

Instead of achieving stability and balance the human population has multiplied with the speed of Darwin's hypothetical birds.

much so that between 8000 B.C. and the birth of Christ the human population increased 25 times. As agriculture grew more sophisticated, with the result that the same amount of land yielded more food, more people could gather in one place, producing the clustering of population that gave rise to cities. The same dynamics have continued up to the present: Agriculture has grown more productive, and the size of the population has continued to increase. But, as we shall soon see, the population has outstripped the food supply.

The population has also increased because people live longer. Until the last century, a newborn infant had a 50-50 chance of living until its first birthday. Today, in the countries with the best maternity systems—a group which, notably, does not include the United States—the odds run 50 to 1 in favor of the child. And once past that first year, the child is likely to live far longer than his forebears. A hunter-gatherer of 10,000 years ago usually died at the age of 35. This was not considered an early age of death well through the medieval period in Europe. At present, though, the average American man lives to 67, while the life expectancy of a woman is 74. These changes have added greatly to the population. Also, a couple once had to produce six to eight children to have just one son survive to adulthood and support them in their old age. Today all six or eight would likely survive and themselves reproduce. And since people have greater lifespans, the population stays larger longer.

Why have the death rates of the human species changed so markedly? Largely because we have been able to stop the ravages of many diseases caused by microorganisms. Public health measures such as the chlorination of drinking water and the inspection of food and milk keep such serious diseases as hepatitis and typhoid in check. Also, antibiotics like penicillin, which have been available only since the late 1940s, save many people from diseases they would have succumbed to formerly. In the tropical regions of the world, widespread use of DDT has increased the human lifespan markedly. The main killer in these areas was malaria, and DDT applied to swamps and stagnant water has all but eliminated the malaria-carrying *Anopheles* mosquito.

We noted in Chapter 14 that any and all species have the reproductive potential to overrun the earth. Normally this does not happen, though, because the effects of predation, territoriality, dominance, and disease keep the population more or less stable. However, in man's case, practical techniques based on scientific knowledge have allowed us to roll death back and to give more people a chance to life.

The uneven pattern of population growth

Although the population of the earth and of almost every nation on the earth is increasing, some populations are growing considerably faster than others. The increase is most marked in the so-called underdeveloped nations of the Third World—Africa, Asia, and Latin America. India and Nigeria have doubling times of 28 years; the Pakistani population reproduces itself in 23 years; and only 19 years from now El Salvador will have twice as many people as it does today. The high-income countries of North America, Europe, and Asia have been more fortunate, because of the widespread use of birth control. Great Britain's doubling time is 140 years, and Japan's is 63 (Figure 15–2). Thus in those nations where poverty is an inescapable fact of life for all but a minority of the citizenry, population is growing the fastest. While more people will be getting a chance at life, the life they'll be getting a chance at is likely to be more meager and destitute than even the impoverished world their parents have known.

Controlling population: death

Stability in a worldwide population means that births equal deaths. Human numbers have increased because births have outpaced deaths,

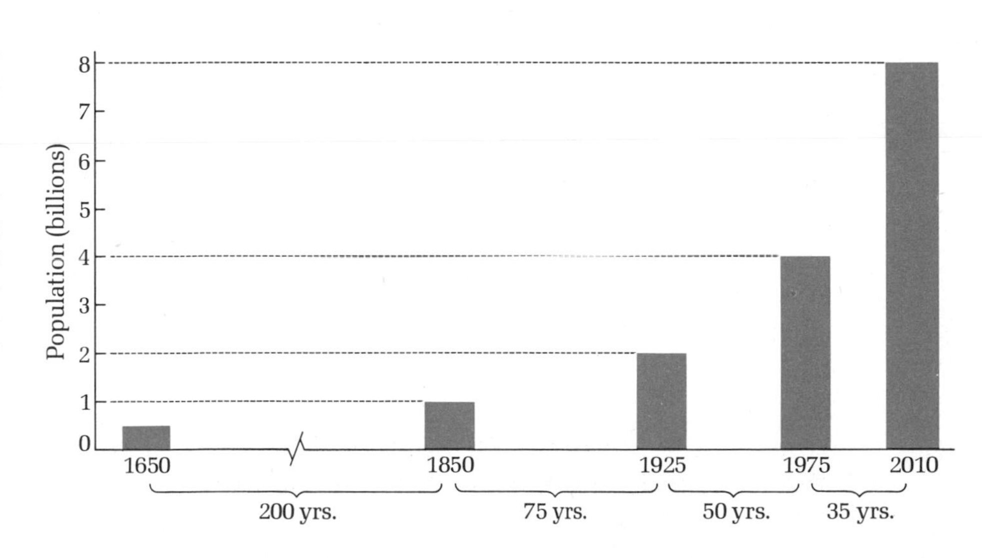

Figure 15–2 The accelerating doubling time of world population. As important as the absolute population total of a nation or region, the doubling time indicates the amount of time required for the population to double, accounting for all births, deaths, and migration. As this graph shows, the world population's doubling time has been consistently shrinking.

particularly over the last 30 years. Obviously the way to stop a population from growing is either to increase the number of deaths or to decrease the number of births. How could it happen that more people might start dying?

One possibility, one that has been a fact of life in many parts of the world for centuries, is famine. Thomas Malthus, the economist whose thinking influenced Darwin, noted that population increases geometrically: 2 → 4 → 8 → 16 → 32 → 64, while the food supply grows arithmetically: 2 → 3 → 4 → 5 → 6 → 7. Thus, at some time, a growing population is bound to outstrip its food supply. The result is famine—mass starvation until the food supply and the population are again in harmony. Famine is well known in human history, occurring regularly for centuries in much of South and East Asia. Oftentimes famine has gone hand in hand with war, the latest example being the Pakistani-Indian conflict that gave birth to the nation of Bangladesh. Because the crops were disrupted, millions starved. As this is being written, a succession of dry years threatens 20 million West Africans with starvation.

It could be argued, with a certain cynical coldness, that famine is nature's way of putting man in his place. But famine does more than kill people. It also disrupts a society completely, as all morality and social convention are subordinated to the need for food. Those who survive, particularly children, are likely to suffer from a wide variety of physical, mental, and emotional problems. And for a nation like the United States with relatively abundant food to make no attempt to stop a famine, maintaining that nature is just taking its course, would be a criminal act of the worst kind. So far, we've tried to help.

And what of war? Aren't starving people who know of the abundance of richer nations likely to try to take what they need? As popular as this scenario is, it's highly overrated. Hungry people don't have the energy to fight or even to organize themselves into any kind of an army. However, a hungry nation can easily be exploited or invaded by a richer and more powerful country, perhaps to rob the poorer country of its natural resources. Famines in nineteenth-century China were commonly exploited

Famine does more than kill people—it disrupts a society completely, as all morality and social convention are subordinated to the need for food.

by the imperialist European powers seeking to extend their control. In our world, where more and more nations are arming themselves with nuclear weapons, any such confrontation is fraught with danger. Nuclear warfare could very possibly stop the population problem by annihilating the human race.

Controlling population: birth

Rather than relying on the horrors of famine and war to kill people once they have been born, it seems more reasonable to reduce births until they equal deaths. But there are problems with mass birth control. The pill, for example, is highly effective, but using it requires care, and even small mistakes make it worthless. IUDs require no care, but enough women reject them that they are inappropriate for controlling births on a massive scale. Tubal ligation makes a woman permanently sterile, but the operation is intricate and costly, far too expensive to be used on large numbers of women at public expense. The best solution at present is vasectomy. But many men object to the operation, partly from misinformation that it will make them impotent, partly because of the legitimate concern that their sterility cannot usually be reversed should they later change their minds. The Indian government, though, has stimulated interest in vasectomies by paying cash bonuses to men who agree to the operation (Figure 15–3).

Campaigns aimed at lowering the birthrate often run into severe cultural barriers. In many countries, children are regarded as assets. More hands make less work, and grown children provide security for the parents when they have grown old. Religious organizations, in the United States as well as elsewhere, have often opposed birth control plans for reasons of dogma. In this country, the Catholic Church lobbied against liberalized birth control and abortion laws, and its influence in Latin America and Italy has made distributing birth control information a crime. Middle-class Europeans and Americans have generally resented any suggestion that they limit family size for the good of all. Shielded by affluence, they see overpopulation and the need for birth control as somebody else's burden—the urban poor and the teeming masses of black Africans, brown Indians, and yellow Chinese. This view is both racist and shortsighted. In his lifetime, one middle-class European, Canadian, Aus-

Figure 15–3 The Indian government, in cooperation with major international health organizations, has instituted an extensive and successful birth control program. The primary purpose, as indicated by this sign, is education—to motivate the Indian citizens to limit family size voluntarily.

tralian, or American uses up more food, creates more pollution, and leaves more garbage than 10 Indians, Africans, or Chinese.

The need to plan ahead

Even if birth control programs were immediately instituted all over the world, and if each childbearing couple bore only two children, thus replacing only themselves, it would still take a long while for the population to stabilize. In 1972, the American birthrate equaled the death rate, but the population still continued to grow. The population contains a large number of juveniles below the age of reproduction. Only when these people have matured and reproduced will the population level off, probably around the year 2000. In the case of a country like India, where 45 percent of the population is immature, leveling cannot occur before the middle of the next century.

Feeding the world

Inseparable from the size of the population is the amount of food needed to feed it. As we saw from Malthus's idea, it seems to be an iron-clad law that a growing population will eventually exceed its food. This is no idle prophecy. For at least half the world's population, to live is to be hungry.

Missing nutrition

What's bad about hunger isn't that it's uncomfortable, but that it affects health severely, and may lead to death. There are two varieties of malnutrition. One is a lack of sufficient Calories; as a result, the body draws on its fat reserves until they are used up. The second is a lack of needed food elements like proteins and vitamins. It's possible to get enough Calories and miss out on many necessary nutrients, a common problem for people whose diets include large amounts of starches and sugars but lack meat or dairy products.

What are the effects of malnutrition? In an adult, it produces a generally lowered state of health. Malnutrition itself doesn't often cause death directly, but the body is so weak that it succumbs easily to infection or

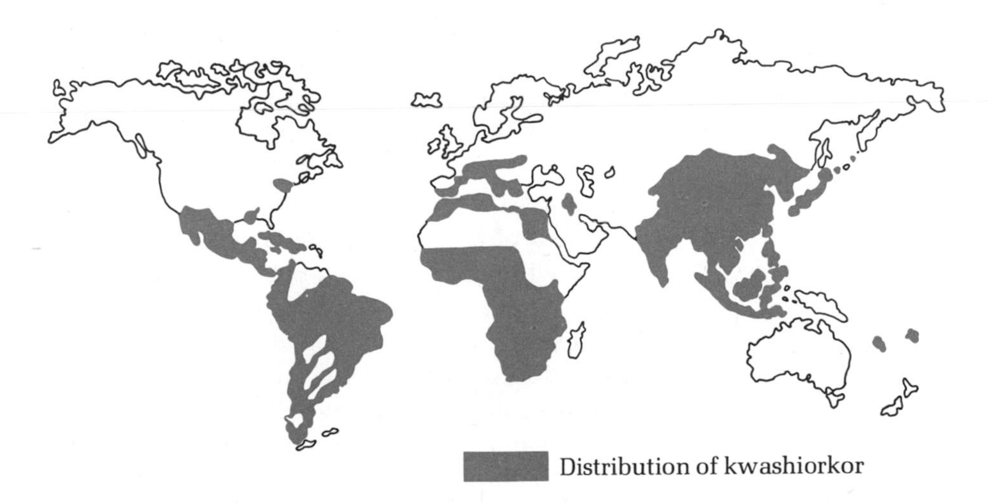

Figure 15–4 The spread of kwashiorkor. This serious malnutrition disease results from protein deprivation. Despite its particular prevalence in underdeveloped countries, kwashiorkor has been reported within the United States, including on some Indian reservations.

injury. But it is among children that malnutrition, particularly a lack of protein, has its most damaging effects. Protein is a principal component of brain tissues, and the brain reaches 80 percent of its adult size in the first 3 years of life. If a growing child gets too little protein in his diet, the protein the brain needs for proper development isn't available. The result is a form of mental retardation, the severity of which depends on the extent of protein deprivation. A diet containing no protein often causes a severe nutritional disease known as kwashiorkor. Although official United States government statistics maintain that kwashiorkor does not exist in this country, it does in fact appear regularly among children on Navaho Indian reservations (Figure 15–4).

Hunger isn't only a reminder that famine may occur at some point in the future. It means that people in the present are forced to live under conditions of poor physical and mental health.

The energetics of shortage

If man's food problem consisted only of getting enough Calories, there would be no hunger. The United States alone could produce enough sugar from cane and beets to supply Calories for ourselves and the rest of the world. The real problem is producing enough protein. One of the best sources of protein is meat, but meat is very expensive, not just in terms of dollars and cents paid to the butcher, but in terms of the energy needed to produce it.

The reason is the energy "lost" from one trophic level to another. As you recall from Chapter 5, a plant fixes only about 1 percent of the energy of the sunlight that falls on it into carbon compounds. The cow that eats the corn gets only 10 percent of the grain's energy, and a man who eats the cow gets only 10 percent of the Calories theoretically available from its carcass. Thus 100 pounds of corn produce 10 pounds of beef to yield 1 pound of human. But that same 100 pounds of corn fed directly to humans will support 10 times as many people. Only the well-to-do nations of America, Europe, and Australia have been able to afford expensive meat protein. Most of the American grain crop, for example, is fed to cattle. In more impoverished nations, people eat their grains directly and live at a lower trophic level. The only meat animals are pigs and chickens, which can scavenge foods humans can't eat, such as the fibrous stalks of rice and corn.

These energetic facts coupled with a rising population put mankind in a bind. To avoid mass starvation, the food supply must increase in both Calories and available protein. Is there any way these two requirements can be met?

More food from old sources

After each world war, both the supporters and the opponents of communism talked of a great red revolution that would sweep the world clean of imperialism, poverty, and starvation. But in the 1950s, some people began to talk of a "Green Revolution" that would be brought about by the peaceful efforts of biological and agricultural scientists. These men and women would produce new high-yield, high-protein grain crops that would provide food for the world's growing billions and stem the rising tide of famine. Prophecies of miracles just around the corner became commonplace. Now, some 20 years later, it appears that the Green Revolution has produced a handful of solid achievements and no miracles.

The International Rice Institute has developed new varieties of rice that may allow tropical Asian countries like the Philippines and Vietnam to triple their rice harvests. Several new strains of wheat developed in Mexico in a research project funded by various American and international foundations and led by Norman Borlaug have increased Mexico's wheat production six times over. These wheats produce about twice as much grain as conventional varieties, and it is possible to plant two or three crops a year instead of just one. Because of Borlaug's work, which won him the Nobel Peace Prize in 1970, Mexico, India, and Pakistan have actually been able to increase their food supplies faster than their populations (Figure 15–5).

Unfortunately, these developments have been nothing more than isolated skirmishes in a war that hunger seems to be winning. For the time being, India, Pakistan, and Mexico will produce enough food, but they may be only delaying starvation, not stopping it. Unless their rates of population growth also slow, the number of people will again outstrip the food supply in the future. As it stands, the world contains more hungry people now than at any time in its history. While we can expect the Green Revolution to help increase the food supply somewhat, there is no prospect of a miracle grain that can feed a family of four from a plot the size of the average backyard.

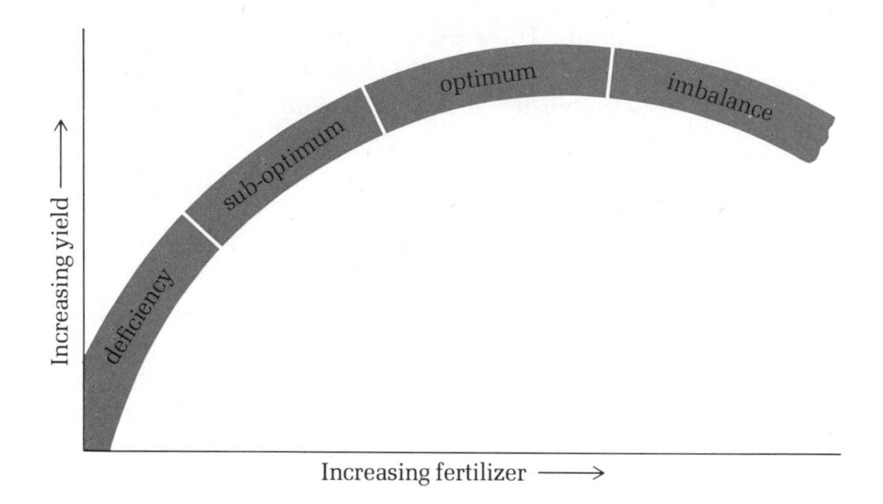

Figure 15–5 The limits of fertilizer use. As more and more people draw on the world's agricultural resources, the need for increased production becomes crucial. Fertilizers have made a substantial contribution, but a point of diminishing returns can be reached, at which agricultural production cannot be increased chemically.

Adding to the land

The key figure in judging the productivity of any crop is its yield per acre. The new crop strains were developed to increase that yield and thus increase the amount of food produced. But couldn't one also increase the food supply by increasing the number of acres under cultivation? We Americans are accustomed to think of this as an easy and simple solution because such a strategy was so much a part of our history. We cut the forests of the East to make farmland, and our greatest food producer, the Midwest, is a plowed and cultivated grassland. Why can't the same strategy be followed elsewhere?

The curious fact is that we're losing the best farmland at a slow but steady rate. Before the days of rapid transportation, cities generally grew up near their food supplies, so that the farms were just outside the city limits. As the cities grew, good quality farmland was paved over with concrete and asphalt and lost to production. When land was plentiful and the population small, such waste could perhaps be tolerated, but that situation no longer holds true. It is estimated that each year the United States loses about 900,000 acres of rural land to tract homes, freeways, parking lots, shopping centers, and so forth (Figure 15–6).

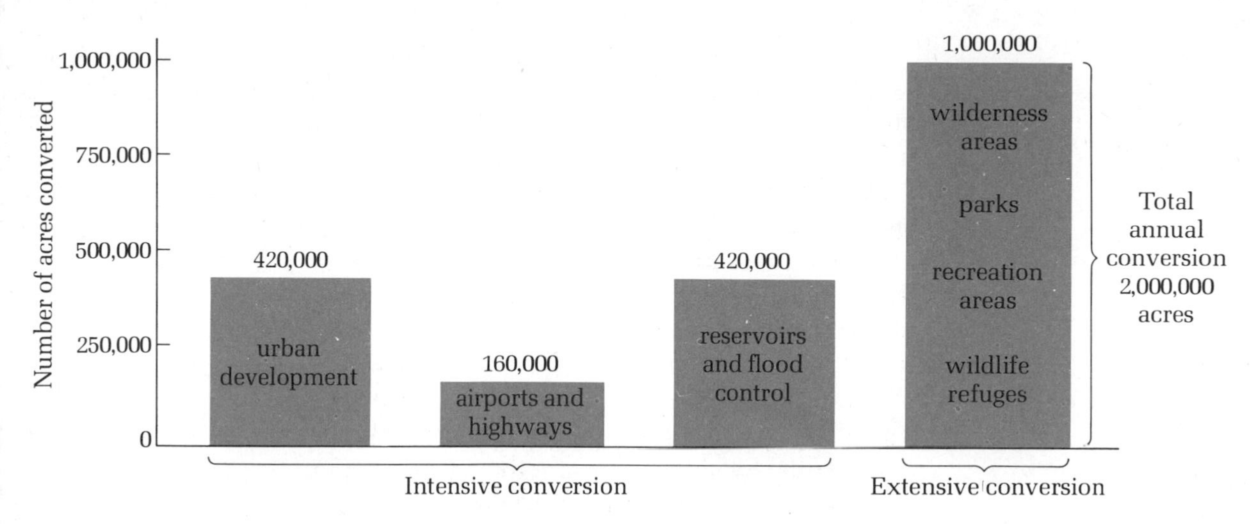

Figure 15–6 Annual loss of agricultural land in the United States. Each year, between 900,000 and 1,000,000 acres of potential agricultural land are converted to meet the needs of our expanding urban and suburban population. An additional 1,000,000 acres are transferred to state and federal jurisdiction for conservation and recreational use.

Can't other areas be planted? Why not clear the tropical forests and supply the deserts with water? Tropical forests can be cleared and planted, but the soil seldom supports planting for more than a season or two. Tropical soils are composed of a hard clay kept soft by the heat, humidity, and absence of sunlight on the forest floor. Many of the nutrients needed for plant life are stored in fungi that cover the soil thickly. When the tree cover is cut, the sun kills the fungi, and the nutrients are lost. Exposed to the sun and rain, the clay soil either erodes away or bakes into a hard crust that cannot be planted. A tropical forest can support a small number of migrant farmers who clear the forest, plant their crops, and then move on when the fields become infertile. But if these farmers become too numerous, as is happening in Central and South America and in the Congo basin of Africa, the forest disappears rapidly, leaving in its place only an unplantable wasteland.

It is estimated that each year the United States loses about 900,000 acres of rural land to tract homes, freeways, parking lots, and shopping centers.

The situation is a bit better in the deserts, where the key to agriculture is water. Much of the agriculture of Israel and California would be impossible without extensive irrigation. Ironically, however, the irrigation that makes agriculture possible may finally put an end to it. The water used for irrigation is usually of lower quality than that used for drinking, and such water contains a fairly high concentration of dissolved salts. In small amounts, these salts are nutrients, but at higher levels they are toxic to the plants. The amount of salt in irrigation water is less than toxic, but natural conditions in desert areas cause it to become concentrated. Since it rarely rains in the desert, the salt left by evaporating water doesn't run off. As a result, the salt accumulates. In addition, a layer of bedrock impermeable to either water or the salts dissolved in it underlies many desert soils. Even if it did rain, the salt couldn't be leached from the soil. Salt accumulation takes time, but it finally takes its toll. Salt contributed to the ruin of 5 million acres in Pakistan, and California's Imperial Valley, one of the most productive vegetable-growing regions of the United States, will probably become unusable by the year 2000 (Figure 15–7). Solutions to the problem of salt accumulation are expensive. Higher-grade water with less salt could be used, or the lower-grade water could be run

Figure 15–7 Like the extensive use of fertilizers, the widespread use of irrigation has severely affected soil quality in several major agricultural areas of the United States. This photograph, taken in the San Joaquin Valley of California, shows the snowlike accumulation of dried salts on furrows of farmland. This almost irreversible damage could take substantial amounts of land out of production by the year 2000.

through a desalination plant before it is piped into the fields. Either plan would make desert agriculture considerably more expensive than it is now.

In sum, then, some acreage may be added to the total of arable land, but not enough to make any lasting difference in the final outcome of man's war with hunger. We're pretty much limited to the land we already have, and too much of that is going into nonproductive urban use.

The prospect of new sources

In the past few years, whenever future famine is prophesied, someone points out to the sea and begins to extoll the oceans as an untapped mine of protein. Most likely, the sea's productivity has been highly overrated, and the food assumed to be there is either nonexistent or unavailable.

The most abundant life zone in the sea lies along the coasts, and this area has been adversely affected by pollutants. The great wide expanse of water in the middle of the oceans is far less populated than the coastal areas. Also, fishing practices have greatly depleted the numbers of many species. Using sophisticated sonar systems that allow them to "see" fish, fishermen have been able to net great numbers of tuna and other fish that run in large schools. The present whale population, only a remnant of what it once was, is being so rapidly depleted by the mechanized whaling fleets of Russia and Japan that many species of whale are facing extinction. Phytoplankton, tiny plants that grow in the surface waters of the open ocean, is not the ultimate human food source it was once thought to be. Plankton grows very thinly, making harvesting uneconomical. Even if the plankton could be harvested, removing it would be unwise because it is the first trophic level in the ocean. Taking away the plankton would accomplish the same thing as cutting our own throats, only more slowly.

The sea could be put to more productive use if we concentrated on harvesting energy at the lower end of the food chain, thus increasing the amount available. Tuna and cod have been popular food fish. If we shifted our efforts to sardines and herring, which are on a lower trophic level, 10 times as much fish protein would be available (Figure 15–8).

There is some possibility that synthetics—simple molecules assembled into complex nutrients by man—could represent an important source of food in the future. But the amount of help available from this source

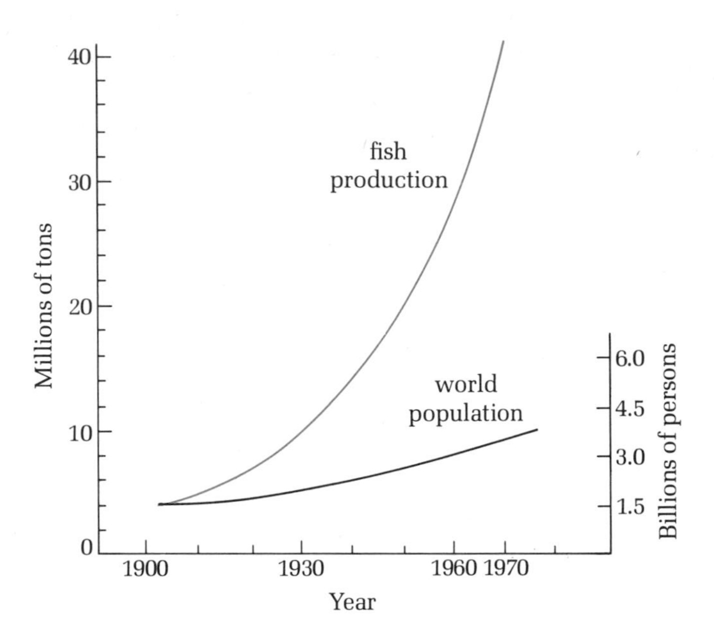

Figure 15–8 Alternative sources of food —particularly protein —might be the key to meeting the world's future nutritional needs. The world's fish catch increased substantially faster than human population through the first 70 years of this century, but overfishing, pollution of coastal waters, and extinction of several food-fish species might limit our opportunity to draw on this valuable resource in the future.

remains very much a matter of speculation. At this time, manufacturing proteins and vitamins necessary to human nutrition is very expensive. There is the chance that the cost will come down, but it would be foolhardy to count on synthetics as anything other than a supplement to the world's needs.

The future

The amount of available food will increase somewhat over the next 10 or 20 years. To a large extent, this will be due to new grains like Borlaug's wheat and to the gradual replacement of traditional peasant practices with more productive methods of farming. But even with the best of luck, the present doubling rate of the human population will in time push the number of people beyond the supply of food. The final solution to the hunger problem is limiting population and limiting it soon.

The prospect of a worldwide brownout

In much the same way that the body needs food to keep running, the technology we Americans have built and the society we live in are founded on the continuing consumption of technological energy. If the supply of energy were to run out, this country would grind to a halt. Because of the natural endowments of the United States, particularly in fossil fuels and water power, we Americans have been able to use great amounts of energy to gas our cars, cool and heat our homes, power a wide variety of household appliances, and to run the thousands of miles of assembly lines central to American manufacturing. We Americans con-

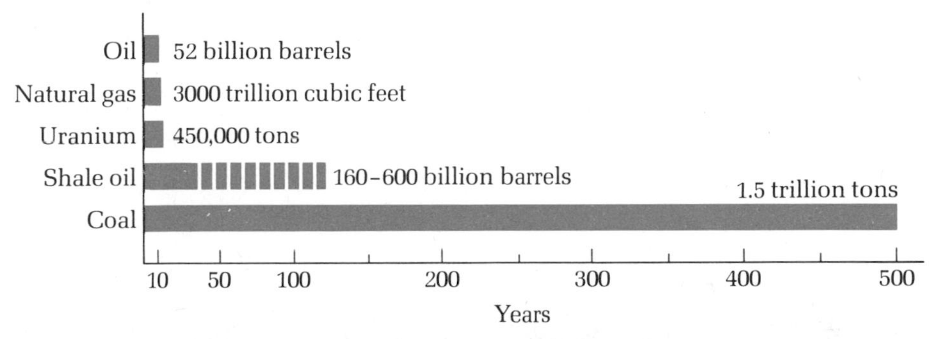

Figure 15–9 United States fuel reserves. At current rates of consumption, our oil, natural gas, uranium, and shale oil reserves will last for less than 50 years. Coal remains our most abundant fuel resource, but there are serious environmental problems attendant to its extraction and burning.

A public meeting to protest the construction of the Trojan nuclear power plant near Portland, Oregon. Despite the expanding need for alternative sources of power, public concern about safety hazards and environmental damage has blocked or slowed the development of nuclear power use.

stitute only 6 percent of the world's population, but we use over 35 percent of the technological energy produced each year in the whole world. But the honeymoon may be over (Figure 15–9).

Dimensions of the dilemma

There are two sources to the problem. The first is that we seem to be coming to the end of the energy resources that have been our traditional mainstay. For the last century there has always been some new supply of ready energy in an untapped coal seam or an undammed river. However, this is no longer the case. Known oil and natural gas reserves may be depleted within 15 years. As things stand now, only 4 percent of the nation's energy comes from water power, and there is little chance to increase this figure. Only so many dams and reservoirs can be built, and that number has been very nearly reached already. While coal is still relatively abundant, most of what remains is low-grade ore. Burning it releases great amounts of sulfur oxides. This coal often fetches such a low price that only strip mining is economically feasible. Thus getting this coal means ruining the land, and burning it means fouling the air.

The other half of the energy problem has to do with population. Not only has the American population grown steadily, but each member of that population uses 5 percent more energy each year. Not only has the total number of people—and thus the overall amount of energy consumed gone up, but the quantity used by each of us has also risen—swelling the total all the more. We keep on consuming more and more energy even when it appears that the fuel gauge is resting on empty.

Because our technology is so dependent on energy, a sudden drop in available fuels could cause severe hardship. What would happen, for example, if the supplies of gasoline and diesel fuel were suddenly exhausted and prices rose astronomically? So far as most private homes are concerned, trips to the grocery store would probably be made on foot instead of by car, and commuters would have to switch over to buses and trains. Such changes would cause inconvenience and trouble, but most likely people could adapt. But imagine what might happen if gasoline and diesel fuel supplies dried up at the time of the grain harvest. Suddenly, not only gasoline would be in short supply, but the prices of grain products like flour and corn meal would skyrocket. Since most livestock are fed grain, beef, pork, and chicken would soon become luxuries known only to the very rich.

Seeking ways out of the bind

Exactly how does one go about forestalling an energy crisis and all the suffering it could cause? Obviously, total energy consumption has to go down. The total amount of energy consumed is the product of two factors:

$$\text{total energy consumption} = \text{population} \times \text{per capita consumption}$$

To bring the total down, you can decrease the population, reduce per capita consumption, or both (Figure 15–10).

We have already looked at the prospects and problems of limiting population. Since the population of this country is going to keep growing at least for the next 30 years—and even longer elsewhere—then something will have to be done about per capita consumption as well. To a certain extent, per capita consumption will come down of its own accord. As the supplies of fuel and power become smaller and smaller, their prices will rise, and people will simply find it too expensive to leave the lights burning all night or to use the car for anything other than serious transportation. Still, this sort of change is not likely to make anything more than a marginal change in the overall consumption of energy. At some point, we might have to distinguish between the "luxury" and "necessity" uses of energy, and dispense with the luxuries. But what sort of criteria does one use to make such a distinction? People differ markedly in what they call an expendable luxury. Some people loathe TV and would banish it from the face of the earth in the blink of an eye, while others would find the loss of the tube nothing less than intolerable. Who is to decide which group is right—the local public utility, a senate committee, or a benevolent dictator? In addition, reducing per capita consumption would have severe economic effects. The lack of power would reduce industrial production.

Seeking new sources of energy

In the past, whenever a principal source of energy seemed about to disappear, some scientific breakthrough produced a ready replacement. Whale oil lit all the lamps of America and Europe in the nineteenth century. Just when it seemed that the whale and its oil were becoming too scarce and too expensive to harvest, the Pennsylvania oil fields were discovered. There was a time when aluminium was considered exotic and expensive. Extracting the metal from bauxite ore took a great amount of electricity, and electricity was costly. Today an improved refinement process for aluminum and cheap power have made aluminum so inexpensive that we even use it for beer and soft drink cans. Given technology's track record, isn't it reasonable to assume that some new source of power will come along to bail us out of our troubles (Figure 15–11)?

Figure 15–10 The energy crisis—before and during. After encouraging increased energy use through pricing advantages and advertising for many years, the public power companies are responding to the energy crunch—warning of wasteful practices and fuel-consuming appliances.

In truth, the alternatives don't look too promising. Atomic power is no panacea. There is always the risk of an accident exposing possible millions of people to radiation poisoning, and the thermal pollution and radioactive wastes of such plants are real ecological threats. Windmills have been used extensively in Holland and the Australian outback, but the wind can be a power source only in those few places where it blows steadily and reliably for most of the year. An electric plant on the Rance estuary in France harnesses the power of the incoming and outgoing tides and another at The Geysers, California, uses steam from the earth's in-

Given technology's track record, isn't it reasonable to assume that some new source of power will come along to bail us out of our troubles?

terior to turn its turbines. Unfortunately, these plants are little more than novelties that can be duplicated only where the earth's crust has the right characteristics and where the coast is the right size and shape.

The most workable option might prove to be solar power. Solar power rests on the *photoelectric effect:* Light falling on certain metals produces an electric current. The metals can be packaged into photoelectric cells, which can be arranged in large light-catching plates. Solar power has all sorts of advantages. Among other things, it gives off no pollutants, and the sun's power is virtually limitless. As conventional sources of power seem to be drying up, governments and industries are taking a more serious look at solar power as an alternative. For example, in 1971, the American government provided no funds for research into solar energy, but by 1973 the amount had risen to $13 million. In the same year, UNESCO, the educational and scientific agency of the United Nations, sponsored an international meeting of scientists interested in the possibilities of solar energy.

With more research money available, those possibilities are being realized more and more. A French pioneer in solar power, Félix Trombe, has constructed a solar furnace capable of reaching 3,500°F and supplanting the coal-fueled blast furnace. The University of Delaware has built a house that gets all of its heat and electricity from solar panels mounted on its roof. Eventually, similar units will supply the power needs of larger buildings such as office complexes. Nor is this the upper limit to solar power. To give some sense of scale, Lake Erie absorbs more energy from the sun in a day than the United States consumes from all sources in a year. However, as we shall see in Chapter 16, there is substantial disagreement about the true potential of solar energy.

Such large-scale schemes, though, are at least 10 or 20 years off. In addition, we must realize that a new source of energy is not in itself a solution. Economists tell us that it takes at least 50 years for a high-energy economy like this one to switch over completely from one fuel to another. The major reason has to do with the lifetime of existing power equipment. Tearing out one kind of equipment to install another is a ter-

Economists tell us that it takes at least 50 years for a high-energy economy like this one to switch over completely from one fuel to another.

ribly expensive proposition, and no corporation is likely to bear such a burden before that original equipment is worn out and worthless. Thus, even if a new power source does become available, a serious energy crisis can develop even as the new power source is waiting in the wings.

Man and the global ecosystem

More than any other creature, humans have left their marks on the earth. We have cut forests, moved mountains, diverted rivers, built cities, planted grasslands, and deposited trash from one end of the globe to the other. How have we affected the living environment about us? Has that effect been to the good or the bad?

Figure 15–11 The Humboldt nuclear power plant in California. Nuclear and solar power are two of the most attractive possible answers to the energy crisis, yet technological and economic factors are still restricting the application of these methods on a large scale.

Succession: one community after another

To get a standard of comparison, let's see first what happens when man's influence is absent. Many of the ecosystems we see about us have resulted from thousands or even millions of years of community evolution, an evolution so complex that the most painstaking research can provide little more than educated guesses about how the community came to be what it is. Periodically, though, nature provides us with an opportunity to see life begin in a new environment—islands rise from the sea, glaciers retreat, landslides clear the side of a mountain. Plants cover the new habitat, and with them come animals. The influences of both change the habitat so that new species of plants and animals appear, oftentimes displacing the original colonists.

We will detail the stages by which a large pond left on a sandy beach by retreating Lake Michigan eventually becomes a forest. At first the pond contains no life at all, but gradually it is colonized by small plants and animals. Debris left by these forms contributes to soil on the bottom of the pond, which becomes populated with bottom-rooted plants. The growing amount of plant debris and collecting silt fills the pond from the edges in, making them boggy. Marsh grasses and rushes convert the bog into a meadow, leaving a temporary pond that fills only after the spring thaw and heavy summer rains. Trees and brush gradually take over the meadow, resulting finally in a forest comprising beech and maple trees. Each stage has its own characteristic animal life, ranging from black bass and bluegills in the bare bottom pond to the great variety of woodland organisms we counted earlier in Chapter 14—from 70 mammals to many millions of mites.

Once the maple and beech forest is established, change stops. Although individual plants and animals die, the community remains remarkably the same as long as there is no major catastrophe like a fire, a flood, or an advancing glacier. Such a final community is called a *climax community*, and the process of change from the first stage to the climax community is an *ecological succession* (Figure 15–12). The details of succession vary considerably from one environment to another. For instance, succession on the slopes of a Hawaiian volcano after a lava flow would feature different plants and animals than the scene on the pond depicted above.

Figure 15–12 An Eastern deciduous forest, an example of a climax community.

However, all successions share certain characteristics. First, at each stage in the succession, the biomass increases, reaching a high point in the climax community. Second, each stage also contains more species than the previous one, and, as is to be expected, the climax community has the most of all. Except for very harsh environments such as the Arctic tundra, climax communities are quite complex, with equally complex and intricate food webs. This complexity lends the climax community its extraordinary staying power. It will last until some catastrophe pushes the community back to an earlier stage of succession.

Man's simplification

Often that catastrophe is the hand of man. While ecological succession, the evolution of communities, favors complexity, man simplifies the communities he comes into contact with.

Part of this simplification is intentional. In essence, agriculture is a simplification, changing a complex food web into one that benefits man directly. Consider the Great Plains, for example. The soil was the key to this ecosystem, for the soil allowed the growth of grasses that trapped sunlight and thus formed the first trophic level. A wide variety of rodents, particularly mice, prairie dogs, and rabbits, lived on the grass or on the insects it harbored. Predators that lived on the rodents came in an equivalently great variety: snakes, owls, hawks, ferrets, badgers, coyotes. The grass also fed the great herbivores, the buffalo and the pronghorn. These in turn provided food for wolves, mountain lions, and Indians. No other grassland, not even the fabled savanna of East Africa, has equaled the biomass of the American plains. White Americans moving in from the East changed this complex, mature community into a much simpler one. The soil was broken with plows to kill the grass and allow grains and vegetables to be planted. Men then ate these foods or fed them to livestock that were subsequently slaughtered for their meat. Denied a food supply, many of the original animal populations were reduced in size. Others were purposely exterminated by poisoning or shooting. And the Indian—from an ecologist's point of view, the top consumer of the system—was pushed into marginal lands unsuitable for agriculture. As a result of all these actions, what had been a complex system became a far simpler community with fewer species and a smaller biomass.

Modern agricultural methods have carried simplification even further. Long-lasting pesticides kill not only unwanted insects, but all insects. Older methods of farming often left relatively large sections of land untouched. Hedgerows, for example, were allowed to grow up between fields, providing considerable cover and food for many animals like quail, pheasant, and foxes. Contemporary "clean" farming methods eliminate all such habitats (Figure 15–13). Natural cycling of wastes is rarely practiced—the men and cattle fed from a cornfield are usually many miles distant—and fertilizers must be used to keep the soil productive. Man continues to eliminate predators, apparently on the grounds that they compete with him or pose a threat to his livestock. By the end of this century, it is possible that no tropical rain forests—the richest biome in terms of both biomass and numbers of species—will remain standing. And although the continental United States contains few rain forests, we Americans have contributed mightily to elimination of forests and grasslands in Vietnam through the widespread military use of herbicides and bulldozers.

Figure 15–13 Simplication of ecosystems through modern agriculture involves the cultivation of single species of animals or plants over an extensive area.

Although man regularly simplifies ecosystems on purpose, he also does so quite by accident, without realizing how many species his actions may affect. We have seen, for example, how pesticides have practically

eliminated several kinds of birds, none of them ever the target of these poisons. We also saw that the Florida Everglades, a water-based ecosystem, are being hurt by drought caused by marsh-draining projects on the edge of the wilderness. Dams erected to stop floods and generate power have so changed the structure of some rivers that native fish species have been killed off.

In our discussion of evolution, we talked about natural selection as the force shaping species. Humans have become something of an artificial selector, a self-styled environmental force. Our cultivated plants and domesticated animals were produced by purposeful breeding. We have declared war on some species directly, while we have pushed others back by denying them suitable living space. Curiously, though, a few animal species have learned how to adapt to man, to find niches within the habitat man dominates. The most notable successes have been the starling, the cockroach, and the brown rat.

Simplicity and vulnerability

Why the ruckus about simplifying ecosystems? Species have passed away before—the dinosaurs, the saber-toothed tiger, the American camel, the wooly mammoth—and the earth doesn't seem any the worse for the loss. Why worry that the world is simpler now than it used to be?

The problem is that in making complex ecosystems simple we also make them more vulnerable to damage. A complex ecosystem can last through anything short of catastrophe by virtue of its complexity. Such a system has achieved a kind of dynamic balance, or equilibrium. One element, say a certain species of tree, can be knocked out by disease, and the remaining organisms will seek a new balance that leaves the community intact. In Eastern forests and woodlots, this happened on a massive scale in the 1950s, when Dutch elm disease killed practically all the remaining elm trees. However, the loss of the elms did not signal the end of the climax forests containing elms. The other trees quickly filled in the ecological niche left by the extinction of the elm, and the community continued to live. The situation was quite different, though, for any property holder whose backyard was a simplified ecosystem that contained elms as the only trees. In this case, Dutch elm disease caused a massive disruption of the ecosystem. Suddenly an area that had provided a shady habitat was denuded and opened to the sun. While the climax community was able to adjust to the change with relatively little problem, the same change totally disrupted the simplified, man-made ecosystem.

Recall the example of the population changes in the snowshoe hare and the lynx. The Arctic forests are simple ecosystems where any given species has only one source of food. The hare feeds on vegetable matter, and the lynx feeds on the hare. If catastrophe strikes either the plant food or the hares, the lynx is in trouble. It has no alternative food source to turn to. Compare the lynx's predicament with that of his close relative, the bobcat, principal predator in much of the South. The bobcat eats plenty of rabbits, but it has many other sources of food as well: mice, rats, shrews, songbirds, quail, pheasants, opossums, immature deer, and larger insects like grasshoppers and crickets. Should one of these species disappear, the bobcat would hardly feel the loss. He still has plenty of alternative foods. In a simple ecosystem, a small change affects the whole community, while complex systems are only slightly affected by the same sort of change. One species may disappear entirely, and the community will go on pretty much the same as before.

History abounds with examples of the price man has paid for simplicity. In certain areas of the world, the population depends on a single food crop. Should anything happen to that crop, the population is faced with serious famine. For centuries, the potato, developed originally in South America, was the staple of the Irish diet. In a little less than 200 years, from 1670 to 1845, the Irish grew from 1 million to 8 million people, the majority of them dependent on the potato for most of their calories. Then, for the 6 years from 1848 to 1854, a kind of parasitic fungus wiped out each and every crop of potatoes before it could be harvested. In the resulting famine, more than 1 million people starved, and another million emigrated to the United States in extreme poverty and poor health. Cyclic famines such as this were common in much of China, dependent on the rice crop, until the early 1960s, when diversification and modernization of agriculture put an end to starvation.

In addition, one must wonder just how much we can tamper with the fabric of life before we endanger ourselves directly. Investigations into how ecosystems work bring out the same points each time—that the balance is stable but delicate; that each part, no matter how small, is necessary to the whole. An ecosystem can only be pushed so far before its balance, achieved by the fine adjustments of centuries of evolution, is lost. Man has already dealt the earth many a sore blow.

The final paradox

Unique among the animals because of our brain, we have been able to comprehend the pathways of life well enough to build tailor-made ecosystems and run them for our own benefit. Now, with the accumulation of experience over the years, we've found that our simplification of the world has made it more susceptible to damage and, perhaps, catastrophe. But, at the same time that the meaning of simplification has grown clear and definite, our technological ability to cause that damage has increased. DDT was developed during World War II and has been used on a worldwide basis only for the past 25 years. Many modern fertilizers came to being in the 1950s. Atomic power plants and nuclear weapons, whose radioactivity is the most lethal force we have yet to meet, are likewise products of the past three decades. With the one hand, man has simplified the world and made it vulnerable, and with the other he has perfected tools capable of dealing a death blow.

The question is: Will he deliver it? How much hope is there for the future? Have we, by our acts, dug our own graves? Or are we standing at the dawn of a new age when man will strike a compromise with other living things so that we all can go on living? These are the questions we shall turn to in the next chapter and look for some expert answers.

Summary

As humans have dominated the world, their numbers have increased with great speed. In the last 10,000 years, the earth's population has gone from 5 million to 3.8 billion. The primary cause of this growth was the development of agriculture, which increased the food supply. In more recent times, public health measures, antibiotics, and pesticides have decreased infant mortality and added to life expectancy. Population

Are we standing at the dawn of a new age when man will strike a compromise with other living things so that we all can go on living?

growth is uneven, hitting a peak in poor tropical countries and leveling off in the richer nations.

Famine and war could restore population to a balance by increasing the number of deaths, but their effect is crude and cruel. The best way to control population is to limit births. However, there is no perfect method, and any campaign of population control must overcome serious cultural, psychological, and even legal barriers. Even if these were surmounted today and the birth rate made to equal the death rate, population growth would continue well into the next century.

Hunger is part of the population problem. Malnutrition comes in two varieties: lack of calories and lack of protein. The world's population has grown so much that people must make use of the lower trophic levels. While the Green Revolution has produced high-yield, high-protein grains, the food supply still lags behind the population. The amount of arable land is actually decreasing as it is taken over by growing cities. Rain forests are unsuitable for planting, and desert agriculture faces the problem of salt accumulation. The sea can be made to produce more food, but it cannot take up all the slack. Synthetic foods on a commercial scale are still far into the future.

As with food, energy is in short supply. Not only is the population growing, but we are using up the last of some fuel reserves. Per capita consumption of energy will probably be reduced as costs rise. Solar power may offer a low-cost, nonpolluting alternative for the future, but finding a new source of power is not the same as shifting a whole technology over to it.

Ecosystems pass through an orderly succession until they reach a climax community characterized by complexity. Man's activity tends to make ecosystems simple again. Simple ecosystems are more vulnerable to damage than are complex ones. Thus man has made the world more liable to damage at the same time that he has developed tools capable of causing that damage.

Questions to consider

1. a. The basic trend in the human population over the past 10,000 years has been an increasing rate of growth. What are some reasons for this increase?

b. Is population growth the same in all nations of the world?
c. What problems are posed by an increasing population?

2. One way to stop population growth would be to let nature have her way and allow famines to run their lethal course. Is this a reasonable idea? Why?

3. What stumbling blocks stand in the way of mass birth control programs?

4. a. What are the varieties of malnutrition?
   b. What are the effects of each kind?

5. a. What is the Green Revolution?
   b. Will the Green Revolution stop hunger before it becomes more widespread?

6. Relatively large amounts of the land are covered by rain forests and deserts. Can these be put into production to increase the world's food supply?

7. Two factors contribute to the present energy crisis. What are they?

8. In a statement on the American energy problem in September, 1973, President Nixon proposed lowering air pollution standards to allow industry to burn high-sulfur coal and oil, granting more permits for offshore oil drilling, and letting strip miners dig for coal on public lands in the West. Would you rate these proposals as sound or unsound? Why?

9. Solar power appears to be a workable solution to the energy problem. However, it is possible that a severe energy shortage could strike even with the technology for solar power available. Why is this the case?

10. a. What is meant by ecological succession?
    b. How does each stage in an ecological succession differ from the previous one?
    c. Why is the final stage in an ecological succession more stable than the first one?

11. a. In what ways does man simplify his environment?
    b. Why is simplification ecologically important?

Glossary

climax community the most stable, complex final stage in a pattern of ecological succession

ecological succession the replacement of one form of community by another over a period of time within a geographical area

photoelectric effect the generation of an electric current by light energy

16 The Future

In the previous chapter, we saw how man's past and present conduct has affected the world we live in. The growing number of humans has put a severe strain on our living space, our food supply, and our energy reserves, and our simplification of the world ecosystem has made it increasingly vulnerable to severe damage. In this chapter, we'll change our time scale and ask the relatively speculative question: What do these facts indicate for the future? In the process, we'll meet some of the people who have thought seriously about the subject.

We have seen that biology is a systematized collection of laws, theories, explanations, and data about living things, collected mostly over the last 200 years. If we're going to get through the next 200 years with our humanity, our health, and our general pleasure in living intact, we'll have to apply these principles to present and future problems with care and understanding. While many of these issues will be decided by government bureaucrats and not by us directly, we have a stake in what the decision-makers decide, and we must be able to evaluate the wisdom of their solutions.

A literature of concern

Like any subject of broad social importance, the question of human survival and well being in the coming centuries has produced a literature of its own. Although most of this writing is less than 20 years old—most of it has been produced since the mid-1960s—several articles, books, and essays have already attained classic status and are continually quoted, collected into anthologies, and reprinted for increasingly larger audiences.

One of the interesting sides of this "literature of the future of man" is that it is generally the product of people whose professional and scientific training might have led them to be concerned with only the narrower aspects of their disciplines. But these botanists, ecologists, animal physiologists, geneticists, bacteriologists, and other biological specialists have questioned the impact of their research on the human community, and have contrasted the biological absolutes we have discussed in this book with the political and economic pressures there are to ignore these absolutes.

It is interesting to note that a whole new discipline—futurology—has recently developed. Basically, futurologists attempt to put all of the social, scientific, economic, and political cards on the proverbial table and to draw some conclusions about what's in store for the human race.

Both futurologists and scientific specialists agree that knowledge is the key to human survival. For example, Glenn Seaborg, the eminent physicist and former chairman of the Atomic Energy Commission, has talked of the importance of education. Sir Julian Huxley, a renowned biologist and the grandson of the champion of Darwin mentioned in Chapter 4, also calls for the gathering of knowledge as the raw material for a humane future.

Let me focus briefly on the human force that will be necessary to carry out much, if not all, of the change demanded to create a livable future. That force is organized knowledge; and it is transmitted by education. It is education, perhaps more than any single factor, that will determine how we survive—the way in which developing nations develop, and the extent that human freedom and dignity flourish in a complex and highly organized world.

Glenn Seaborg in a speech to the American Association for the Advancement of Science

We must realize that our aim is not mere quantity, whether of people or goods or anything else, but quality—quality of human beings and of the lives they lead. Once we have grasped this, things begin to fall into place.

Such a view has important implications for science and education. It implies that the most important sciences today for the modern world are not physics and chemistry and their applications in technology, but evolutionary biology and ecology and their applications in scientific conservation.

Ecology is the science of the relations of living things with each other and with their environment. It of course includes human ecology, which deals with the relations of man with his fellow men and with the world's resources, both material resources and psychological or enjoyment resources. So I would plead for much more emphasis in education on biology in general and ecological biology and human ecology in particular.

Sir Julian Huxley in "The World Population Problem"

However, the noted British-born economist Kenneth Boulding is obviously skeptical about the number of "sure things" we have to go on. Even so, he does admit that "we have to worry about the future, simply because the greatest dilemma of mankind is that all knowledge is about the past and all decisions are about the future."

In October of 1970, I was in Mauritius, the home of the Dodo. I suppose that is a good place to start talking about survival, because the Dodo didn't make it. It also illustrates the extraordinary difficulties of saying anything about survival because we do not know much about the survival functions of species of any kind, whether biological, intellectual or even academic . . . as a matter of fact, one of the reasons for the popularity of astrology in the aquarian age (the aquarian age is one in which everyone is all wet) is that celestial mechanics is almost the only system that has any real futurology, outside of a little bit in developmental biology.

There are only two things we know about the future. One is where and when eclipses will take place and the other is that a kitten will never grow up into a rhinoceros. Outside of these two things, I am very skeptical about futurology.

Kenneth Boulding in "The Dodo Didn't Make It"

Evolution toward the future

One of the major biological links—one of the few we've got—between the past and future is the theory of evolution. It gives us a predictable way of viewing a species' potential for change, with the environment serving as the mold.

In Chapter 4, we saw how mankind has been adapting and evolving over the past 2 million years. There are some readily identifiable trends: The human brain has enlarged, human posture has become increasingly erect, body hair has decreased, and the proportion of muscular weight to total body weight has decreased. So can we, as some physical anthropologists have idly wondered, extend this line of evolution into the future? Will *Homo sapiens*, circa A.D. 33,000, be completely bald, with a huge cerebrum and a fairly light, tall skeletal frame?

We really have no idea and no way of knowing. Our evolutionary fate will depend on the environmental conditions and the selectivity they impose on the human gene pool. Remember that the steady growth of the human brain leveled off 150 thousand years ago, and we still haven't figured out why. But, in any case, the important problems of the future do not really concern the physical appearance of the human race several thousand years from now.

The renowned biologist René Dubos has made a critical point about human evolution: Our environment is now making extraordinary demands on our gene pool, but evolution often takes hundreds of centuries to select for the "fittest."

Clearly our adaptive potentialities are not unlimited. Even now they may be exceeded by some of the stresses created by contemporary technological developments. In the course of his prehistoric evolution, man was repeatedly exposed to seasonal famine, inclement weather, infectious processes, physical fatigue, and many forms of fear. This evolutionary experience has generated in his genetic constitution the potentiality to adapt to many different kinds of stresses. But he now faces dangers that have no precedent in his evolutionary past. He probably does not possess the responses that will be necessary to adapt to many of the new environmental threats created by modern technology: the toxic effects of chemical pollutants and synthetic substances, the physiological and mental aberrations resulting from the mechanizations of life, the artificial and violent stimuli that are ubiquitous in the technological world.

René Dubos in "Man's Participatory Evolution"

We know further that the average urban dweller in the United States now suffers from dangerously high levels of stress, inhales thousands of pounds of noxious substances during his lifetime (see Chapter 6), frequently has to drink impure water, and is carrying an excessive amount of DDT in his fatty tissues. However, in Chapter 4, we saw how the peppered moths adapted to industrial pollution in a matter of two or three decades. Can't man adapt equally fast? The answer is a definite no. To begin with, the adaptation of the moths involved only one trait—body and wing color—and, in evolutionary terms, it was a very minor change. Major changes—and the ability to breathe carbon monoxide safely, for example, would be a major change indeed—take hundreds of centuries. One million years were needed for the human brain to triple in size, and that is the fastest known case of large-scale adaptation. There's no way that the slow pace of evolution can even begin to match the speed with which the environment is deteriorating. Nor do we have any assurance that the needed genes are present in the human gene pool, as was the black-phase trait in the peppered moth. John Q. Public of 2075 will be very similar physiologically to his 1975 counterpart—only he'll probably be in much poorer health.

Can we direct our own evolution, perhaps through our expanding knowledge of genetics? After all, a foremost scientist, James Bonner of the California Institute of Technology, recently speculated that man might eventually be able to manipulate the gene and choose his future. But, before we seize on this attractive possibility as the solution to the dilemma, let's look at the pros and cons.

The changing scope of biomedical science

In a fascinating little book published in 1970, the Hungarian physicist Dennis Gabor speculated on the possible range of scientific, technological, and social innovations during the next 50 years. In his discussion, he itemizes 21 future feats of bioengineering, and provides the dates estimated for their achievement by groups of experts working at the Institute for the Future at Wesleyan University in Connecticut. He lists his "innovations of the future" in what he considers the decreasing order of their desirability.

1. *Economical mass-administered contraceptives*	Est. 1983
2. *Cures for cancer*	Est. 1990
3. *Accurate predictions of the effects of drugs*	Est. 2000
4. *Mutation-free forms of diagnostic devices using ultrasonic radiation*	Est. "near future"
5. *Immunization against most bacterial and viral diseases*	Est. 1980
6. *Transplants without rejection of tissues*	Est. 1983
7. *Artificial hearts*	Est. 1980
8. *Natural "renewal" of worn-out organs with the body's own renewal processes*	Est. 2010
9. *Repair of nerve cells*	Est. 2000
10. *Postponement of aging*	Est. 2015
11. *Early detection of abnormalities in the human fetus by readings of its chromosomes*	Est. 1980
12. *Determination of the sex of children with 90% accuracy*	Est. 1980
13. *Fertilization of human ovum outside the mother, followed by implantation into the uterus*	Est. 1990
14. *Human clone producing a series of identical humans*	Est. 1985
15. *Development of fetus outside a human uterus*	Est. 2015

16. *Molecular (chemical) manipulation of genes*	Est. 2020
17. *Drugs to enhance sensory perception and learning*	Est. 1985
18. *The perfect tranquilizing drug*	No estimate
19. *Drugs to raise intelligence permanently*	Est. 2020
20. *Personality-altering drugs with predictable effects*	Est. 1980
21. *Creation of artificial life*	Est. 1980

Adapted from Dennis Gabor, *Innovations: Scientific, Technological, and Social*

Gabor's priorities are very illuminating, particularly since he once said we cannot predict the future, but we can invent it. How many of Gabor's innovations strike you as being desirable? Why do you think he placed the most dramatic possibility, the creation of life in a "test tube" last? What about the idea of clones of identical humans, produced in just the same way as the identical toads in Chapter 10?

Here's another possibility that relates directly to Gabor's innovations 17 and 19. David Krech, a well-known psychologist, has been doing a great deal of research into the chemistry of thinking, or psychoneurobiochemistry. He and his colleagues have found that rats raised in particularly stimulating environments tend to have larger brains, larger brain cells, and more activity of two important brain enzymes than rats raised in a duller environment. Would it be possible to alter the human brain chemically with such enzymes to increase its potential for thought and memory? Even if we can, do we want to? If it's feasible, is it also desirable?

The problem of weighing desirability against technological feasibility is central. Garrett Hardin, a well-known biologist at the University of California at Santa Barbara, has rather dryly summarized the drive of Western man's "religious" preoccupation with scientific discovery and application: "1. Anything we can dream of, we can invent; 2. Anything we can invent, we must use." This may sound farfetched, but recall the arguments of the proponents of the supersonic transport, or, moving back a couple of decades, the national agony over the development and use of the atomic bomb. We are becoming frightened by our technological potential.

> *In 1961, an Italian researcher fertilized a human ovum with sperm and kept it alive for 29 days until it was the size of a pea. Noting that it was starting to grow into a monstrosity, he terminated the experiment—but only after he had given it conditional baptism and extreme unction. His work caused such a religious furor that subsequent experiments along these lines have not been publicized at all.*
>
> E. Fuller Torrey in "Ethical Issues in Future Medicine," *Toward Century 21*

Granting that the issue of feasibility and desirability is resolved satisfactorily, we must recognize that our knowledge is limited. The research currently being conducted into the simulated replication of DNA and the possibilities of test-tube babies is, of course, fascinating stuff, in the tradition of the best science fiction, and will certainly lead to discoveries that will be applied to the cure and prevention of the inherited diseases described in Chapter 8.

But, as far as direct manipulation of the gene itself—to control heredity on a mass scale—is concerned, we face four major objections. The Nobel Prize–winning geneticist Joshua Lederberg makes a strong case for not going off the deep end about big genetic breakthroughs along these lines in the immediate future. Lederberg might even disagree with Bonner about the potential for meaningful control of evolution in the coming years. But it is interesting to note that Lederberg has been directly responsible for a good deal of our "new" perspective on the gene, both through his pioneering research and his writings in the popular press.

The second problem is a moral one. We already have the technological and medical information necessary to prevent the birth of a child already conceived and to "permit" the death of elderly or terminally ill patients. But we have not yet resolved the moral problems behind abortion and euthanasia. And the moral problems posed by genetic manipulation make those of abortion and euthanasia seem trivial in comparison.

The third problem is the "who decides" of the issue. Would you like to put your confidence in a bureaucracy or an elected commissioner of genetic control? How about a gigantic think-tank, The Institute for Genetic Policy, that might decide which traits should be emphasized in coming generations and select accordingly? What would happen to our diversity and vigor as a people? What might happen to *your* individuality?

There are two specific limitations, both very, very important, in the human utility of biological advances such as DNA replication. The most important of these is that, at the present time, when it comes to an application in man, we don't really have any very useful application to make of the first sample of DNA, much less than of any copies that we might make of it . . . DNA in the form in which it is isolated in the test tube has no known biological activity in man.

[Secondly,] we have no insight, at the present time, as to how we would begin to approach the question of substituting one DNA molecule already present in a sperm or an egg with another DNA molecule that we had outside of it.

Joshua Lederberg in "Human Implications of Biological Discovery," *Toward Century 21*

The fourth question is an ironic one. Look again at Gabor's list. Would the control of conception and development, extrauterine growth of offspring, and other alterations of the present-day concept of reproduction clash with our sense of our own humanity? Would the freedom to choose the traits, such as sex, intelligence, physical skills, and so on, of our offspring make us more or less human? Would we be reduced to breeding machines, or would we be elevated to a superior form of biological system with complete control over its destiny?

We tend, of course, to think of scientific advances as enhancing man's position in nature. To be free of terrifying diseases and plagues, to have leisure time to pursue our intellectual and cultural interests—these are all benefits of the scientific revolution. But with its political and moral dilemmas, scientific advance is fast becoming a two-edged sword.

Too many too fast

Before you start losing sleep over the idea of a government project to produce a clone of 4,000 identical Henry Kissingers or 2,000 Billie Jean Kings, let's turn our attention to some of the people who have suggested that the dilemmas posed by advances in medical science pale in comparison with the true problem of the twentieth century—the population explosion.

A few years ago, Garrett Hardin wrote a brief article for the American Association for the Advancement of Science. Initially, it was probably read by several thousand scientists here and abroad, and perhaps in earlier times would have only become part of the literature of a select scientific community. As it turned out, "The Tragedy of the Commons" has achieved a very important status in the scientific literature of the 1960s—and a far wider audience than scientists alone.

The metaphor of Hardin's title refers to the common pasture ground, or commons, of small, rural communities in medieval England. The commons, as its name indicates, was a shared resource open to all and unregulated in its use. It was at once the property of everyone and of no one. According to Hardin, the earth's ecosystem is a commons, and its uncontrolled exploitation cannot continue indefinitely. Each person's use of the collective resources of the community reduces everyone else's share proportionately. The major problem confronting us, Hardin says, is the freedom to breed and the population growth that results. He maintains that we are sacrificing all of our other freedoms to maintain this one.

> *There is no technical solution to the problem. An implicit and almost universal assumption of discussions published in professional and semipopular scientific journals is that the problem under discussion has a technical solution. A technical solution may be defined as one that requires a change only in the techniques of the natural sciences, demanding little or nothing in the way of change in human values or ideas of morality . . .*
>
> *My thesis is that the "population problem" . . . is a member of this ("no technical solution problems") class . . . the population problem cannot be solved in a technical way.*
>
> Garrett Hardin in "The Tragedy of the Commons"

"Every new enclosure of the commons involves the infringement of somebody's personal liberty." How much longer will this infringement continue? As we saw in Chapter 15, the fight against starvation has largely been a losing battle. Yet, population growth on a worldwide scale continues.

One of the foremost advocates of a complete reevaluation of our unlimited population growth and the mentality of "the bigger, the better" is Paul Ehrlich, an ecologist who several years ago wrote a brief and pungent book entitled *The Population Bomb.* In addition to stating the potential threat we face from excessive population growth, the book forced its readers to recognize that the population problem is not only a future problem—it is with us right now.

Contrast the facts and figures of population growth we saw in Chapter 15 with Ehrlich's interesting observation that in the year 1966, the world population increased by 70 million people. However, we produced no additional food to feed them, thus reducing every earthly resident's share of the world's food supply by 2 percent! Ehrlich estimates that we will have to triple the world's food supply between now and the turn of the century to match the needs of the expected world population at that time.

Think of what it means for the population of an underdeveloped country to double in 25 years. In order just to keep living standards at the present inadequate level, the food available for the people must be doubled. Every structure and road must be duplicated. The amount of power must be doubled. The capacity of the transport system must be doubled. The number of trained doctors, nurses, teachers, and administrators must be doubled. This would be a fantastically difficult job in the United States—a rich country with a fine agricultural system, immense industries, and rich natural resources. Think of what it means to a country with none of these.

Paul Ehrlich in *The Population Bomb*

Here's the catch: This is all supposed to occur while we are continually taking more and more land out of agricultural production, tampering with our water supplies through industrial pollution and the administration of pesticides, and throwing one monkey wrench after another into the world ecosystem with damaging agricultural processes. For example, as we saw in Chapter 15, monoculture, the cultivation of a single crop over a large surface area, is a dangerous, simplifying technique. Yet, it is very efficient, if enough pesticides and other controls are used. How can we reconcile the Russian roulette of such methods with mankind's driving—and growing—need for food?

Food and other resources should not be the only standard for evaluating the desirability of further population growth. Think how many choices you may personally make each year on the basis of population. Did crowds keep you from enjoying a recent vacation? Was finding a parking space or a seat on a city bus a problem this morning? How about finding a job or a pleasant place to live now and after graduation?

Ehrlich and other people concerned with the problem have called for a national population policy coordinated with an international effort. One possibility might be to administer universal antifertility substances through public water supplies. If a couple were to have a child, they would receive an antidote, and their fertility would thus be restored. Another idea, again a drastic one, would be the public licensing of childbirth. It would be as illegal to have an unlicensed child as it is to drive an unlicensed car.

Both worldwide plague and thermonuclear war are made more probable as population growth continues. These, along with famine, make up the trio of potential "death rate solutions" to the population problem—solutions in which the birth rate–death rate imbalance is redressed by a rise in the death rate rather than by a lowering of the birth rate. Make no mistake about it, the imbalance will be redressed [*emphasis Ehrlich's*]. *The shape of the population growth curve is one familiar to the biologist . . . a population grows rapidly in the presence of abundant resources, finally runs out of food or some other necessity, and crashes to a low level or extinction.*

The situation was recently summarized very succinctly: "It is the top of the ninth inning. Man, always a threat at the plate, has been hitting Nature hard. It is important to remember, however, that Nature bats last."

Paul Ehrlich in "Eco-Catastrophe!"

Mass birth control and birth licensing can be criticized severely on any number of moral and political grounds. They are presented here not to advocate them, but to show that the problem is sufficiently drastic that equally drastic solutions are being seriously entertained. And, no matter what the flaws in these particular proposals, the problem of population growth remains.

As we noted in Chapter 9, we already possess the birth control techniques to limit population quite effectively. But, as Hardin has pointed out, technology has little to do with this dilemma—rather, we need to question the growth atmosphere in which we find ourselves; too many people want too many children. We are still economically, politically, and socially oriented toward childbirth. Is our "right" to reproduce without limit to supersede all of our other rights? Can the earth, as a whole, "afford" the children that we, as individuals, can afford to raise?

Energy resources: some further questions

The energy crisis of the 1970s has certainly sharpened people's awareness of how potentially scarce our resources are and how quickly they might be depleted. In 1973, the governor of Oregon called for a ban on the unnecessary use of electricity for outdoor advertisements. Our need for imported oil is already affecting our foreign policy, and brownouts and fuel shortages are becoming commonplace in many parts of the country.

Many of the questions of energy use come back to us as individuals, just as the population crisis does. Do you consider gas mileage an important factor in the purchase of a car? In fact, have you questioned whether you actually need to purchase a car at all? How about electrical home appliances? Are the excessive energy resources used in the production of much of our modern packaging worth the price we are paying?

As we saw in Chapter 15, the current crisis centers on our depletion of fossil-fuel reserves. The noted biochemist and writer Isaac Asimov has estimated that our total energy use is doubling every 15 years. Interestingly, his solution to the problem would include pursuing nuclear energy as a major power source. Nuclear energy has advantages—it involves a relatively inexhaustible raw material, it does not involve burning (thus eliminating an air pollution problem), and we probably possess most of the technology needed to make it possible. But the problems of accident, thermal pollution, and radioactive damage to the human gene pool remain unsolved.

It is estimated that there is enough gas, oil, and coal here and there in the earth to keep us going 7,500 years at our present rate of energy use. Unfortunately, not all of this fuel can be dug or drilled out of the earth. Some of it is so deep, or spread out so thin, it would take more energy to get the fuel than that fuel would yield when burned. The fuel we could get out at an energy-profit would only last us about 1,000 years at the present rate of use . . . if the population continues to increase and our needs continue to double every 15 years, all the gas and oil, and coal will be gone in 135 years . . . if mankind wishes to maintain the present level of civilization, we must find some alternate sources of energy, sources that will produce no pollution and are plentiful enough to keep us going and supply all our needs.

Isaac Asimov in "The Power Crisis that Threatens the World"

Even solar energy, which is really rather promising in terms of its potential yield, its safety, and its lack of pollution, is of limited feasibility at present. The complexity of the debate about alternative energy resources is exemplified by the positions of the respected science journalist Irving Bengelsdorf, who dismisses solar energy as financially improbable, and John Holdren, a Berkeley nuclear physicist, who finds substantial evidence of its future importance.

The enormous amount of sun power intercepted by earth is about 100,000 times greater than the entire world's presently installed electrical power–generating capacity. But it is terribly diffuse and difficult to concentrate. The capital investment for a solar power installation is prohibitive.

Irving Bengelsdorf in "Are We Running Out of Fuel?"

In the year 1970, mankind consumed an amount of energy equal only to the amount of solar energy that strikes earth's outer atmosphere in fifteen minutes. Because solar energy is clear, free and abundant, why did we not use it to solve our energy problems long ago? There are two main reasons. First, the solar energy reaching the surface of the earth is dilute: to acquire enough for large projects it must be collected over a large area . . . second, solar energy is variable: on cloudy days not much gets through, none at night, and in winter less is available than in summer . . . but these problems are not overwhelming . . . one of the great built-in advantages of solar energy is that transmission and distribution are free . . . two Arizona astronomers, Drs. Alden Meinel and Marjorie Meinel, have proposed a system in which the solar to electrical conversion efficiency would be 25 to 30 percent, making the generation of electricity at even large central power plants a possibility.

John Holdren in "Defusing Old Smoky by Plugging into Nature"

What now?

In Chapter 6, we encountered the concept of synergism, which means that two or more things together can have effects greater than the sum of all of them. Synergism is a value-free concept. Of itself, it's neither good nor bad, its meaning depends on the context. The synergistic aspects of air pollution described in that earlier chapter are obviously harmful.

But synergism can also have beneficial aspects. It can mean that research in one scientific or technical area can have applications in other areas, beyond those already anticipated. The space program, for example, has yielded many innovations for medicine and for individual comfort and convenience as well as a great deal of scientific data that may provide the basis for the major innovations of the future.

But these multifaceted aspects of discovery and progress require extraordinary coordination and cooperation among people, scientific groups, and nations. Coordinated effort is certainly going to prove the key to solving many of our current environmental and biological problems.

Contradictory and paradoxical policies have been a significant problem. For example, if we look at the American tax laws, it is obvious that child rearing is officially endorsed by our government, in that income tax deductions are permitted for each dependent child. On the other hand, our government has spent a good deal of money on public education programs concerning birth control, and has subsidized private groups seeking the stabilization of our population. There was also a well-publicized squabble between Secretary of Commerce Maurice Stans and Environmental Protection Agency head William Ruckelshaus in 1972 over the economic implications of pollution control. Our democratic system and our varied society demand that we provide accommodations for a wide variety of perspectives, many of them conflicting. But the contradictions may prove fatal for us.

Besides coordination of effort, there is the question of how much scientific and technical attention we are willing to pay to the problems of the future. A few years ago, John Platt, a biophysicist at the University of Michigan, wrote an article that pinpointed our collision course with disaster and itemized the individual catastrophes that might stand between us and survival into the next century. His conclusions about the need for concentrated attention to the problems ahead are worth noting.

> *What can we do? I think that nothing less than the application of the full intelligence of our society is likely to be adequate. These problems will require the humane and constructive efforts of everyone involved. But I think they will also require something very similar to the mobilization of scientists for solving crisis problems in wartime. I believe we are going to need large numbers of scientists forming something like research teams or task forces for social research and development. We need full-time interdisciplinary teams combining men of different specialties, natural scientists, social scientists, doctors, engineers, teachers, lawyers, and many other trained and inventive minds, who can put together our stores of knowledge and powerful new ideas into improved technical methods, organizational designs, or "social inventions" that have a chance of being adopted soon enough and widely enough to be effective. Even a great mobilization of scientists may not be enough. There is no guarantee that these problems can be solved, or solved in time, no matter what we do. But for problems of this scale and urgency, this kind of focusing of our brains and knowledge may be the only chance we have.*
>
> John Platt in "What We Must Do"

John Fischer, critic and writer, has proposed a solution that relates directly to our educational institutions. His plan for Survival U is a fascinating one: A school where each department—indeed, each instructor—is obliged to demonstrate the utility of his area of specialization to the future of mankind. Survival of the human race would be the rationale of the institution, just as the development of an educated clergy or a scholarly upper class was the goal of higher education in centuries past. According to Fischer, this type of immediate attention and mobilization is the only way of keeping the concern of people such as we have met in this chapter from falling on deaf ears.

> *Warnings could be quoted from a long list of other social scientists, biologists, and physicists, among them such distinguished thinkers as René Dubos, Buckminster Fuller, Loren Eiseley, George Wald, and Barry Commoner. They are not hopeless. Most of them believe that we still have a chance to bring our weapons, our population growth, and the destruction of our environment under control before it is too late. But the time is short, and so far there is no evidence that enough people are taking them seriously.*
>
> John Fischer in "Survival U: Prospectus for a Really Relevant University"

The urgency of planning for the future—so that innovations do not come flying at us and catch us unprepared—is pointed out by Barry Commoner, an eminent ecologist and author, and by Nigel Calder, an English journalist and science writer. To a greater extent than any civilization before us, we are going to be able to predict and shape our future; it will not simply happen to us.

> *If we are to survive, we need to become aware of the damaging effects of technological innovations, determine their economic and social costs, balance these against the expected benefits, make the facts broadly available to the public, and take the action needed to achieve an acceptable balance of benefits and hazards. Obviously, all this should be done* before *we become massively committed to a new technology. One of our most urgent needs is to establish within the scientific community some means of estimating and reporting on the expected benefits and hazards of proposed environmental inter-*

ventions in advance. Such advance consideration could have averted many of our present difficulties with detergents, insecticides, and radioactive contaminants.

Barry Commoner in "To Survive on the Earth"

Besides the different styles of science and its management in different countries, there are wide disparities in the sophistication of science policymaking. Some "advanced" countries have scarcely reached the starting post, where you decide that research is worth patronizing on a significant scale. Even the leading scientific nations are just at the point of learning how to encourage technological innovation in all economic and administrative activities, and how to look ahead to anticipate the consequences, for good or ill, of those innovations. The organization of serious public debate about the uses of science has scarcely begun. Still less is there any ordering of moral and political attitudes that might give some coherence to the argument, or a pattern for new policies.

Nigel Calder in *Technopolis: Social Control of the Uses of Science*

Planning is also important because the pace of our technology, as Aldous Huxley, the well-known author, points out, is so rapid that it will be almost impossible for our cultural institutions to keep up. We need to plan for the future's problems as well as for its benefits—and to know which is which.

Some day, let us hope, rulers and ruled will break out of the cultural prison in which they are now confined. Some day. And may that day come soon! For, thanks to our rapidly advancing technology, we have very little time at our disposal. The river of change flows ever faster, and somewhere downstream, perhaps only a few years ahead, we shall come to the rapids, shall hear, louder and ever louder, the roaring of a cataract.

Aldous Huxley in "The Politics of Population/Second Edition"

A final word

The scope of the problems facing us in the coming decades—and the urgency of the solutions suggested—might strike you as terribly pessimistic and discouraging. But any form of planning for the future is based on the assumption that we do indeed have a future, and, unlike any generation before us, we have the information and the skill to make of it as we may.

Our current problems are based on one basic misconception—that mankind is somehow outside the natural way of things and that we can "legislate" our own variations of natural laws. One of the major goals of this book has been to help us put ourselves back into our natural context.

If we are indeed living satisfying and healthy lives a century from now, it will be because enough people recognized this simple principle in time. And, hopefully you'll be one of them.

> *There may yet be, in the untrammeled tenth of America, enough nature, unsecond-guessed by technological arrogance, to build a good future on . . . where we consider our beginnings and our beyondings, where we learn to absorb, and to respect and love and remember.*
>
> *The wild places are where we began. When they end, so do we. We had better not speed their passing. Man's talent can keep them if he lets it. Something happened and can still happen on a summer island, to substantiate all this hypothesis. Drive near (you can't, happily, drive to places like this), park, and ask for a sarsaparilla, then think about the island and about the other places there ought to be that are like it enough to count, this year, next year, and forever after.*
>
> David Brower in the Foreword to *Summer Island* by Elliot Porter

Summary

Out of the concern about the future of mankind that has emerged in the past two decades an important literature has developed. The need to

plan for the future requires the accumulation of masses of information and strict attention to the laws of biological science.

We cannot rely on the adaptive potential of mankind to wait out the threat from environmental deterioration, because the pace of evolution is extremely slow. Our world is changing rapidly while we remain biologically the same.

Biological and medical advances raise the two-edged sword of extraordinary progress toward the curing of disease and the alleviation of pain on the one hand, coupled with the moral and practical dilemmas of adjusting to the new concept of human life that they make possible on the other.

Tempting though it is to speculate about the exciting technical developments of the future, many authorities feel that the significant question to be answered is how we will deal with our expanding population and our shrinking resources. We already are scientifically ready to control worldwide population growth, but social, cultural, political, and economic factors still equate large populations with strength. We have so severely damaged our earthly resources that we face a serious starvation problem today, with little chance of supporting the much larger populations of the future.

Our depletion of fossil fuels has become so critical that we must turn our attention to the development of alternative methods of energy production. In all probability, either atomic or solar power will be the major energy source of the future, if the many economic, technological, and ecological problems associated with them can be worked out.

In order to assure a satisfying future for ourselves and the coming generations, we need to coordinate our research efforts, turning our attention toward the broad questions of survival. Changes in our educational systems and our planning procedures will be necessary if we are to meet the challenges ahead.

Sources

Asimov, Isaac. "The Power Crisis that Threatens the World." *Boys Life*, November 23, 1971, pp. 33–35.

Bengelsdorf, Irving. "Are We Running Out of Fuel?" *National Wildlife*, February–March 1971, pp. 4–8.

Boulding, Kenneth. "The Dodo Didn't Make It: Survival and Betterment." *Bulletin of the Atomic Scientists*, May 1971, pp. 19–20.

Brower, David. Foreword in Elliot Porter, *Summer Island: Penobscot County*. San Francisco: The Sierra Club, 1966.

Calder, Nigel. *Technopolis: Social Control of the Uses of Science*. New York: Simon and Schuster, 1969, p. 23.

Commoner, Barry. *Science and Survival*. New York: Viking, 1963, pp. 122–123.

Dubos, René. "Man Over Adapting." *Psychology Today*, February 1971, p. 51.

Ehrlich, Paul. "Eco-Catastrophe!" *Ramparts*, September 1969, p. 28.

———. *The Population Bomb*. New York: Ballantine, 1971, pp. 22–23.

Fischer, John. "Survival U: Prospectus for a Really Relevant University," *Harper's*, September 1969, p. 14.

Gabor, Dennis. *Innovations: Scientific, Technological, and Social*. New York: Oxford University Press, 1970, pp. 67–75.

Hardin, Garrett. "The Tragedy of the Commons." *Science*, December 13, 1968, pp. 1243–1248.

Holdren, John. "Defusing Old Smoky by Plugging into Nature." *Sierra Club Bulletin*, 1971, pp. 24–26.

Huxley, Aldous. "The Politics of Population/Second Edition." *The Center Magazine* (The Center for the Study of Democratic Institutions), March 1969, p. 17.

Huxley, Sir Julian. "The World Population Problem." In Glen and Rhoda Love, *Ecological Crisis*. New York: Harcourt, 1970, p. 78.

Lederberg, Joshua. "Human Implications of Biological Discovery." In C. S. Wallia, ed., *Toward Century 21*. New York: Basic Books, 1970, pp. 47–48.

Platt, John. "What We Must Do." *Science*, November 28, 1969, pp. 1115–1121.

Seaborg, Glenn. Speech to the annual meeting of the American Association for the Advancement of Science, June 1973. In *Science*, July 6, 1973, p. 18.

Torrey, E. Fuller. "Ethical Issues in Future Medicine." In C. S. Wallia, ed., *Toward Century 21*. New York: Basic Books, 1970, p. 31.

A further sample of the literature of concern

Carson, Rachel. *Silent Spring*. New York: Fawcett World Library, 1970, paperback.
Probably the book which started the public environmental concern of the 1960s; a poetic, skillful, and pungent presentation of the damaging effects of careless pesticide use.

Commoner, Barry. *The Closing Circle*. New York: Bantam, 1972, paperback.
An ecologist's view of the biological imperatives of man's future.

Cousteau, Jacques-Yves and Philippe Diole. *Life and Death in a Coral Sea*. Garden City: Doubleday, 1971, illustrated.
Two marine biologists explore the threat to one of nature's most fragile environments.

de Bell, Garrett. *The Environmental Handbook*. New York: Ballantine Books, 1970, paperback.
Prepared for the first Environmental Teach-In, this is a worthwhile collection of articles, combined with suggestions for community and individual action to protect the environment.

Ehrlich, Paul and Anne Ehrlich. *Population, Resources, and Environment: Issues in Ecology*. San Francisco: W. H. Freeman, 1970.

A detailed and highly readable analysis of the limited and the flexible factors in the current environmental crisis.

Lappé, Frances. *Diet for a Small Planet.* New York: Ballantine Books, 1971, paperback.
A popular and delightful exploration of our dietary habits and how they can be brought into harmony with the needs of our "shrinking" world.

Lauwerys, J. A. *Man's Impact on Nature.* Garden City: Natural History Press 1969, illustrated.
A well-illustrated perspective on the cultural, technological, and physical aspects of man's effect on his environment.

Love, Glen and Rhoda Love, eds. *Ecological Crisis: Readings for Survival.* New York: Harcourt Brace Jovanovich, 1970.
An excellent anthology of the classics of recent environmental literature.

McHarg, Ian. *Design with Nature.* Garden City: Natural History Press, 1971, paperback.
A beautiful photographic presentation of the concept of ecological design, the planning of buildings and cities in harmony with nature.

Paddock, William and Paul Paddock. *Famine—1975!* Boston: Little, Brown, 1970, paperback.
A powerful and controversial argument for immediate attention to the growth of the world's population and its dangerous drain on the world's food resources.

Shepard, Paul, and Daniel McKinley. *The Subversive Science.* Boston: Houghton Mifflin, 1971, paperback.
One of the earlier, and one of the best, anthologies on ecology and its perspective on man's potential survival.

Smith, Robert. *The Ecology of Man: An Ecosystem Approach.* New York: Harper & Row, 1972, paperback.
A well-organized sampling of the recent literature on ecology and environmental issues, covering many of the topics discussed in previous chapters.

Taylor, Gordon Rattray. *The Biological Time Bomb.* New York: New American Library, 1969, paperback.
A rather flashy but thoroughly engrossing tour of the new developments in biological science and their potential effects on human life and values.

Index

Page numbers in boldface type indicate glossary entries.

Chapter 7 *Fig. 7-3:* After Michael Neushul, *Botany,* Santa Barbara, Calif., Hamilton, 1974. *Fig. 7-5:* Richard H. Gross.

Chapter 8 *Fig. 8-2:* A. Richard H. Gross. *Fig. 8-3:* Richard H. Gross. *Fig. 8-8:* Drawing, T. S. Painter, "A New Method for the Study of Chromosome Aberrations and the Plotting of Chromosome Maps in *Drosophila melanogaster,*" *Genetics,* vol. 19, 1934, pp. 175-188; list, T. H. Morgan, "The Constitution of the Hereditary Material," *Proceedings of the American Philosophical Society,* vol. 54, 1915, pp. 143-153. *Fig. 8-10:* Courtesy of J. I. Bryant. *Fig. 8-17:* From Sickle Cell Anemia Research and Education, San Francisco.

Chapter 9 *Fig. 9-1:* Courtesy Carolina Biological Supply House. *Fig. 9-3:* Photo by Keith Gillett, Australia. *Fig. 9-5:* B. Richard H. Gross. *Fig. 9-6:* Richard H. Gross. *Fig. 9-9:* Richard H. Gross. *p. 236:* Richard H. Gross. *Fig. 9-11:* Richard H. Gross. *Fig. 9-13:* B. Richard H. Gross. *Fig. 9-15:* Ward's Natural Science Establishment, Inc. *Fig. 9-16:* Richard H. Gross. *Fig. 9-19:* New York Racing Association. *Fig. 9-22:* After a photograph from Ward's Natural Science Establishment, Inc. *Fig. 9-24:* From W. H. Masters and V. E. Johnson, "The Sexual Response Cycles of the Human Male and Female: Comparative Anatomy and Physiology," in *Human Sexual Behavior,* B. Lieberman, ed., New York, Wiley, 1971, p. 8. *Fig. 9-25:* From Kenneth Jones et al., *Sex,* 2nd ed., New York, Harper & Row, 1973, p. 45. *Fig. 9-26:* Adapted from Kenneth Jones et al., *Sex,* 2nd ed., New York, Harper & Row, 1973, pp. 54-55.

Chapter 10 *p. 272:* Richard H. Gross. *Fig. 10-3:* Adapted from John Moore, *Heredity and Development,* 2nd ed., New York, Oxford, 1972, pp. 232-233. *Fig. 10-4:* Drawn from a photograph in Roberts Rugh and Landrum Shettles, *From Conception to Birth,* New York, Harper & Row, 1971, p. 30. *Fig. 10-6:* With permission from Rugh and Shettles: *From Conception to Birth: the Drama of Life's Beginnings,* Harper & Row, 1971. *Fig. 10-8:* Adapted from Edward Wilson et al., *Life on Earth,* Stamford, Conn., Sinauer, 1973, p. 338. *Fig. 10-9:* From S. Hörstadius, *Biological Reviews,* vol. 14, 1939, pp. 132-179. *Fig. 10-10:* Adapted from Edward Wilson et al., *Life on Earth,* Stamford, Conn., Sinauer, 1973, p. 310. *Fig. 10-11:* Giant chromosomes, Richard H. Gross. *Fig. 10-13:* Adapted from *Newsweek,* April 16, 1973, p. 58. *Fig. 10-14:* Charles Moore, Black Star.

Chapter 11 *Fig. 11-2:* B. Richard H. Gross. *Fig. 11-3:* Richard H. Gross. *Fig. 11-5:* California Academy of Sciences, Steinhart Aquarium. *Fig. 11-6:* Beetle, Richard H. Gross; dragonfly, Richard H. Gross; crab, Steve H. Kratka. *Fig. 11-9:* Richard H. Gross. *Fig. 11-12:* Richard H. Gross. *Fig. 11-13:* B. From F. D. Ommanney, *The Fishes,* New York, Time-Life Books, 1968, p. 37. C. Richard H. Gross. *Fig. 11-14:* U.S. Fish and Wildlife Service by David B. Marshall. *p. 317:* Catfish, Richard H. Gross. *Fig. 11-17:* Wide World Photos. *Fig. 11-18:* Yerkes Primate Research Center. *Fig. 11-20:* Steve H. Kratka.

Chapter 12 *Fig. 12-1:* Courtesy Carolina Biological Supply Company. *Fig. 12-4:* B. Richard H. Gross. *Fig. 12-10:* B. Richard H. Gross. *Fig. 12-17:* Drawn from a photograph in James Olds, "Pleasure Centers in the Brain," *Scientific American,* October 1956, p. 112. *Fig. 12-19:* Richard H. Gross. *Fig. 12-20:* Adapted from A. Nason, *Modern Biology,* New York, Wiley, 1965.

Chapter 13 *p. 372:* Richard H. Gross. *Fig. 13-2:* From Otto Lowenstein, *The Senses,* Gretna, La., Pelican, 1966, p. 46. *Fig. 13-3:* Photos by Thomas Eisner. *Fig. 13-4:* R. A. Mendez, © Animals Animals 1973. *Fig. 13-11:* Andreas Feininger, Time-LIFE Picture Agency. *Fig. 13-12:* Richard H. Gross. *Fig. 13-15:* Richard H. Gross. *Fig. 13-16:* From Edward Wilson et al., *Life on Earth,* Stamford, Conn., Sinauer, 1973, p. 563. *Fig. 13-22:* From Frank Brown et al., *The Biological Clock,* New York, Academic Press, 1970, p. 34.

Chapter 14 *Fig. 14-1:* B. Smith, © Animals Animals 1972. *p. 417:* Alligator, gar, and sparrowhawk, Richard H. Gross; osprey, Steve H. Kratka. *Fig. 14-2:* U.S. Department of Commerce, National Marine Fisheries Service. *Fig. 14-3:* T. Fuller, © Animals Animals 1973. *Fig. 14-4:* Steve H. Kratka. *Fig. 14-5:* Virgil N. Argo. *p. 423:* Orchid, Richard H. Gross. *Fig. 14-6:* Birds, Steve H. Kratka; fish, California Academy of Sciences, Steinhart Aquarium; lions, Steve H. Kratka. *Fig. 14-7:* T. Kojo, © Animals Animals 1974. *Fig. 14-8:* C. Baldwin, © Animals Animals 1974. *Fig. 14-10:* Harry F. Harlow, University of Wisconsin Primate Laboratory. *Fig. 14-11:* Walter Chandoha. *Fig. 14-12:* Adapted from Robert L. Smith, *Ecology and Field Biology,* New York, Harper & Row, 1966, p. 41. *Fig. 14-13:* Adapted from A. S. Bodenheimer, *Monographiae Biologicae,* vol. 6, no. 1, 1958, p. 276. *Fig. 14-14:* California Academy of Sciences. *Fig. 14-15:* From D. A. MacLulich, University of Toronto Studies, Biological Series No. 43, 1937. *Fig. 14-16:* From A. S. Leopold, Wisconsin Conservation Bulletin No. 321, 1943. *Fig. 14-17:* B. Gallagher, © Animals Animals 1972. *Fig. 14-18:* P. Caulfield, © Animals Animals 1971.

Chapter 15 *Fig. 15-1:* After J. D. Durand, "Population Problems," *Proceedings of the American Philosophical Society,* vol. 3, no. 3, 1967. *Fig. 15-2:* From Tadd Fisher, *Our Overcrowded World,* New York, Parents' Magazine Press, 1969, p. 23. *Fig. 15-3:* International Labour Office. *Fig. 15-4:* From Tadd Fisher, *Our Overcrowded World,* New York, Parents' Magazine Press, 1969, p. 202. *Fig. 15-5:* Adapted from George Borgstrom, *Too Many: The Biological Limitations of Our Earth,* New York, Macmillan, 1969. *Fig. 15-6:* Data from U.S. Department of Agriculture. *Fig. 15-7:* California State Department of Water Resources. *Fig. 15-8:* From Fisheries of North America. *Fig. 15-9:* Data from U.S. Geological Survey. *p. 471:* Photo by Jan Fardell, Courtesy Friends of the Earth. *Fig. 15-10:* Courtesy Pacific Gas and Electric Company. *Fig. 15-11:* Photo by John Armstrong, Courtesy Friends of the Earth. *Fig. 15-12:* Richard H. Gross. *Fig. 15-13:* Steve H. Kratka. *p. 481:* Steve H. Kratka.

Chapter 16 *pp. 493-494:* Adapted from Dennis Gabor, *Innovations: Scientific, Technological, and Social,* © 1970 Oxford University Press, by permission of The Clarendon Press, Oxford.